JSRAE Technical Book Series

Compressors for Air Conditioning and Refrigeration

Japan Society of Refrigerating and Air Conditioning Engineers

Preface

Dear reader, in your hands is a book that we have all expected for many years. This is a comprehensive monography about refrigeration compressors, specifically positive displacement types. Some can say: *Only refrigeration compressors?* However, after reading this book everyone will witness a diversity of designs, approaches, and manufacturing as a living example of human creativity embodied in compressor designers and other engineers responsible for these beautiful machines.

This book brings a broad set of concise information needed to deeply understand the principles and function of fluid mover in a vapor compression refrigeration cycle. It covers it's kinematic, main manufacturing issues, and more. This book also discusses lubrication, an introduction to lubricants and their interaction with refrigerant in compression. Descriptions of motors and their control strengthens the usefulness of the text. Towards the end, this book presents measuring technology to determine characteristics of compressor operation with descriptions of the instruments. This book is an excellent text for every refrigeration engineer!

This collective effort of a strong group of engineers gathered around Compressor Technology Committee of the Japan Society of Refrigerating and Air Conditioning Engineers presents a significant contribution to science and applications, important for both academics and practicing engineers. I am sure that it will be frequently used as a reference, source of information, and even a textbook. This is a monument to heroic efforts of the writers and numerous engineers who conceived the principles of operation and put endless hours to make these compressor designs operational, reliable, efficient, compact, and cost effective.

I have fully enjoyed reading this book and hope you will too.

Pega Hrnjak, in Urbana IL January 2018

Contents

Chapter 1 Introduction ···1

  1.1 Refrigeration Cycle ···1

    1.1.1 What is refrigeration? ···1

    1.1.2 What is a cycle? ···1

    1.1.3 Vapor compression refrigeration cycle ···1

  1.2 Role of Compressors ···2

  1.3 History of Compressors (Early Period) ···2

  1.4 Classification of Compressors ··3

Chapter 2 Basic Theory ···6

  2.1 Compression Theory ···6

    2.1.1 Reversed Carnot cycle ···6

    2.1.2 *P-h* diagram ···6

    2.1.3 Adiabatic compression ···9

    2.1.4 Equation of energy ··· 12

    2.1.5 Leakage ·· 12

  2.2 Dynamic-mechanical Analysis ·· 16

    2.2.1 Motion of linkage ··· 17

    2.2.2 Equation of energy and mechanical efficiency ·· 18

    2.2.3 Excitation forces and vibration of a compressor body ····································· 18

  2.3 Lubrication Theory ··· 19

    2.3.1 Basic Reynolds equation ··· 19

    2.3.2 Modified Reynolds equation based on average flow model ······························· 19

    2.3.3 Contact theory ··· 20

    2.3.4 Journal bearing ·· 21

    2.3.5 Thrust bearing ··· 24

    2.3.6 Rolling bearing ·· 26

  2.4 Compressor Efficiency ·· 27

  2.5 Refrigerants ··· 29

    2.5.1 History of refrigerants ·· 29

    2.5.2 Required characteristics of refrigerant ·· 29

    2.5.3 Types of refrigerant ··· 30

    2.5.4 Designation of refrigerants ·· 31

    2.5.5 Comparison of commonly used refrigerants ··· 31

    2.5.6 Applications of various refrigerants ·· 32

    2.5.7 Refrigerant safety ··· 32

    2.5.8 Safety measures ··· 33

Chapter 3 Reciprocating Compressors ·· 35

  3.1 History of Reciprocating Compressors ··· 35

  3.2 Operating Principle and Characteristics ·· 35

    3.2.1 Compression action ··· 35

3.2.2 Theoretical stroke volume ··· 37

3.3 Structure of Compressors ··· 38

   3.3.1 Motion conversion mechanism ·· 38

   3.3.2 Suspension mechanism ·· 38

   3.3.3 Lubrication system ·· 39

   3.3.4 Valve system ·· 40

   3.3.5 Piston ring ·· 41

   3.3.6 Discharge muffler ··· 42

   3.3.7 Shaft seal mechanism ·· 42

   3.3.8 Capacity control mechanism ·· 44

3.4 Analysis of Kinematics ·· 45

   3.4.1 Kinematic analysis ·· 45

   3.4.2 Equation of energy and mechanical efficiency ·· 47

   3.4.3 Excitation forces and vibration of compressor body ·································· 47

3.5 Application and Performance ··· 49

3.6 Other Mechanisms of Reciprocating Compressors ·· 50

   3.6.1 Ball-joint mechanism ·· 50

   3.6.2 Two-cylinder mechanism ·· 50

   3.6.3 Two-stage compression ··· 50

   3.6.4 Linear motor-driven mechanism ·· 51

Chapter 4 Rotary Compressors ··· 52

4.1 History of Rotary Compressors ·· 52

4.2 Operating Principle and Characteristics ·· 53

   4.2.1 Compression action ··· 53

   4.2.2 Displacement volume ·· 54

4.3 Structure and Design of Compressor Components ·· 54

   4.3.1 Basic structure ·· 54

   4.3.2 Crankshaft ··· 55

   4.3.3 Bearings ··· 56

   4.3.4 Rolling piston ··· 56

   4.3.5 Vane ·· 57

   4.3.6 Cylinder ··· 57

   4.3.7 Discharge valve ·· 57

4.4 Dynamic-mechanical Analysis ··· 58

   4.4.1 Mechanism and movements ·· 58

   4.4.2 Constraint forces and equations of motion ··· 59

   4.4.3 Equation of energy and mechanical efficiency ·· 61

   4.4.4 Mechanical excitation forces and vibration of a compressor body ·············· 62

4.5 Applications and Performance ··· 62

   4.5.1 High-temperature application rotary compressors ······································ 62

   4.5.2 Low-temperature application rotary compressors ······································ 66

   4.5.3 Medium-temperature application rotary compressors ·································· 67

4.6 Other Types of Rotary Compressor Mechanisms ··· 67

   4.6.1 Two-cylinder type ··· 67

4.6.2 Swing compressors ⋯⋯⋯⋯⋯⋯⋯⋯⋯⋯⋯⋯⋯⋯⋯⋯⋯⋯⋯⋯⋯⋯⋯⋯⋯ 68

4.6.3 Liquid injection ⋯⋯⋯⋯⋯⋯⋯⋯⋯⋯⋯⋯⋯⋯⋯⋯⋯⋯⋯⋯⋯⋯⋯⋯⋯⋯ 70

4.6.4 Capacity control ⋯⋯⋯⋯⋯⋯⋯⋯⋯⋯⋯⋯⋯⋯⋯⋯⋯⋯⋯⋯⋯⋯⋯⋯⋯⋯ 70

4.6.5 Two-stage compressors ⋯⋯⋯⋯⋯⋯⋯⋯⋯⋯⋯⋯⋯⋯⋯⋯⋯⋯⋯⋯⋯⋯⋯ 73

4.6.6 Two-stage compressors with an intermediate pressure casing ⋯⋯⋯⋯⋯ 74

4.7 Lubrication and Oil Separation ⋯⋯⋯⋯⋯⋯⋯⋯⋯⋯⋯⋯⋯⋯⋯⋯⋯⋯⋯⋯ 74

4.7.1 Lubrication system for vertical rotary compressors ⋯⋯⋯⋯⋯⋯⋯⋯⋯ 74

4.7.2 Lubrication system for horizontal rotary compressors ⋯⋯⋯⋯⋯⋯⋯ 76

4.7.3 Oil separation in rotary compressors ⋯⋯⋯⋯⋯⋯⋯⋯⋯⋯⋯⋯⋯⋯⋯ 76

4.8 Noise and Vibration of Rotary Compressors ⋯⋯⋯⋯⋯⋯⋯⋯⋯⋯⋯⋯⋯ 77

4.8.1 Excitation forces and vibration reduction ⋯⋯⋯⋯⋯⋯⋯⋯⋯⋯⋯⋯⋯ 77

4.8.2 Vibration reduction through torque control ⋯⋯⋯⋯⋯⋯⋯⋯⋯⋯⋯⋯ 78

4.8.3 Noise of rotary compressors ⋯⋯⋯⋯⋯⋯⋯⋯⋯⋯⋯⋯⋯⋯⋯⋯⋯⋯⋯ 79

4.9 Production Technology ⋯⋯⋯⋯⋯⋯⋯⋯⋯⋯⋯⋯⋯⋯⋯⋯⋯⋯⋯⋯⋯⋯⋯ 81

4.9.1 High-precision machining ⋯⋯⋯⋯⋯⋯⋯⋯⋯⋯⋯⋯⋯⋯⋯⋯⋯⋯⋯⋯ 81

4.9.2 High-precision assembling ⋯⋯⋯⋯⋯⋯⋯⋯⋯⋯⋯⋯⋯⋯⋯⋯⋯⋯⋯⋯ 82

Chapter 5 Scroll Compressors ⋯⋯⋯⋯⋯⋯⋯⋯⋯⋯⋯⋯⋯⋯⋯⋯⋯⋯⋯⋯⋯⋯ 85

5.1 History of Scroll Compressors ⋯⋯⋯⋯⋯⋯⋯⋯⋯⋯⋯⋯⋯⋯⋯⋯⋯⋯⋯⋯ 85

5.2 Operating Principle and Characteristics ⋯⋯⋯⋯⋯⋯⋯⋯⋯⋯⋯⋯⋯⋯⋯⋯ 85

5.2.1 Operating principle ⋯⋯⋯⋯⋯⋯⋯⋯⋯⋯⋯⋯⋯⋯⋯⋯⋯⋯⋯⋯⋯⋯⋯ 85

5.2.2 Scroll wrap geometry ⋯⋯⋯⋯⋯⋯⋯⋯⋯⋯⋯⋯⋯⋯⋯⋯⋯⋯⋯⋯⋯⋯ 86

5.2.3 New scroll wrap ⋯⋯⋯⋯⋯⋯⋯⋯⋯⋯⋯⋯⋯⋯⋯⋯⋯⋯⋯⋯⋯⋯⋯⋯ 88

5.3 Structure and Design of Compressor Components ⋯⋯⋯⋯⋯⋯⋯⋯⋯⋯⋯ 89

5.3.1 Compression mechanism ⋯⋯⋯⋯⋯⋯⋯⋯⋯⋯⋯⋯⋯⋯⋯⋯⋯⋯⋯⋯ 89

5.3.2 Forces and moments working on the drive mechanism ⋯⋯⋯⋯⋯⋯⋯ 90

5.3.3 Sealing mechanism ⋯⋯⋯⋯⋯⋯⋯⋯⋯⋯⋯⋯⋯⋯⋯⋯⋯⋯⋯⋯⋯⋯ 91

5.3.4 Structures of scroll compressor ⋯⋯⋯⋯⋯⋯⋯⋯⋯⋯⋯⋯⋯⋯⋯⋯⋯ 92

5.4 Kinematic Analysis ⋯⋯⋯⋯⋯⋯⋯⋯⋯⋯⋯⋯⋯⋯⋯⋯⋯⋯⋯⋯⋯⋯⋯⋯ 93

5.4.1 Working forces and equations of motion ⋯⋯⋯⋯⋯⋯⋯⋯⋯⋯⋯⋯ 93

5.4.2 Equation of energy and mechanical efficiency ⋯⋯⋯⋯⋯⋯⋯⋯⋯⋯ 95

5.5 Other Scroll Compressor Mechanisms ⋯⋯⋯⋯⋯⋯⋯⋯⋯⋯⋯⋯⋯⋯⋯⋯ 96

5.5.1 Capacity control ⋯⋯⋯⋯⋯⋯⋯⋯⋯⋯⋯⋯⋯⋯⋯⋯⋯⋯⋯⋯⋯⋯⋯ 96

5.5.2 Injection ⋯⋯⋯⋯⋯⋯⋯⋯⋯⋯⋯⋯⋯⋯⋯⋯⋯⋯⋯⋯⋯⋯⋯⋯⋯⋯ 96

5.6 Uses and Features ⋯⋯⋯⋯⋯⋯⋯⋯⋯⋯⋯⋯⋯⋯⋯⋯⋯⋯⋯⋯⋯⋯⋯⋯⋯ 96

5.7 Production Technology ⋯⋯⋯⋯⋯⋯⋯⋯⋯⋯⋯⋯⋯⋯⋯⋯⋯⋯⋯⋯⋯⋯⋯ 98

5.7.1 Scroll machining ⋯⋯⋯⋯⋯⋯⋯⋯⋯⋯⋯⋯⋯⋯⋯⋯⋯⋯⋯⋯⋯⋯⋯ 98

5.7.2 Machining accuracy control ⋯⋯⋯⋯⋯⋯⋯⋯⋯⋯⋯⋯⋯⋯⋯⋯⋯⋯ 99

5.8 Selection Criteria of Scroll Compressor ⋯⋯⋯⋯⋯⋯⋯⋯⋯⋯⋯⋯⋯⋯⋯ 99

Chapter 6 Twin Screw Compressors ⋯⋯⋯⋯⋯⋯⋯⋯⋯⋯⋯⋯⋯⋯⋯⋯⋯⋯⋯ 101

6.1 History of Twin Screw Compressors ⋯⋯⋯⋯⋯⋯⋯⋯⋯⋯⋯⋯⋯⋯⋯⋯ 101

6.2 Working Principle and Basic Structure ⋯⋯⋯⋯⋯⋯⋯⋯⋯⋯⋯⋯⋯⋯⋯ 102

6.2.1 Working principle ⋯⋯⋯⋯⋯⋯⋯⋯⋯⋯⋯⋯⋯⋯⋯⋯⋯⋯⋯⋯⋯⋯ 102

6.2.2 Basic structure ⋯⋯⋯⋯⋯⋯⋯⋯⋯⋯⋯⋯⋯⋯⋯⋯⋯⋯⋯⋯⋯⋯⋯ 103

6.2.3 Oil-free type and oil-injected type ·· 104

6.2.4 Semi-hermetic type and open type ·· 105

6.3 Rotor Profiles ··· 106

6.3.1 Basics and history of rotor profiles ·· 106

6.3.2 Rotor profile modification ·· 107

6.3.3 Number of lobes ··· 108

6.3.4 Wrap angle ··· 109

6.3.5 L/D_m and center distance ·· 109

6.3.6 Sealing line and internal leakage passage ·· 110

6.3.7 Displacement volume ·· 111

6.4 Structure and Design of Compressor Components ·· 111

6.4.1 Screw rotors ·· 111

6.4.2 Rotor casing ··· 112

6.4.3 Inlet housing ·· 112

6.4.4 Outlet housing ·· 112

6.4.5 Suction port ··· 112

6.4.6 Discharge port ·· 112

6.4.7 Other ports ·· 113

6.4.8 Bearings ·· 113

6.4.9 Shaft seals ··· 115

6.4.10 Oil supply system ·· 117

6.4.11 Capacity control mechanism ·· 119

6.4.12 Balance piston ·· 122

6.5 Dynamic-mechanical Analysis ·· 122

6.5.1 Motion of the mechanism, constraint forces and motion equations ························· 122

6.5.2 Equation of energy and mechanical efficiency ··· 124

6.5.3 Mechanical excitation force and vibration of compressor body ······························· 124

6.5.4 Lateral and torsional critical speeds ·· 125

6.6 Operating Characteristics and Applications ·· 126

6.6.1 Performance characteristics ·· 126

6.6.2 Capacity and pressure range ··· 130

6.6.3 Single-stage compression and multiple-stage compression ······································· 131

6.6.4 Main applications ·· 131

6.7 Noise and Vibration ··· 133

6.7.1 Vibration characteristics ··· 133

6.7.2 Noise characteristics ··· 135

6.8 Production Technology ·· 136

6.8.1 Screw rotor machining ·· 136

6.8.2 Machining accuracy control ··· 139

Chapter 7 Single Screw Compressors ··· 142

7.1 History of Single Screw Compressors ··· 142

7.2 Basic Mechanism and Operating Principle ·· 142

7.2.1 Compression mechanism ·· 142

7.2.2 Operating principle ·· 143

7.2.3 Basic structure of single screw compressors ·················143

7.2.4 Semi-hermetic type and open type ·················145

7.3 Structure and Design of Compressor Components ·················145

7.3.1 Shaft ·················145

7.3.2 Screw rotor ·················145

7.3.3 Gate rotor ·················145

7.3.4 Discharge port ·················146

7.3.5 Bearings ·················146

7.3.6 Shaft sealing feature ·················146

7.3.7 Casing ·················146

7.3.8 Capacity control mechanism ·················146

7.3.9 Oil separator ·················148

7.4 Rotor Tooth Profile ·················148

7.4.1 Rotor tooth profile ·················148

7.4.2 Number of rotor teeth ·················148

7.4.3 Displacement volume ·················149

7.4.4 New gate rotor tooth profile ·················149

7.5 Performance and Noise ·················151

7.5.1 Efficiency characteristics ·················151

7.5.2 Noise characteristics ·················151

7.6 Other Mechanisms with Single Screw Compressors ·················152

7.6.1 Cooling system by oil injection ·················152

7.6.2 Cooling system by liquid refrigerant injection ·················152

7.6.3 Economizer cycle ·················152

Chapter 8 Automotive Air Conditioning Compressors ·················154

8.1 History of Automotive Air Conditioning Compressors ·················154

8.2 Automotive Air Conditioning Systems ·················155

8.2.1 Characteristics of automotive air conditioning systems ·················155

8.2.2 Types of automotive air conditioning compressors ·················157

8.2.3 Power transmission mechanism ·················157

8.3 Axial Compressors ·················158

8.3.1 Wobble plate compressors ·················158

8.3.2 Swash plate compressors ·················162

8.3.3 Variable capacity compressors ·················165

8.3.4 Noise and vibration of axial compressors ·················174

8.4 Scroll Compressors ·················175

8.4.1 History of scroll compressors ·················175

8.4.2 Structure and characteristics of scroll compressors for automotive air conditioners ···········176

8.4.3 Thrust bearing structure ·················178

8.4.4 Capacity control mechanism ·················179

8.4.5 Torque fluctuation and belt life ·················179

8.5 Rotary Vane Compressors ·················180

8.5.1 Operating principle ·················180

8.5.2 Displacement volume ·················181

8.5.3 Basic structure ···182

8.5.4 Dynamic-mechanical analysis ···184

Chapter 9　Refrigeration Oil ···190

9.1 Types of Refrigeration Oil ···190

9.1.1 MO refrigeration oil ···190

9.1.2 AB refrigeration oil ··191

9.1.3 PAG refrigeration oil ···191

9.1.4 PVE refrigeration oil ···191

9.1.5 POE refrigeration oil ···192

9.2 Interaction with Refrigerant ··192

9.2.1 Miscibility ··193

9.2.2 Solution properties ··194

9.2.3 Electrical insulation ···195

9.2.4 Lubricity ··197

9.2.5 Stability ··199

9.2.6 Compatibility with organic materials ···201

9.3 Refrigeration Oil Selection Method and Considerations for Use ·······················201

9.3.1 Refrigeration oil selection method ··201

9.3.2 Considerations for use of refrigeration oil ···202

Chapter 10　Motors and Inverters ···205

10.1 Motor Structure and Performance ···205

10.1.1 Motor classifications by drive power waveform ···206

10.1.2 Motor structure and components ···208

10.1.3 Motor component fabrication ···210

10.1.4 Alternative stator fabrication method ··213

10.2 Motor Design ···213

10.2.1 Operating principle of induction motors ···213

10.2.2 Operating principle of permanent-magnet synchronous motors ··················214

10.2.3 Motor design specifications (specification requirements, output-torque relationship) ······215

10.3 Motor Materials and Evaluation Methods ···217

10.3.1 Magnet wires ··218

10.3.2 Varnish ··219

10.3.3 Insulating papers (films) ··221

10.3.4 Electrical steel sheets ···222

10.3.5 Permanent magnets ··224

10.4 Inverter Structure and Control ··226

10.4.1 Inverter classifications and characteristics ···226

10.4.2 Inverter structure and components ··227

10.4.3 Principle of voltage-based inverters ···231

10.4.4 Principle of motor control ···232

10.5 Electrical Characteristic Measurements (with Inverter Control) ·······················235

10.5.1 Types of loss and measurement methods ···235

10.5.2 Types of noises and measurement methods ···236

10.6 Standards and Regulations ··237
    10.6.1 Power supply voltages, standards and regulations in major countries ························237

Chapter 11　Testing ···241
  11.1 Performance Tests ···241
    11.1.1 Main measurement items ···241
    11.1.2 Test methods ···242
  11.2 Noise Tests ···244
    11.2.1 Measurement conditions and equipment ···245
    11.2.2 Measurement methods ··246
  11.3 Vibration Tests ···246
    11.3.1 Measurement conditions and equipment ···246
    11.3.2 Measurement methods ··247
  11.4 Reliability Tests ··248
    11.4.1 Types of reliability tests ···248
    11.4.2 Reliability test implementation ···249
    11.4.3 Reliability data analysis ··249
    11.4.4 Reliability evaluation ···249
  11.5 Pressure Tests ···249
    11.5.1 Pressure resistance test ···249
    11.5.2 Strength test ··250
    11.5.3 Airtightness test ···250
  11.6 Other Tests ···250

Chapter 12　Measurement Technologies ···251
  12.1 Measurements of Basic Physical Quantities ··251
    12.1.1 Temperature measurement ···251
    12.1.2 Thermocouples and resistance thermometer sensors ·······································252
    12.1.3 Pressure measurement ··253
    12.1.4 Flow rate measurement ···255
  12.2 Other Measurement Technologies and Applications ···257

Chapter 13　About This Book ···262

Index ···265

Chapter 1 Introduction

1.1 Refrigeration Cycle

1.1.1 What is refrigeration?

"Refrigeration" does not only mean cooling and freezing a material. In science and technology, refrigeration generally refers to a wider variety of processes where a material is cooled below the range of temperatures that spontaneously occur in naturally existing materials, such as air or water ("environmental temperature"). As heat always moves from a hotter material to a cooler one, cooling a material below the environmental temperature and maintaining it at such level requires a man-made equipment and an amount of energy. On the other hand, cooling a heated material down to the environmental temperature does not require any special equipment or energy, and therefore such cooling process is clearly excluded from the definition of refrigeration[1]. The amount of heat removed from a material in order to cool it has to be dumped into something else (for example, released into the air or taken away by coolant water), which causes heating of the heat-dumped object. In that sense, refrigeration and heating are both ends of the same heat movement process, and whether to call it "refrigeration" or "heating" depends on for what purpose the heat movement is initiated[2].

Refrigeration is implemented in various aspects of our lives, such as the air conditioning of houses, commercial buildings and vehicles, the dehumidification and drying, the processing, distribution and storage of food, as well as various applications in industrial and medical fields. Refrigeration is a truly essential technology to our life.

1.1.2 What is a cycle?

The state of a material changes depending on its temperature and pressure. The working medium that is used in a process of refrigeration, or the "refrigerant" (refer to Section 2.5), also changes its state depending on its temperature and pressure. A quantity that helps to express the current state of a material is called "quantity of state". Examples of the quantity of state are pressure, temperature, density, internal energy, entropy and enthalpy. The quantity of state of an object will change when heat is moved to or from it in order for refrigeration to take place. Quantity of state is determined solely by the current state of the object, independently of the path by which the object has reached that state. On the other hand, work and heat are not quantity of state as they are dependent on the path of the change. Quantities of state that have changed as a result of heat movement can be brought back to the original quantities by way of other processes. Such sequence of processes, where the starting and the ending quantities of state are identical, is called a "cycle" and the cycle can be repeated continuously.

1.1.3 Vapor compression refrigeration cycle

Figure 1.1 shows a typical vapor compression refrigeration cycle. A simple vapor compression refrigeration cycle consists of a compressor, a condenser, an expansion valve and an evaporator. Liquid refrigerant at the exit of the condenser is depressurized as it passes through the expansion valve and becomes partially vaporized. As a result of this, an amount of heat, or "heat of vaporization", moves away from the refrigerant. Such movement of heat will cause the refrigerant to be cooled to its saturation temperature corresponding the pressure at the inlet of the evaporator. At the evaporator, the refrigerant becomes fully vaporized as its liquid portion absorbs heat from the outside. The fully vaporized refrigerant is then sucked into the compressor and compressed to become a high-pressure and high-temperature gas. The compressed gas gives off its heat as it passes through the condenser and changes back into liquid, before it enters the expansion valve. Thus the refrigerant

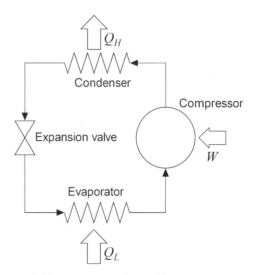

Fig. 1.1 Vapor compression refrigeration cycle

can be utilized repeatedly in a refrigeration cycle, to cool a space or an object in a continuous manner. Heat dissipated from the refrigerant at the condenser can be used to warm a room or heat an object. The refrigeration cycle, which absorbs low-temperature heat through the evaporator and gives off high-temperature heat through the condenser, is often referred to as "heat pump" in that it pumps up heat from a lower level to a higher level. The vapor-compression refrigeration cycle has a particularly good efficiency as it utilizes latent heat accompanying the phase change of refrigerant. Because of this, it is widely used in refrigerators and air conditioners.

1.2 Role of Compressors

A refrigerant has a specific relationship between the temperature and the pressure at which it vaporizes or condenses. Such mutually corresponding temperature and pressure are relatively referred to as "saturation temperature" and "saturation pressure". At a given pressure, the liquid refrigerant vaporizes when it receives heat from a heat source whose temperature is higher than its saturation temperature corresponding to the given pressure. On the other hand, the vapor refrigerant condenses, or changes back into liquid, when it gives off heat to a heat sink whose temperature is lower than its saturation temperature corresponding to that pressure. To change the refrigerant vapor back into liquid in the refrigeration cycle shown in Fig. 1.1, it is necessary to cool the refrigerant by using a heat sink having a temperature lower than its saturation temperature corresponding the condenser's pressure. In general, the saturation temperature of a refrigerant increases with pressure. When a refrigerant is pressurized enough so that its saturation temperature rises above that of the air or water surrounding it, the refrigerant can be condensed by being cooled by the air or water. For example, when isobutane, a refrigerant commonly used in household refrigerators, is pressurized to 0.53 MPa abs., its saturation temperature becomes 40°C, and it condenses as it gives off heat to the room air. As described above, in order to raise the saturation temperature of a refrigerant above that of the available heat sink so that heat radiation and condensation occurs, it is necessary to increase the refrigerant pressure. This is the role of the compressor in a refrigeration cycle - to increase the temperature and pressure of the refrigerant. As the refrigerant is sucked into the compressor immediately after it has vaporized from liquid inside the evaporator, pressure inside the evaporator is maintained at the saturation pressure corresponding to the temperature of the vaporizing refrigerant. As the pressure inside the

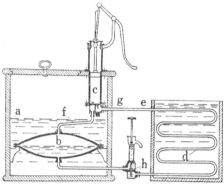

a: Container   d: Condenser
b: Evaporator   f: Suction line
c: Pump   h: Expansion valve

Fig. 1.2 Perkin's ether refrigerator[3]

evaporator is controlled by the amount of refrigerant entering through the expansion valve, the amount of refrigerant evaporating by absorbing heat (refrigeration load) from the outside and the amount of refrigerant sucked into the compressor, the compressor also helps to maintain the pressure inside the evaporator (and consequently the evaporating temperature) at the required level.

1.3 History of Compressors (Early Period)*

In 1834, Jacob Perkins of the United Kingdom acquired a patent for a vapor-compression refrigeration system for the first time in history. His invention, shown in Fig. 1.2, was a cycle using diethyl ether as refrigerant[3]. Lowe of the United States built the world's first carbon dioxide gas compressor in 1866, and then in 1886 Windhausen of Germany succeeded to put the technology into practical use[4, 5]. Meanwhile, David Boyle of the United States invented an ammonia compressor in 1873, and Raoul Pictet of Switzland successfully fabricated a sulfer dioxide gas compressor in 1874. Figure 1.3 shows a compressor production factory around 1876[6]. Compared with humans depicted in the picture, it can be seen that compressors at that time were very large in size. Figure 1.4 shows a vertical multiple-cylinder reciprocating ammonia compressor in 1870s. It was driven by a large steam engine with a crankshaft in common. The operation speed was in the range of 120 to 180 rpm. These compressors used a crosshead, and the piston rods were fixed to their respective pistons.

Around 1900, belt-driven compressors powered by electric motors emerged. Some compressors were capacity controlled by way of a manual bypass valve that con-

---

* Depending on the source literatures used, date of the events may be different in a range of few years.

nected the discharge and the suction pipes as shown in Fig. 1.5, according to the refrigeration load at the evaporator[6]. Compressors were made progressively smaller from 1900 through to 1925, while the operation speed was increased to approximately 300 to 600 rpm. The refrigerant seal portion was moved from the piston rod to the rotary shaft, and electrical motor driving became more common. Figure 1.6 shows a compressor production factory in 1920s[6]. The types of refrigerant commonly used at the time were ammonia, sulfur dioxide, and carbon dioxide, with isobutane and propane being used in limited applications. Most of these were gradually excluded from use due to hazardous risks such as excessively high pressure, toxicity or flammability. By around 1920, ammonia had become the dominant refrigerant for large refrigeration systems[4]. Then in 1930, fluorine-based synthetic refrigerants were developed and put in use[4].

As being described in the next section, a variety of compressor mechanisms besides reciprocating compressors have been developed and implemented; rotary vane compressors were put into use in 1920s, screw compressors in 1930s, rolling piston compressors in 1960s, and scroll compressors in 1980s. The history of each type of compressor is described in detail in the respective section of this textbook.

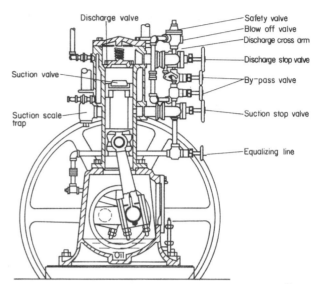

Fig. 1.5 Compressor with manual cylinder bypass[6]

1.4 Classification of Compressors

Compressors that are used for vapor-compression refrigeration systems can be classified, according to their compression mechanism, into two types; positive displacement type and kinetic type. In a positive displacement type compressor, refrigerant gas is sucked into an enclosed space and volume of the space is geometrically reduced so that the refrigerant pressure increases. The kinetic type compressor is also called a turbo compressor, and the centrifugal compressor among the turbo compressor is typically used for refrigeration system. In a centrifugal compressor, an impeller raises the refrigerant gas pressure by centrifugal force and at the same time accelerates the gas by rotation. This kinetic energy is converted into pressure as the refrigerant gas passes through the diffuser located at the exit of the impeller.

The most common types of positive-displacement compressors are reciprocating and rotary compressors, which

Fig. 1.3 1865 compressor manufacturing shop[6]

Fig. 1.4 Vertical compressor driven by compound Corliss engine[6]

Fig. 1.6 Compressor shop of the late 1920s[6]

are classified as follows:

(1) Reciprocating type:

    Crank-piston compressor

    Scotch yoke-piston compressor

    Swash plate-piston compressor

    Free-piston compressor (linear mechanism)

(2) Rotary type:

    Rotary vane (sliding vane) compressor

    Rolling piston compressor

    Scroll compressor

    Screw compressor

Commonly used types of compressors and their applications are listed in Table 1.1 [7].

Compressors can also be classified, according to how the casing that houses the compression mechanism is sealed, into the open type, the hermetic type and the semi-hermetic type. In an open type compressor, the driving shaft penetrates the casing and is externally driven by an outer motor or engine. A mechanical seal or lip seal is used to seal the section where the driving shaft penetrates the casing. Open type compressors are typically used where an external power source is available, as in the case of compressors for automotive air conditioner, or where copper wire for electric motor winding is at a risk of chemical attack by ammonia as refrigerant. Open type compressors are also widely used for large industrial refrigeration systems. Both hermetic and semi-hermetic compressors have an electrical motor enclosed in the compressor casing. Hermetic compressors have the casing completely closed off and sealed by welding, while semi-enclosed compressors have a section of the casing secured by mechanical fasteners like bolts, so that it can be opened by undoing the fasteners. Almost all household refrigerators and domestic air conditioners use hermetic compressors.

In the following chapters, this textbook focuses on the positive-displacement type compressors.

References

1) S. Tezuka: "Refrigerating Machine", Kyoritsu Shuppan Co., Ltd., Tokyo, p. 1 (1983). (in Japanese)

2) Roy J. Dossat: "PRINCIPLE OF REFRIGERATION", John Wiley & Sons, Inc., USA, p. 97 (1981).

3) J. Nagaoka: "Refrigerating Engineering", Corona Publishing Co., Ltd., Tokyo, p. 2 (1976). (in Japanese)

4) Japan Society of Refrigerating and Air Conditioning Engineers: "History of Refrigeration in Japan", Japan Society of Refrigerating and Air Conditioning Engineers, Tokyo (1998). (in Japanese)

5) K. Tojo: "History of A/C and Refrigerating Compressors", Proc. of the 2005 JSRAE Annual Conf., C318-1-4, Tokyo (2005). (in Japanese)

6) A. B. Newton: "The refrigeration compressor - the steps to maturity", Int. J. of Refrigeration, 4 (5), pp. 246-254 (1981).

7) Japan Society of Refrigerating and Air Conditioning Engineers: "Refrigerating and Air Conditioning Technologies", Japan Society of Refrigerating and Air Conditioning Engineers, Tokyo, p. 32 (Partially modified) (2008). (in Japanese)

Chapter 1 Introduction

Table 1.1 Classification of compressor[7]

Classification			Shape	Sealing structure	Main applications (Past applications)	Driving capacity range [kW]	Feature
Positive displacement type	Reciprocating type	Piston-crank		Open	Refrigeration, Heat pump, Automotive air conditioner	0.4 ~ 120	Easy to use Variety of models Low cost Unsuitable for large capacity
				Semi-hermetic	Refrigeration, Air conditioner, Heat pump	0.75 ~ 45	
				Hermetic	Refrigerator, Air conditioner	0.1 ~ 15	
		Piston-swash plate		Open	Automotive air conditioner	0.75 ~ 5	Only for automotive air conditioner Capacity control
	Rotary type	Rolling piston		Hermetic	Air conditioner, Small freezer, Showcase, Water heater, (Refrigerator)	0.1 ~ 5.5	Small capacity Rotational speed is increasing.
		Rotary vane		Open	Automotive air conditioner	0.75 ~ 2.2	Small size
				Hermetic	(Refrigerator), (Air conditioner)	0.6 ~ 5.5	
	Scroll type			Open	Automotive air conditioner	0.75 ~ 2.2	
				Semi-hermetic	Automotive air conditioner for electric vehicle, Refrigeration		
				Hermetic	Air conditioner, Refrigeration, Water heater	0.75 ~ 30	To be small capacity Rotational speed is increasing.
	Screw type	Twin rotor		Open	(Bus air conditioner)	Around 6	This type is often used for refrigeration and heat pump because it is suitable for high pressure ratio than centrifugal type. Being hermetic is accelerated for small capacity one.
					Refrigeration, Air conditioner, Automotive air conditioner, Heat pump	20 ~ 1800	
				Semi-hermetic	Refrigeration, Air conditioner, Heat pump	30 ~ 300	
		Single rotor		Open	Refrigeration, Air conditioner, Heat pump	100 ~ 1100	
				Semi-hermetic	Refrigeration, Air conditioner, Heat pump	22 ~ 90	
Kinetic type	Centrifugal type (Turbo type)			Open	Refrigeration, Air conditioner	90 ~ 10000	Suitable for large capacity Unsuitable for high pressure ratio
				Hermetic			

5

Chapter 2 Basic Theory

2.1 Compression Theory

2.1.1 Reversed Carnot cycle

The Carnot cycle is an ideal heat engine that operates between two different temperature levels, in the sequence of isothermal expansion → adiabatic expansion → isothermal compression → adiabatic compression. Figure 2.1 shows the *T-s* diagram of the Carnot cycle. In the Carnot cycle, the working fluid isothermally expands from point 3 to point 2 at temperature T_H while receiving heat q_H from a high-temperature heat source. Then the fluid adiabatically expands from point 2 to point 1 to reach temperature T_L, after which it is isothermally compressed from point 1 to point 4 while rejecting heat q_L to a low-temperature heat sink. Finally, the fluid is adiabatically compressed from point 4 to point 3 to return to the original state. The entire cycle is a reversible process and is the most effective heat engine that operates between two different temperature levels. As described above, the Carnot cycle can also be run in the reversed sequence of adiabatic compression → isothermal compression → adiabatic expansion → isothermal expansion. In the reversed cycle, shown in the *T-s* diagram of Fig. 2.2, the working fluid receives heat while isothermally expanding from point 4 to point 1 on the low-temperature side, and it rejects heat while being isothermally compressed from point 2 to point 3 on the high-temperature side, thus generating refrigeration effect. This is called the reversed Carnot cycle, or the Carnot refrigeration cycle. Heat q_L that is received per unit mass of the working fluid from the low-temperature heat source can be determined by integrating $\delta q = Tds$ and is expressed as area [1-1'-4'-4], while the amount of work w required to drive the cycle can be expressed as area [1-2-3-4]. Hence the ratio of refrigeration capacity to the power given is:

$$\varepsilon_L = \frac{q_L}{w} = \frac{T_L}{T_H - T_L} \qquad (2.1)$$

This is called the coefficient of performance, or COP. On the other hand, heat q_H rejected per unit mass of the working fluid to the high-temperature heat sink can be expressed as area [2-1'-4'-3], so that the coefficient of performance for heating, such as when warming an indoor space, is:

$$\varepsilon_H = \frac{q_H}{w} = \frac{T_H}{T_H - T_L} \qquad (2.2)$$

which gives $\varepsilon_H = \varepsilon_L + 1$.

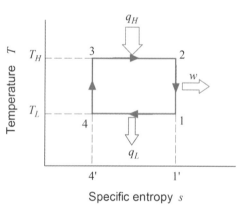

Fig. 2.1 Carnot cycle

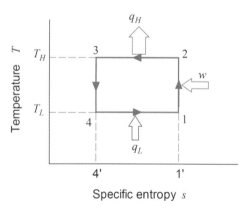

Fig. 2.2 Reversed Carnot cycle

2.1.2 *P-h* diagram

The vapor compression refrigeration cycle can be considered as a result of efforts to create an actual cycle that is as close to the reversed Carnot cycle as possible. In the actual vapor compression refrigeration cycle, the isothermal compression in the reversed Carnot cycle is replaced by isobaric process through the condenser, the isothermal expansion by isobaric process through the evaporator, and the adiabatic expansion by isenthalpic process through the throttle valve, respectively. As the amount of energy conversion that occurs at each component device of a refrigeration cycle can be expressed as enthalpy variation, a *P-h* diagram, where the vertical axis represents absolute pressure[*] P and the horizontal axis specific enthalpy[**] h, is typically used for describing a refrigeration cycle. Figure 2.3 shows an example of *P-h* diagram. The diagram includes horizontal isobaric lines (a), vertical iso-specific enthalpy lines (b), saturated liquid curve (c), saturated vapor curve (d), iso-quality curves (e), isothermal lines (f), iso-specific entropy lines (g), iso-specific volume lines (h), and the critical point (i). The state and properties of refrigerant at any pressure and temperature can be read on the diagram. The diagram is also convenient for getting the variation of energy corresponding to the change of refrigerant condition in the form of enthalpy difference.

(1) Refrigeration cycle on the *P-h* diagram

Figure 2.4 shows the *P-h* diagram of a refrigeration cycle. Each process of the cycle represented in the diagram is described below.

Compression process (1 → 2)

The compression of a refrigerant compressor can be modeled as an adiabatic compression process. As the specific entropy remains constant in the adiabatic compression process, the process is expressed by an iso-specific entropy line on the *P-h* diagram. The state of refrigerant at point 2, which has just come out of the compressor, is superheated gas. The amount of specific enthalpy difference before and after compression is the amount of adiabatic compression work (refer to Section 2.1.3).

Process through the condenser (2 → 2' → 3' → 3)

The change of refrigerant state through the condenser is a constant-pressure heat radiation process, which is represented by a horizontal isobaric line in the *P-h* diagram. Refrigerant after coming out of the compressor as superheated gas (point 2) is cooled by heat radiation to become saturated gas at point 2'. Then, further heat radiation starts to cause the refrigerant to condense so that quality of refrigerant decreases. As a result, saturated liquid is obtained at point 3'. This condensation process occurs at a constant temperature for a pure refrigerant[***]. Heat associated with such constant-temperature phase transition is called "latent heat". When saturated liquid is further cooled, the refrigerant temperature starts to decrease again so that subcooled liquid is obtained at point 3. When movement of heat associates with a temperature change in a medium, as in the case of cooling from point 2 to point 2' or that from point 3' to point 3, the heat is called "sensible heat". For the specific enthalpy change from point 2 to point 3 through the condenser, heat q_H is radiated per unit mass of refrigerant.

Expansion process (3 → 4)

The expansion process that occurs as the refrigerant goes through a throttle valve (expansion valve) is an iso-specific enthalpy process since the process does not involve any exchange of energy with the outside. The expansion valve serves to adjust the refrigerant flow rate through the evaporator so that the refrigerant pressure decreases to a level corresponding to the evaporating temperature. With vaporizing a part of the refrigerant, the resulting movement of heat away from the refrigerant decreases its temperature to the saturation level corresponding to its evaporating pressure. As a result, the refrigerant becomes a two-phase state at the exit of the expansion valve (before entering the evaporator). The expansion through the expansion valve causes specific entropy of refrigerant to increase and the process becomes irreversible one, which makes the vapor-compression refrigeration cycle an irreversible cycle. The above-described entropy increase can be considered as a result that the kinetic energy of the refrigerant, which has been ac-

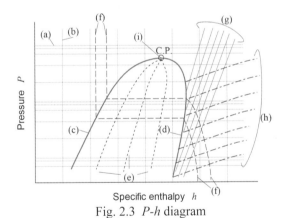

Fig. 2.3 *P-h* diagram

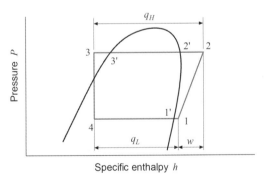

Fig. 2.4 Refrigeration cycle on *P-h* diagram

---

[*]Pressure that is zero-referenced against a perfect vacuum. Except where otherwise specified, all the pressure values cited in this textbook are absolute pressures. On the other hand, pressure that is zero-referenced against ambient air pressure is called "gauge pressure", which is the value indicated by most commonly available pressure gauges.
[**]The adjective "specific" indicates that it is a per-unit-mass value.

[***]HFC410A, one of the most commonly used refrigerants for air-conditioning systems, is a mixed refrigerant that has a saturated temperature difference of approximately 0.1°C between the gas state and the liquid state. Refrigerants with this type of characteristics are called "quasi-azeotropic mixed refrigerants" (refer to Section 2.5.3).

celerated by pressure difference across the expansion valve, is dissipated and lost due to turbulence. This type of loss is called "throttling loss". In such a refrigeration cycle as a carbon dioxide cycle that involves a significant throttling loss at the expansion valve, use of an expander in place of the expansion valve can significantly improve the theoretical efficiency of the cycle[1].

Process through the evaporator (4 → 1' → 1)

The change of refrigerant state through the evaporator is, like the one through the condenser, a constant-pressure transition, which is represented by a horizontal isobaric line on the P-h diagram. The two-phase refrigerant (point 4) that comes out of the expansion valve and enters the evaporator absorbs heat from the outside, so that the liquid phase of refrigerant evaporates and quality of refrigerant increases. The refrigerant becomes saturated gas at point 1'. Then it is further heated to become superheated gas at point 1, where it is sucked into the compressor. For the specific enthalpy change from point 4 to point 1 through the evaporator, heat q_L is received per unit mass of refrigerant. This heat absorbing effect per unit mass of refrigerant is the refrigerating effect.

Figure 2.5 shows a refrigeration cycle represented by a T-s diagram. The adiabatic compression from point 1 to point 2 is shown as an iso-specific entropy line parallel to the vertical axis of the diagram, while transition through the condenser from point 2 to point 3 is along an iso-pressure line. The diagram also indicates temperature changes in the superheated and subcooled regions. As the expansion process, from point 3 to point 4, is an iso-enthalpy transition and is irreversible, specific entropy increases during this process. Transition through the evaporator, from point 4 to point 1, is also along an iso-pressure line.

On the T-s diagram, the amount of energy conversion through each component of the cycle is represented as an area. For example, the heat quantity absorbed through the evaporator is represented as area [4-1'-1-1s-4s] and the compression work through the compressor as area [1-2-2'-3'-3-3s-4s-4-1']. The heat quantity rejected through the condenser, or area [1s-2-2'-3'-3-3s], is equal to the sum of the above two. T-s diagrams are commonly used for representing temperature changes through heat exchangers or to discuss performance of different refrigeration cycles by comparing their graph shapes, while they are not very convenient for calculating graph areas. Due to this drawback, P-h diagrams are used more commonly for capacity and power calculations of refrigeration cycles.

(2) COP and the effects of evaporating and condensing temperatures

As will be explained in the next Section 2.1.3, the adiabatic compression work per unit mass of refrigerant through the compressor can be expressed as specific enthalpy variation before and after compression, (h_2-h_1). The heat quantity absorbed through the evaporator and the one rejected through the condenser are expressed, per unit mass of refrigerant, as (h_1-h_4) and (h_2-h_3) respectively. As mentioned above, one of the advantages of using the P-h diagram is that the amount of energy conversion involving the respective process of a refrigeration cycle can be expressed in a simple and easy-to-understand manner. The coefficient of performance (hereinafter "COP"), which is the ratio of absorbed heat q_L to compression work w in a refrigeration system (COP ε_L) or that of rejected heat q_H to compression work w in a heating system (COP ε_H), can be expressed by the following equations:

$$\varepsilon_L = \frac{q_L}{w} = \frac{h_1 - h_4}{h_2 - h_1} \quad (2.3)$$

$$\varepsilon_H = \frac{q_H}{w} = \frac{h_2 - h_3}{h_2 - h_1} = \frac{w + q_L}{w} = 1 + \varepsilon_L \quad (2.4)$$

Also, compression power L and heat transfer rate Q corresponding to mass flow rate G of compressor can be obtained by multiplying the respective specific enthalpy differences by the mass flow rate.

Figure 2.6 is a P-h diagram representing the state of a refrigeration cycle under different evaporating temperature. As shown in the figure, when the evaporating temperature is decreased from T_L to T_L', the amount of compression work per unit mass of refrigerant increases and the refrigeration effect decreases, resulting in a significant reduction in the COP obtained by Eq. (2.3). With the reduction of the evaporating pressure, the specific volume of the refrigerant sucked into the compressor increases (and the density

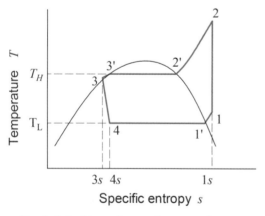

Fig. 2.5 Refrigeration cycle on T-s diagram

decreases) so that a smaller mass of refrigerant is sucked into and compressed by the compressor, resulting in much larger decrease of refrigeration effect of the cycle. Decrease of the evaporating pressure also causes the compression ratio to increase, which results in a significant increase in the discharge temperature in the case of refrigerant having a higher ratio of specific heats. In addition, with the reduction of the evaporating pressure, the compression work per unit mass of refrigerant increases and the mass flow rate decreases. As a result, as the evaporating pressure decreases, the compression power shows a maximum value and then starts to decrease. Evaporating pressure decrease does not result in excessive increase in the compressor load unlike the case of condensing pressure increase as described below.

Figure 2.7 is a p-h diagram representing the state of a refrigeration cycle under different condensing pressure. As shown in the figure, when the condensing temperature increases from T_H to T_H', the amount of compression work per unit mass of refrigerant increases and the refrigeration effect decreases, resulting in a significant reduction in the COP obtained by Eq. (2.3). Note that with the increase of the condensing pressure, the refrigerant mass flow rate does not change and therefore the compression power increase definitely.

2.1.3 Adiabatic compression
(1) Absolute work and industrial work

Figure 2.8 shows the P-v (pressure-specific volume) diagram of a closed system (for example, a system in which fluid is confined in a space totally enclosed by a cylinder and a piston) where fluid is compressed from point 1 to point 2. The amount of work required to compress a unit mass fluid from state 1 to state 2 is expressed as area [1-2-2'-1'], or:

$$w_a = -\int_1^2 P dv \qquad (2.5)$$

As in the above case, the amount of work that is done to the outside by a system or that done to a system from the outside is called "absolute work".

On the other hand, the P-v diagram of an open system, for example where the compressor runs a cycle of suction → compression → discharge, is shown in Fig. 2.9. In this case, in addition to the absolute compression work expressed by Eq. (2.5), the system requires to have the absolute discharge work, expressed as area [2-3-0-2'], done to itself by the outside. The system also does the absolute suction work to the outside, expressed as area [1-4-0-1']. As in this example, the amount of work that is done to an open system by the outside is called "industrial work". In

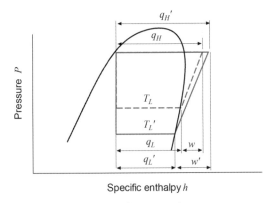

Fig. 2.6 Influence of evaporating temperature

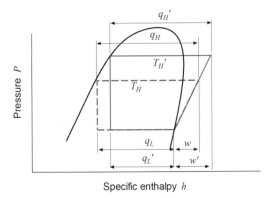

Fig. 2.7 Influence of condensing temperature

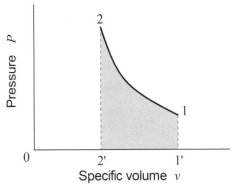

Fig. 2.8 Absolute work

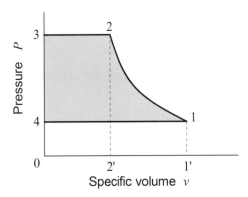

Fig. 2.9 Industrial work

Compressors for Air Conditioning and Refrigeration

the case of a compressor, it is the amount of per-unit-mass work required to achieve the compressor function, which can be expressed as:

$$w_t = -\int_1^2 P dv + P_2 v_2 - P_1 v_1 = \int_1^2 v dP \qquad (2.6)$$

(2) First law of thermodynamics

The first law of thermodynamics is a law of conservation of energy. In a closed system, the first law of thermodynamics relative to an infinitesimal change per unit mass can be expressed as:

$$du = \delta q - P dv \qquad (2.7)$$

where u is specific internal energy. For a constant-volume transition, the above expression is replaced by:

$$du = \delta q \qquad (2.8)$$

Then, the amount of change in the specific internal energy can be expressed by the following equation, with the constant-volume specific heat as c_v:

$$du = c_v dT \qquad (2.9)$$

In the case of an open system, it is necessary to take into account the amount of energy that is brought into or taken away from the system by boundary-crossing fluid. On the condition of a steady flow where kinetic and potential energies can be ignored, use of the specific enthalpy that represents the energy of incoming fluid flow[*] leads to:

$$h = u + Pv \qquad (2.10)$$

$$dh = du + P dv + v dP \qquad (2.11)$$

In a steady-flow open system, the first law of thermodynamics can be expressed as:

$$dh = \delta q + v dP \qquad (2.12)$$

For a constant-pressure process, the above equation is replaced by:

$$dh = \delta q \qquad (2.13)$$

Then, the amount of change in specific enthalpy can be

---

*When fluid enters an open system crossing the system boundary, pressure at the system entrance must be overcome by upstream flow pressure so as to squeeze the fluid in through the entrance. The amount of this work per unit mass is Pv.

---

expressed by the following equation, with the constant-pressure specific heat as c_p:

$$dh = c_p dT \qquad (2.14)$$

Combination of Eqs. (2.7), (2.9), (2.12) and (2.14) gives:

$$\delta q = c_v dT + P dv = c_p dT - v dP \qquad (2.15)$$

Also, based on the following equation of state of an ideal gas:

$$Pv = RT \qquad (2.16)$$

where R is the gas constant. The derivative of Eq. (2.16) is:

$$d(Pv) = P dv + v dP = R dT \qquad (2.17)$$

Combination of Eqs. (2.15) and (2.17) gives:

$$c_p - c_v = R \qquad (2.18)$$

Furthermore, with the ratio of specific heats

$$\kappa = c_p / c_v \qquad (2.19)$$

and Eq. (2.18), the followings are obtained:

$$c_v = \frac{R}{\kappa - 1} \qquad (2.20)$$

$$c_p = \frac{\kappa R}{\kappa - 1} \qquad (2.21)$$

(3) Adiabatic compression work

Compression inside the compressor, which occurs instantaneously, can be considered as an adiabatic process. When Eq. (2.12) is integrated with $\delta q = 0$, the following is obtained:

$$h_2 - h_1 = \int_1^2 v dP \qquad (2.22)$$

Equation (2.22) means that the amount of industrial work expressed by Eq. (2.6) is equal to the specific enthalpy difference before and after compression. The amount of adiabatic compression work needed by the compressor per unit mass of refrigerant can be determined as the amount of change in specific enthalpy, which can be easily read along the iso-specific entropy lines on the P-h diagram, but this is

Chapter 2 Basic Theory

on the assumption that the compression process occurs adiabatically. When heat radiation from the compressor causes decrease in discharge temperature, it is not appropriate to determine the amount of compression work as the specific enthalpy difference before and after compression.

By substituting Eq. (2.17) into Eq. (2.15) with $\delta q=0$, the following is obtained:

$$c_p\left(Pdv+vdP\right)-RvdP=0 \tag{2.23}$$

Furthermore, by substituting Eqs. (2.18) and (2.19) into Eq.(2.23), the following is obtained:

$$c_p Pdv+c_v vdP=0 \tag{2.24}$$

$$\kappa\frac{dv}{v}+\frac{dP}{P}=0 \tag{2.25}$$

By integrating Eq. (2.25), the following relationship between pressure and specific volume in the adiabatic compression process is obtained:

$$Pv^\kappa = \text{const.} \tag{2.26}$$

Combination of the above Eq. (2.26) and the equation of state of an ideal gas, Eq. (2.16), gives:

$$Tv^{\kappa-1} = \text{const.} \tag{2.27}$$

$$TP^{-\frac{\kappa-1}{\kappa}} = \text{const.} \tag{2.28}$$

By calculating the compression work expressed by Eq. (2.6) with Eq. (2.26), the following per-unit-mass adiabatic compression work can be obtained as follows.

$$
\begin{aligned}
w_t &= \int_1^2 vdP = P_1^{\frac{1}{\kappa}}v_1\int_1^2 P^{-\frac{1}{\kappa}}dP \\
&= P_1^{\frac{1}{\kappa}}v_1\frac{\kappa}{\kappa-1}\left[P^{\frac{\kappa-1}{\kappa}}\right]_1^2 \\
&= P_1^{\frac{1}{\kappa}}v_1\frac{\kappa}{\kappa-1}\left(P_2^{\frac{\kappa-1}{k}}-P_1^{\frac{\kappa-1}{k}}\right) \\
&= P_1v_1\frac{\kappa}{\kappa-1}\left\{\left(\frac{P_2}{P_1}\right)^{\frac{\kappa-1}{k}}-1\right\}
\end{aligned}
\tag{2.29}
$$

In addition, the adiabatic compression work by Eq. (2.22) is expressed as follows by using Eqs. (2.14) and (2.21).

$$w_t = C_p\left(T_2-T_1\right)=\frac{R\kappa}{\kappa-1}\left(T_2-T_1\right) \tag{2.30}$$

The above equation can be transformed to the form of Eq. (2.29) by using Eqs. (2.16) and (2.28).

The actual compression process cannot be a completely adiabatic one and involves some exchanges of heat. In some cases it is expressed as a polytropic change using a polytropic index n instead of the ratio of specific heats κ in the above Eqs. (2.26) and (2.29). However, the unambiguous adiabatic compression work expressed by Eq. (2.29) is more commonly used for definition of efficiencies being described later in this textbook.

Adiabatic compression power L of the compressor can be obtained by multiplying the per-unit-mass adiabatic compression work of Eq. (2.29) by compressor mass flow rate G:

$$L=Gw_t=G\left(h_2-h_1\right) \tag{2.31}$$

If cylinder volume V is used instead of specific volume v for pressure-volume diagram or in Eq. (2.29), the obtained work would be the amount of work per compression cycle, which can be multiplied by the number of compression cycles per unit time N to obtain the compression power.

$$L=N\int_1^2 VdP \tag{2.32}$$

The power value expressed by Eq. (2.32) represents the power for a volume flow of $NV_1=Q_1$.

(4) How to obtain κ

The value for ratio of specific heats κ in Eq. (2.26) is different depending on the type of refrigerant used. Unlike the ratio of specific heats of an ideal gas, that of an actual refrigerant is not perfectly constant, therefore caution is required to apply a value of ratio of specific heats, which is obtainable from a physical properties table or physical property calculation software, to the calculation of adiabatic compression work. Especially in the case of fluorine-based synthetic refrigerants, the ratio of specific heats listed in a physical properties table is significantly different from the κ value that satisfies Eq. (2.26).[*] Therefore, it is not appropriate to directly apply a κ value from a physical properties table to calculations based on Eqs. (2.26) and (2.29). In order to obtain a κ value being used for Eqs. (2.26) and (2.29), define an iso-entropy process over the typical operation range, then provide a κ value that will satisfy

*Concerning the compression process of R410A, the κ value that satisfies Eq. (2.26) is approximately 1.07 while the specific heat ratio typically shown in a physical properties table is approximately 1.3 to 1.5.

Eq. (2.26) using the pressures and specific volumes at the ends of such iso-entropy process. However, a κ value that satisfies Eq. (2.26) would be different from the value that satisfies Eq. (2.27) or (2.28). It is necessary to obtain a κ value appropriate for the intended purpose.

2.1.4 Equation of energy

The ideal compression process in a compressor can be analyzed using a physical property software or Eq. (2.26), on the assumption that it is an iso-entropy process. When leakage is involved, the amount of change in specific volume caused by the leakage may be included in the calculation of Eq. (2.26). On the other hand, the change of state inside an actual compressor is influenced by such factors as leakage or heat transfer and it is necessary to analyze not only the compression process but also the suction and discharge processes in the same way. Figure 2.10 shows the incoming and outgoing movements of fluid and heat, with the compression chamber defined as a single control volume. The figure shows that fluid with specific enthalpy h_i and mass flow rate G_i enters the control volume of pressure P and volume V, and fluid with specific enthalpy h_o and mass flow rate G_o exits from the same. The control volume changes its volume at the rate dV/dt and receives heat Q from the outside. Under the assumption that physical properties of fluid, such as specific volume v and specific enthalpy h, are uniform inside the control volume, the equation of energy for the fluid with mass M in the control volume is expressed as follows:

$$\frac{d(Mu)}{dt} = G_i h_i - G_o h_o + Q - P\frac{dV}{dt} \quad (2.33)$$

As most compressor engineers are familiar with enthalpy rather than internal energy, the following energy equation using specific enthalpy is presented by substituting Eq. (2.10) into Eq. (2.33).

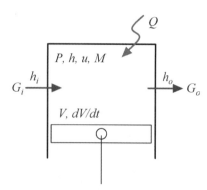

Fig. 2.10 Control volume

$$M\frac{dh}{dt} = G_i(h_i - h) - G_o(h_o - h) + Q + V\frac{dP}{dt} \quad (2.34)$$

Assuming that the specific enthalpy of the fluid exiting from the control volume is equal to the specific enthalpy of fluid in the control volume, the above Eq. (2.34) can be translated to:

$$M\frac{dh}{dt} = G_i(h_i - h) + Q + V\frac{dP}{dt} \quad (2.35)$$

Examining a compression process by applying the energy equation includes advantages such that thermal effects caused by heat transfer and leakage can be considered; that calculation of one cycle of the compressor is possible without distinguishing suction, compression and discharge processes; that various leakage or heat transfer models can be applied; and that the approach can be applied to a wide variety of refrigerants by combining with the physical property calculation algorithm of refrigerants.

Equations (2.33) to (2.35) all assume that the compression chamber is a single control volume in which refrigerant gas is uniform. However, in cases where non-uniformity of refrigerant condition inside the compression chamber must be taken into account, more detailed analysis should be conducted by applying computational fluid dynamics (CFD).

2.1.5 Leakage

As the performance of a positive-displacement compressor is significantly affected by its internal leakage, it is very important to correctly estimate how much internal leakage occurs. The most dominant factor that influences internal leakage is the gap size, which varies depending on the machining and assembly accuracy of components as well as the degree of heat- or pressure-induced deformations. The following paragraphs explain the basics of how to estimate the leakage for a given gap size.

(1) Ideal fluid

When refrigerant gas that is a compressible fluid flows through a suction valve of the compressor, the flow may be treated approximately as an incompressible fluid flow because pressure difference across the suction valve is not so large. Incompressible fluid with no or negligible viscosity is called as ideal fluid. As shown in Fig. 2.11 where there is a pressure difference across a nozzle (= "gap"), the relationship between the fluid upstream of the

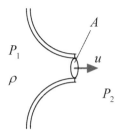

Fig. 2.11 Nozzle flow (Incompressible flow)

nozzle and one that has been accelerated through the nozzle can be expressed by the following Bernoulli's equation:

$$\frac{u^2}{2} + \frac{P}{\rho} = \text{const.} \quad (2.36)$$

where u is the flow velocity. The Bernoulli's equation can be derived from the law of energy conservation, it can also be obtained by applying the equation of motion to the small-fluid element on a streamline. As the fluid velocity upstream of the nozzle is generally negligible as compared with one through the nozzle in Eq. (3.36), the mass flow rate of the fluid passing through the nozzle with cross section area A can be expressed by the following equation:

$$G = \rho A u = A\sqrt{2(P_1 - P_2)\rho} \quad (2.37)$$

In practice, the actual flow rate influenced by flow losses through the opening is estimated by multiplying the flow rate obtained from Eq. (2.37) by a coefficient of flow C (0<C<1)

$$G = C\rho A u = CA\sqrt{2(P_1 - P_2)\rho} \quad (2.38)$$

(2) Viscous fluid

In the case of a flow of oil or one passing through a very narrow clearance, the flow is significantly affected by the viscosity of the fluid. In the case of a flow under small pressure difference, the effect of viscosity becomes relatively large. When fluid flows a circular pipe of diameter d shown in Fig. 2.12 (a) and where friction caused by the fluid's viscosity has a dominant effect, the relationship between the pressure difference ΔP for length l and the average flow velocity u through the pipe can be expressed by the following equation:

$$\Delta P = \lambda \frac{l}{d} \frac{\rho u^2}{2} \quad (2.39)$$

where λ is coefficient of pipe friction, which varies depend-

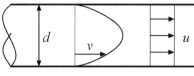

(a) Flow in circular pipe

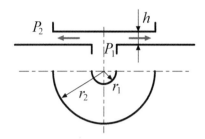

(b) Flow in radial channel

Fig. 2.12 Viscus flow

ing on the flow condition inside the pipe and the surface roughness of the inner wall. Where the pipe's inner wall is smooth, the pipe friction coefficient will be a function of the Reynolds number Re ($= ud/v$; v is kinematic viscosity of the fluid). In the case of a laminar flow, the following equation applies:

$$\lambda = 64/\text{Re} \quad (\text{Re} < 2300) \quad (2.40)$$

In the case of a turbulent flow, various experimental formulas are provided and the most widely used one is the Blasius equation shown below:

$$\lambda = 0.3164\,\text{Re}^{-1/4} \quad (3\times10^3 < \text{Re} < 10^5) \quad (2.41)$$

Though Eq. (2.39) is effective for circular pipe with diameter d, it is conveniently applied to non-circular passages by using an equivalent diameter, i.e. hydraulic diameter, expressed by the following equation:

$$d = 4m = 4\frac{A}{L} \quad (2.42)$$

where m is called as hydraulic mean depth and equal to the fluid-passage cross section area A divided by the perimeter length (wetted perimeter length) L.

In the case of a laminar flow passing through a narrow gap between a pair of parallel flat plates, the mass flow rate per width B is:

$$G = \frac{\Delta P h^3}{12 v l} B \quad (2.43)$$

where h is the gap height and l is the gap length. Similarly, the mass flow rate of a laminar flow passing through a radially shaped passage with entrance radius r_1, exit radius r_2 and height h as shown in Fig. 2.12 (b) is[2]:

$$G = \frac{\Delta P \pi h^3}{6 \nu \ln(r_2/r_1)} \quad (2.44)$$

(3) Compressible fluid

When refrigerant gas is treated as compressible fluid and expands adiabatically through a nozzle shown in Fig. 2.13, the energy equation is expressed as:

$$h_1 = h_2 + \frac{u^2}{2} \quad (2.45)$$

which gives a nozzle exit flow velocity of:

$$u = \sqrt{2\frac{\kappa}{\kappa-1}\frac{P_1}{\rho_1}\left\{1-\left(\frac{P_2}{P_1}\right)^{\frac{(\kappa-1)}{\kappa}}\right\}} \quad (2.46)$$

and a mass flow rate of:

$$G = \rho_2 A u = A\sqrt{2\frac{\kappa}{\kappa-1}P_1\rho_1\left\{\left(\frac{P_2}{P_1}\right)^{\frac{2}{\kappa}}-\left(\frac{P_2}{P_1}\right)^{\frac{(\kappa+1)}{\kappa}}\right\}} \quad (2.47)$$

The sound velocity of compressible fluid is expressed as:

$$u_c = \sqrt{2\frac{\kappa}{\kappa+1}\frac{P_1}{\rho_1}} \quad (2.48)$$

As pressure propagates at the sound velocity, if the flow velocity expressed by Eq. (2.46) exceeds the sound velocity expressed by Eq. (2.48), the nozzle exit pressure does not match the downstream pressure and then the flow rate does not change even if the downstream pressure is decreased.

This is a critical flow state. In this textbook, such state is called as a "choked state", thus distinguishing it from the critical state of refrigerant. A choked state coincides with a condition where the mass flow rate expressed by Eq. (2.47) reaches its maximum value as the downstream pressure is decreased. A critical pressure ratio that creates the choked state can be expressed by the following equation:

$$\frac{P_c}{P_1} = \left(\frac{2}{\kappa+1}\right)^{\frac{\kappa}{\kappa-1}} \quad (2.49)$$

When the downstream pressure is below the choke pressure, the mass flow rate is expressed as follows:

$$G_c = A\sqrt{2\frac{\kappa}{\kappa+1}P_1\rho_1\left(\frac{2}{\kappa+1}\right)^{\frac{2}{\kappa-1}}} \quad (2.50)$$

The value for adiabatic compression index κ for Eqs. (2.46) - (2.50) can be obtained using the κ calculation method described in Section 2.1.3 (4). If it is difficult to determine a precise adiabatic compression index, for example in the case of a leakage flow of supercritical carbon dioxide coming out of a gas cooler, flow rate can be obtained from the calculation of flow velocity based on Eq. (2.45) utilizing a specific enthalpy variation that corresponds the iso-entropy transition through the nozzle with a help of a physical property calculation software.

(4) Fanno flow[3]

As leakage flow inside a compressor passes through a narrow gap, the flow is often analyzed as an adiabatic flow through a passage with fluid friction (Fanno flow). In a flow channel consisting of a frictionless entrance nozzle and a frictional strait passage as shown in Fig.2.14, the relationship between the Mach number M_t at the entrance cross section t of the frictional passage and the passage length l can be expressed by the following equation when the flow is choked at the exit cross section e of the passage:

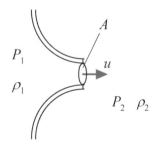

Fig. 2.13 Nozzle flow (Compressible flow)

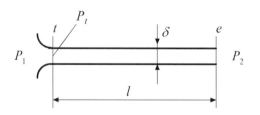

Fig. 2.14 Frictional flow channel

$$\lambda \frac{l}{2\delta} = \frac{1-M_t^2}{\kappa M_t^2} + \frac{\kappa+1}{2\kappa} \ln\left\{\frac{(\kappa+1)M_t^2}{2+(\kappa-1)M_t^2}\right\} \qquad (2.51)$$

where 2δ is the equivalent circular diameter of Eq. (2.42) when the gap is very narrow. The pipe friction coefficient λ, being a function of the Reynolds number and the inner wall surface roughness, is expressed as follows for a laminar flow through a narrow passage[4]:

$$\lambda = \frac{96}{\mathrm{Re}} \qquad (2.52)$$

The Reynolds number can be obtained by the following equation using mass flow rate G, coefficient of viscosity μ and passage width B perpendicular to page:

$$\mathrm{Re} = \frac{2G}{\mu B} \qquad (2.53)$$

The pressure ratio before and after the nozzle and one between the entrance and the exit of the passage in Fig. 2.14 can be expressed by the following equations:

$$\frac{P_1}{P_t} = \left(1 + \frac{\kappa-1}{2}M_t^2\right)^{\frac{\kappa}{k-1}} \qquad (2.54)$$

$$\frac{P_t}{P_e} = \frac{1}{M_t}\sqrt{\frac{\kappa+1}{2+(\kappa-1)M_t^2}} \qquad (2.55)$$

If the critical pressure ratio $\xi = (P_1/P_t)\cdot(P_t/P_e)$ is smaller than the actual pressure ratio P_1/P_2, a choked state will result, and then exit temperature T_e, exit flow velocity u_e and the mass flow rate G can be obtained from the following equation using exit Mach number $M_e=1$, $P_e=P_1/\xi$. Here R is the gas constant:

$$T_e = \frac{2T_1}{(\kappa-1)M_e^2+2} \qquad (2.56)$$

$$u_e = M_e\sqrt{\kappa R T_e} \qquad (2.57)$$

$$G = \frac{\delta B u_e P_e}{R T_e} \qquad (2.58)$$

If the flow does not reach the choked state at the passage exit, the flow condition of channel is calculated under a condition that the frictional passage is imaginarily extended till the imaginary exit becomes the choked state. At first, by giving an appropriate Mach number M_t at the passage

entrance, a critical passage length l^* is obtained using Eq. (2.51) and then the nozzle pressure ratio and the passage entrance/exit pressure ratio are calculated using Eqs. (2.54) and (2.55). Next, by applying Eq. (2.51) to the imaginary passage between the actual passage exit and the imaginary passage exit with the choked state, Mach number M_e at the actual passage exit is obtained using the following equation:

$$\lambda \frac{l^*-l}{2\delta} = \frac{1-M_e^2}{\kappa M_e^2} + \frac{\kappa+1}{2\kappa} \ln\left\{\frac{(\kappa+1)M_e^2}{2+(\kappa-1)M_e^2}\right\} \qquad (2.59)$$

This M_e is substituted into the left hand side of Eq. (2.55) instead of M_t and the imaginary entrance/exit pressure ratio P_e/P^* of the imaginary flow passage is obtained, then the entrance/exit pressure ratio of the actual passage is given as follows:

$$\frac{P_1}{P_e} = \frac{P_1}{P_t}\frac{P_t}{P^*}\bigg/\frac{P_e}{P^*} \qquad (2.60)$$

The above calculation is iterated by giving different Mach numbers M_t for the friction passage entrance until the pressure ratio of Eq. (2.60) matches the actual pressure ratio P_1/P_2. The exit temperature, the exit flow velocity and the mass flow rate can be determined using Eqs. (2.56) to (2.58).

(5) Gas-liquid two-phase flow

Leakage inside a compressor often becomes a two-phase flow composed of refrigerant gas and lubricating oil. Especially when estimating the seal effect of the lubricating oil, it is important to consider the leakage flow as a gas-liquid two-phase flow.

At a homogeneous flow model that regards a gas-liquid two-phase flow as a single phase flow with apparent physical properties, an apparent specific volume (or density) and an apparent coefficient of viscosity are appropriately used for the two-phase flow.

Specific volume: With the gas-to-liquid phase velocity ratio (= slip ratio) S, an apparent specific volume for a two-phase flow[5] can be expressed as follows:

$$v_{tp} = \left\{xv_g + (1-x)Sv_\ell\right\}\left(x + \frac{1-x}{S}\right) \qquad (2.61)$$

For the homogeneous flow with $S=1$, Eq. (2.61) yields to:

$$v_{tp} = v_\ell + x(v_g - v_\ell) \qquad (2.62)$$

where v is specific volume, x is vapor quality and subscripts tp, g and l represent "two-phase", "gas-phase" and "liquid

15

phase" respectively.

Viscosity: With the gas-phase and liquid-phase viscosity, μ_g and μ_l, respectively, the apparent viscosity μ_{tp} for a two-phase flow can be expressed as follows by several researches:

· Cicchitti[6]

$$\mu_{tp} = \mu_l(1-x) + \mu_g x \tag{2.63}$$

· Dukler[7]

$$\mu_{tp} = \mu_l(1-\beta) + \mu_g \beta \tag{2.64}$$

where β is the volume flow ratio, which, for the homogeneous flow model, can be expressed by:

$$\beta = \frac{x v_g}{x v_g + (1-x) v_\ell} \tag{2.65}$$

· McAdams[8]

$$\frac{1}{\mu_{tp}} = \frac{1-x}{\mu_l} + \frac{x}{\mu_g} \tag{2.66}$$

· Beattie - Whalley[9]

$$\mu_{tp} = \mu_g \beta + \mu_l (1-\beta)(1+2.5\beta) \tag{2.67}$$

· Lin[10]

$$\mu_{tp} = \frac{\mu_l \mu_g}{\mu_g + x^{1.4}(\mu_l - \mu_g)} \tag{2.68}$$

Though unified view for these correlations has not been fully established, Dukler's Eq. (2.64) and McAdams' Eq. (2.66) are used more commonly than others.

On the other hand, for a separated two-phase flow model, which assumes that the gas and the liquid phases move in separate stratified flows, Lockhart-Martinelli two-phase flow model[11] is used for estimation of the pressure loss. Lockhart and Martinelli expressed the two-phase frictional pressure drop, $(-dp_f/dz)$, using imaginary friction loss values, $(-dP_f/dz)_{g0}$ and $(-dp_f/dz)_{l0}$, which were expected when single phase of gas or liquid flowed fulfilling the whole area of the flow passage:

$$-\frac{dP_f}{dz} = \phi_g^2 \left(-\frac{dP_f}{dz}\right)_{g0} = \phi_l^2 \left(-\frac{dP_f}{dz}\right)_{l0} \tag{2.69}$$

where ϕ_g and ϕ_l are two-phase friction multipliers. They introduced the following Lockhart-Martinelli parameter X defined by Eq. (2.70), and showed a relationship with the two-phase friction multipliers.

$$X^2 = \frac{(-dP_f/dz)_{l0}}{(-dP_f/dz)_{g0}} \tag{2.70}$$

For the two phase friction multipliers, the next Chisholm's equation[12] is favorably used.

$$\phi_g^2 = 1 + CX + X^2 = \phi_l^2 X^2 \tag{2.71}$$

It is reported[13] that coefficient C often becomes smaller for narrow gaps than for ordinary circular pipes.

In addition, the following equations are proposed for the gas-to-liquid phase slip ratio by some researchers:

· Zivi[14]

$$S = \left(\frac{\rho_\ell}{\rho_g}\right)^{1/3} \tag{2.72}$$

· Chisholm[15]

$$S = \left\{1 - x\left(1 - \frac{\rho_\ell}{\rho_g}\right)\right\}^{1/2} \tag{2.73}$$

· Smith[16]

$$S = K + (1-K)\left[\frac{\frac{\rho_\ell}{\rho_g} + K\left(\frac{1-x}{x}\right)}{1 + K\left(\frac{1-x}{x}\right)}\right]^{\frac{1}{2}} \tag{2.74}$$

where Smith recommends $K=0.4$ based on experiment results.

2.2 Dynamic-mechanical Analysis

The basic structure of a compressor is a mechanical linkage, where each linked component has its own mass and moves while transmitting force. As the linkage moves, transfer of masses occurs in the system and a constraint force generates as a reaction force at a kinematic pair where each component meets and interacts with one another. Coulomb's friction force emerges in proportion to the constraint force, which contributes to the mechanical loss. And furthermore, the constraint force is transmitted to the outside through the compressor body in the form of vibration. An amount of mechanical loss is one of the important factors defining the compressor performance, as is compressor body vibration. To develop a high-performance compressor, it is essential to conduct a correct dynamic-mechanical analysis of these factors.

2.2.1 Motion of linkage

The dominant factor deciding a constraint force at a kinematic pair in the link mechanism is the compression gas force, while a force of inertia caused by the link movement also plays an important role. To correctly analyze the inertia force, equations of motion for the link mechanism must be derived.

The equation of motion should be derived based on d'Alembert's principle, that states that "the motion of a body occurs like that the sum of all forces acting on the body including the inertia force becomes balanced"[17]. When a body of mass m makes a linear motion of displacement x due to the effect of force F, as shown in Fig. 2.15 (a), the inertia force ($-m\ddot{x}$) appears in the direction defined by coordinate x. As a result, an equation of motion is:

$$-m\ddot{x} + F = 0 \qquad (2.75)$$

In the case of a rotational motion as shown in Fig. 2.15 (b), the inertia torque ($-I\ddot{\theta}$) appears in the direction of rotational coordinate θ. And then, an equation of motion is derived so that the sum of all the torques including the inertial torque becomes zero:

$$-I\ddot{\theta} + N = 0 \qquad (2.76)$$

where I is moment of inertia of the rotating body and N is a driving torque. Any motion of link mechanism can be analyzed by applying the basic equations of motion, Eqs. (2.75) and (2.76), to the mechanism.

Figure 2.16 (a) shows a most simplified compressor model. This model represents a compressor as a simple linear motion system so that the most basic theory can be easily understood, though commonly used compressors include a rotational mechanism as an essential element. Inside the compressor body, a motor core of mass m_1 makes a linear movement x_1 under the action of motor magnetic force F, and a piston of mass m_2 is working together with a linear movement x_2. As the actual motor core and the piston are linked by a pin, let's assume here that the motion of these components is constrained to each other by the following relationship:

$$x_2 = f(x_1) \qquad (2.77)$$

In Fig. 2.16 (a), the piston compresses gas inside the cylinder and gas force generated by pressure difference across the piston is represented by F_p. The gas force also acts on the cylinder head in the opposite direction. And furthermore, frictional force f works on the piston wall and the cylinder wall in opposite directions.

To analyze the movement of the link mechanism, equations of motion about the mechanism must be derived. At that time, as shown in Fig. 2.16 (b), the mechanism is sepa-

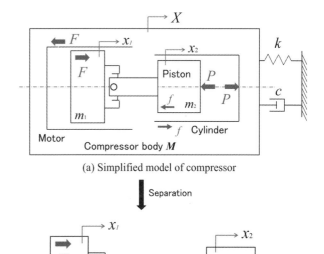

Fig. 2.16 Simplified model of compressor and method of dynamic analysis

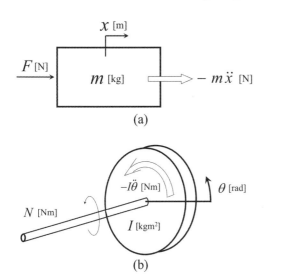

Fig. 2.15 Basic model of linear and rotational motions

rated into two parts and constraint force C is inevitably introduced at the connecting point of the parts. The piston part receives a constraint force as a driving force C from the motor core in the rightward direction, while the motor core receives its reactive force C in the opposite direction. Inertia forces caused by the link motion, $(-m_1\ddot{x}_1)$ and $(-m_2\ddot{x}_2)$, appear in the directions defined by coordinates x_1 and x_2 respectively. Equations of the motion should be derived so that the sum of all the forces working on the linked components including these inertia forces becomes zero.

$$-m_1\ddot{x}_1 + F - C = 0 \qquad (2.78)$$

$$-m_2\ddot{x}_2 + C - F_p - f = 0 \qquad (2.79)$$

From Eq. (2.79), the following constraint force can be obtained:

$$C = m_2\ddot{x}_2 + F_p + f \qquad (2.80)$$

By substituting the above equation into Eq. (2.78), a motion equation that defines the link movement can be derived, as follows:

$$m_1\ddot{x}_1 + m_2\ddot{x}_2 = F - F_p - f \qquad (2.81)$$

In consideration of the constraint condition of Eq. (2.77), the solution of this motion equation can be numerically analyzed with the characteristics of motor magnetic force F and gas compression force F_p, and x_1 and x_2 about the link motion are made clear. By substituting the solution into Eq. (2.80), the constraint force C can be obtained.

Commonly used compressors include a rotational mechanism, of which dynamic-mechanical analysis should be conducted applying the rotational motion Eq. (2.76). The rotational radius of vanes of a rotary vane compressor varies during vane rotation. To analyze such motion on a cylindrical coordinate system, the Coriolis force must also be considered.

2.2.2 Equation of energy and mechanical efficiency

An equation of energy can be obtained by multiplying the motion Eq. (2.79) by velocity and integrating the resulting equation over one cycle. As the integral for the inertia term becomes zero in a steady-state operation, the energy equation can be derived as:

$$E_{motor} = E_{gas} + E_{fric} \qquad (2.82)$$

where

$$E_{motor} = \int_{1-rev} F\,\dot{x}_1 dt$$

$$E_{gas} = \int_{1-rev} P\,\dot{x}_2 dt \,,\; E_{fric} = \int_{1-rev} f\,\dot{x}_2 dt$$

The above Eq. (2.82) states that all energy supplied by the motor is consumed in gas compression and friction.

Mechanical efficiency η_m represents the ratio of how much of the motor-supplied energy is effectively used as gas-compression energy, and it is expressed by the following equation:

$$\eta_m \equiv \frac{E_{gas}}{E_{motor}} \qquad (2.83)$$

Combined with Eq. (2,82), the above equation can be rewritten as:

$$\eta_m = \frac{E_{motor} - E_{fric}}{E_{motor}} \qquad (2.83)$$

The smaller the friction loss is, the higher the mechanical efficiency will be. In a project of compressor development, various engineering efforts are made to minimize friction loss.

2.2.3 Excitation forces and vibration of a compressor body

Forces working on the compressor body, as already illustrated in Fig. 2.16 (a), include motor magnetic reaction force F, gas force F_p working on the cylinder head and friction force f on the cylinder wall. In the figure, the vibration-response displacement of the compressor body is expressed as X. The excitation force of the compressor body working in the direction defined by coordinate X is given as $(-F + F_p + f)$. Combined with Eq. (2.81), this excitation force can be expressed as:

$$-F + F_p + f = -m_1\ddot{x}_1 - m_2\ddot{x}_2 \qquad (2.84)$$

Here, the inertia force alone remains as the excitation source. This is called an "unbalanced inertia force". Forces such as motor magnetic force, gas force and friction force are all internal forces and do not externally appear as an excitation source. Under the governance of motion Eq. (2.81), the difference between the driving force F and load $(F_p + f)$ becomes as an unbalanced inertia force, which shakes and vibrates the compressor body. This forms the basis of engineering approaches that "decreasing the amount of load variation helps minimizing the unbalanced inertia force, which contributes to reduction of the compressor vibration" or that "if the driving force is made to respond faster to the load variation, the unbalanced inertia force is minimized and as a result compressor vibration is reduced".

As the linked components with their own masses keep moving and change their positions inside the compressor body, a center of gravity of the compressor changes its position over time. In a strict sense, the vibration of the compressor body is a non-linear parametric excitation system. However, it is common to ignore such movement of the center of gravity due to it being negligibly small, as is its resulting vibration, so as to use a linearized equation of motion. Assuming that the compressor body is supported by a spring with spring constant k and a damper with damping coefficient c, an equation of motion that determines the magnitude of compressor body vibration caused by the unbalanced inertia force can be, with d'Alembert's principle applied, expressed as follows:

$$-M\ddot{X} - c\dot{X} - kX - m_1\ddot{x}_1 - m_2\ddot{x}_2 = 0 \qquad (2.85)$$

where M is the mass of compressor body. By dividing the above equation with M, the motion equation can be finally derived as:

$$\ddot{X} + 2\omega_n\zeta\dot{X} + \omega_n^2 X = \frac{-1}{M}(m_1\ddot{x}_1 + m_2\ddot{x}_2) \qquad (2.86)$$

where ω_n is the natural frequency and ζ the damping ratio, each of which can be expressed by the following equation:

$$\omega_n \equiv \sqrt{\frac{k}{M}}, \ \zeta \equiv \frac{c}{2\sqrt{Mk}} \ or \ \frac{c}{2M\omega_n} \qquad (2.87)$$

In the above analysis, the compressor vibration is simply treated as single-degree-of-freedom system. As commonly experienced vibration has six degrees of freedom in general, six equations of motion like the above one will be derived in an actual situation. It is more common to express such equations in the form of a determinant.[18]

2.3 Lubrication Theory

Reynolds derived, by applying the Navier-Stokes equation[19] that describes the motion of a viscous fluid to narrow-gap flows, that a wedge-shaped oil film inside the bear-

ing generates hydrodynamic pressure. This is the theory of fluid lubrication, and its fundamental equation is the (basic) Reynolds equation[19], which forms the basis of journal and thrust bearing designs. Furthermore, the above equation has been extended, for the analysis of mixed lubrication condition, into the modified Reynolds equation[20, 21]. The following sections provide an overview of the basic and the modified Reynolds equations, the rough surface contact theory[22], and about different types of bearings, that are journal, rolling and thrust bearings.

2.3.1 Basic Reynolds equation

It is often necessary to regard the flow of lubricating oil (density ρ) through a gap inside a commonly used bearing, as shown in Fig. 2.17, as a fluid field in a two-dimensional (x, y) laminar flow. Assuming that any pressure gradient along oil film thickness direction axis z can be ignored, the basic two-dimensional Reynolds equation can be derived at as follows:

$$\frac{\partial}{\partial x}\left(\frac{\rho h^3}{12\eta}\frac{\partial P}{\partial x}\right) + \frac{\partial}{\partial y}\left(\frac{\rho h^3}{12\eta}\frac{\partial P}{\partial y}\right)$$
$$= \frac{u_1 + u_2}{2}\frac{\partial(\rho h)}{\partial x} + \frac{v_1 + v_2}{2}\frac{\partial(\rho h)}{\partial y} + \frac{\rho h}{2}\frac{\partial}{\partial x}(u_1 + u_2) \qquad (2.88)$$
$$+ \frac{\rho h}{2}\frac{\partial}{\partial y}(v_1 + v_2) + \frac{\partial}{\partial t}(\rho h)$$

where, P is pressure, η is viscosity coefficient. h is the height of the bearing surface gap, and the boundary speeds at $z = 0$ and height h are given as u_1, v_1 and u_2, v_2 respectively. Integrating pressure P obtained from the Reynolds equation over the entire bearing surface gives the oil film force (bearing load capacity).

Also, the viscous friction force working on the bearing can be obtained by integrating the following shearing stress over the entire bearing surface.

$$\tau_{zx} = -\frac{\eta(u_1 - u_2)}{h} \pm \frac{h}{2}\frac{dP}{dx}$$
$$\tau_{zy} = -\frac{\eta(v_1 - v_2)}{h} \pm \frac{h}{2}\frac{dP}{dy} \qquad (2.89)$$

2.3.2 Modified Reynolds equation based on average flow model

Where, as is often the case with high-load slow-rotation applications, the lubricating oil film thickness and the height of surface roughness are approximately the same, such impeding factors as direct contact between projections,

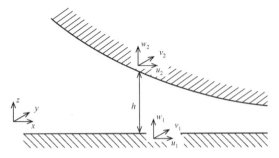

Fig 2.17 General model of bearing

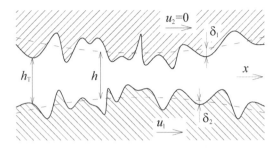

Fig. 2.18 Contacts between rough surfaces

generation of oil film by microscopic-scale wedge effect around projections and oil-flow blocking by the projections are expected. It is essential to include those effects of surface roughness into evaluation of fluid lubrication. Average flow models, based on probabilistic/statistical approach of analyzing bearing surface roughness, such as that of Patir and Cheng, are well known. Their governing equation is called the modified Reynolds equation, which is expressed at as follows:

$$\frac{\partial}{\partial x}\left(\phi_x \frac{h^3}{12\eta}\frac{\partial \overline{P}}{\partial x}\right) + \frac{\partial}{\partial y}\left(\phi_y \frac{h^3}{12\eta}\frac{\partial \overline{P}}{\partial y}\right)$$
$$= \frac{u_1+u_2}{2}\frac{\partial h_T}{\partial x} + \frac{u_1-u_2}{2}\sigma\frac{\partial \phi_s}{\partial x} + \frac{v_1+v_2}{2}\frac{\partial h_T}{\partial y} \quad (2.90)$$
$$+ \frac{v_1-v_2}{2}\sigma\frac{\partial \phi_s}{\partial y} + \frac{\partial h_T}{\partial t}$$

where, $\overline{P}$ is average pressure, ϕ_x and ϕ_y are pressure-flow coefficients, ϕ_s is shear-flow coefficient and σ is the standard deviation of surface roughness. In a commonly experienced case of contact between rough surfaces shown in Fig. 2.18, σ is given as $\sigma = \sqrt{\sigma_1^2 + \sigma_2^2}$, where σ_1 is the standard deviation of roughness of the lower surface and σ_2 is that of the upper surface. The sum of the average distance, that is the apparent oil film thickness h between the two surfaces, σ_1 and σ_2 corresponds to the local oil film thickness h_T (= $h + \delta_1 + \delta_2$). With the ratio of apparent oil film thickness h to the standard deviation of surface roughness σ given as H_r (= h/σ), the local oil film thickness h_T can be given as follows according to the magnitude of H_r.

$$h_T = h \qquad \text{at } H_r \geq 3$$
$$h_T = \frac{3\sigma}{256}\{35 + Z(128 + Z(140 + Z^2(-70 + Z^2(28 - 5Z^2))))\}$$
$$\text{at } H_r < 3 \quad (2.91)$$

where $Z = H_r/3$.

The average shear stress taking the surface roughness into account can be obtained by the following equation:

$$\overline{\tau}_x = \frac{\eta(u_2 - u_1)}{h_T}(\phi_f \pm \phi_{fs}) \pm \phi_{fp}\frac{h_T}{2}\frac{\partial \overline{p}}{\partial x}$$
$$\overline{\tau}_y = \frac{\eta(v_2 - v_1)}{h_T}(\phi_f \pm \phi_{fs}) \pm \phi_{fp}\frac{h_T}{2}\frac{\partial \overline{p}}{\partial y} \quad (2.92)$$

The first term of the right side of the equation is the shear stress generated by movement of the wall, where additional shear stresses induced by surface projections are compensated by ϕ_f and ϕ_{fs}. The second term is the shear stress arising from the pressure difference-induced flow, where, similarly to the first term, surface projection effect is compensated by ϕ_{fp}. The plus sign corresponds to $z = h_T$ and the minus sign to $z = 0$.

2.3.3 Contact theory

Even though two surfaces are apparently in contact with each other, they are actually in contact only at a number of contact points as shown in Fig. 2.19. The sum of the areas of these contact points is called "real contact area". At these contact points, the outer surface layer of the material is being destroyed by plastic flow so as to expose the inner metal for direct contact, where the materials may adhere to each other. When such two surfaces are moved against each other, the adhesion areas is sheared. The amount of force required for the motion is the solid friction force. To determine the amount of solid friction force, it is necessary to determine the real contact area.

The Greenwood-Williamson contact theory[22] regards surface projections like the ones shown in Fig. 2.20 (a) as spherical projections with different height and curvature

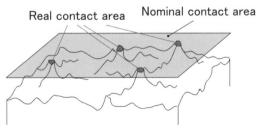

Fig. 2.19 Real contact area

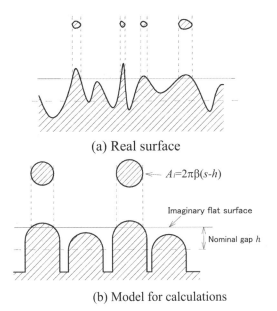

Fig. 2.20 Greenwood-Williamson model of solid contacts

radius β, to calculate the load capacity by considering whether the projection contact is elastic or plastic so as to determine the real contact area. To simplify the theory, the modified Greenwood-Williamson model, which treats that plastic deformation is always taking place at the contact point of surface projections, is used more commonly.

When the projection end point curvature radius β, apparent gap height h and projection height s are given, the average contact area of individual projections, A_{ave}, is calculated by the following equation:

$$A_{ave} = \int_h^\infty A_i \cdot \phi(s) \frac{ds}{\sigma} \\ = 2\pi\beta \int_h^\infty (s-h) \cdot \phi(s) \frac{ds}{\sigma} \qquad (2.93)$$

As the projection heights of roughness on a machined surface exhibit a Gaussian distribution in general, a distribution function ϕ of the projection heights can be expressed by the following equation:

$$\phi(z) = \frac{1}{\sqrt{2\pi}} e^{-\frac{1}{2}z^2} \qquad (2.94)$$

Assuming that the number of projections included in the apparent contact surface is N_{asp}, real contact area A can be finally determined by the following equation:

$$A = 2N_{asp}\pi\beta/\sigma \int_h^\infty (s-h) \cdot \phi(s) ds \qquad (2.95)$$

2.3.4 Journal bearing

A journal bearing is a type of sliding bearing that supports a radial load working perpendicularly to its shaft. The term "journal" refers to the part of the rotating shaft supported by the bearing. Load is supported by the fluid film between the shaft and the bearing. Journal bearings are generally superior in load capacity and damping performance and have a long service life. Though journal bearings can work with either gas or liquid, journal bearings in compressors mainly use liquid refrigerating machine oil as the working fluid.

(1) Petroff's equation

One of the most basic journal bearing theories is the Petroff's equation, that evaluates coefficient of friction μ. Petroff's theory assumes that, as shown in Fig. 2.21, the shaft with radius R is positioned at the center of the bearing. When the frictional force caused by oil film viscosity is divided by bearing load W, friction coefficient μ is obtained.

$$\mu = \frac{\eta \pi^2 (2R)^2 NL}{Wc} \qquad (2.96)$$

This is the Petroff's equation. Here, the following Sommerfeld number S, representing the bearing operating status, is introduced;

$$S \equiv \frac{\eta N}{P_m}\left(\frac{R}{c}\right)^2 \qquad (2.97)$$

where, P_m is the average bearing surface pressure $W/(2RL)$ under bearing load W. Combined with Sommerfeld number S, the Petroff's equation can be expressed as follows:

$$\frac{R}{c}\mu = 2\pi^2 S \qquad (2.98)$$

The left side of this equation, $(R/c)\mu$, is the frictional coefficient rate, which increases linearly and proportionally to Sommerfeld number S.

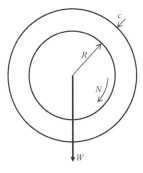

Fig. 2.21 Journal bearing with concentric shaft center

(2) Infinitely long bearing theory

Where the shaft load is small and the shaft rotational speed is high, that is, where the Sommerfeld number S is high, the shaft will be positioned very close to the exact center of the bearing and then the Petroff's equation can be applied effectively. More commonly, however, it is necessary to consider shaft eccentricity inside the bearing. As shown in Fig. 2.22 (a), load W works to cause the shaft center O to deviate downward from the bearing center O' with eccentricity e. When the shaft is rotating clockwise in Fig. 2.22 (a), oil is drawn into the right-hand side of the wedge-shaped space formed between the shaft and the bearing, so that the very high pressure generated inside the oil film works to cause the shaft center O to deviate upward and leftward.

Various characteristics of journal bearings in such case can be calculated by solving the Reynolds equation with bearing gap and boundary conditions given. At first, the problem to determine the flow of oil and the pressure of oil film between the shaft and the bearing, which can significantly affect the shaft eccentricity, can be solved in an equivalent manner to a plane bearing problem shown in Fig. 2.22 (b) by assuming that the centrifugal effect of curvature on the oil film is sufficiently smaller than the oil film pressure generated and is therefore negligible. This approach conceptually spreads out the oil film presenting in the range of $\theta = 0$ to $\theta = 2\pi$ in Fig. 2.22 (a) without any curvature effect, until the shaft circumference matches a virtual single flat plate, which moves in the rightward direction (along x axis) at the shaft circumferential speed U as shown in Fig. 2.22 (b). At the same time, the bearing surface is conceptually spread as a virtual surface so that the distance between the shaft and the bearing surfaces is equal to the oil film thickness h shown in Fig. 2.22 (a):

$$h(\theta) = c(1 + \varepsilon \cos\theta) \quad (2.99)$$

where ε, defined by $\varepsilon \equiv e/c$, represents eccentricity ratio of the shaft center.

Assuming that the oil is incompressible, that oil viscosity remains constant without being affected by temperature and pressure, and that the bearing's axial (y axis direction) length L is sufficiently large, oil flow can be analyzed single-dimensionally and then the Reynolds Eq. (2.88) is reduced to as follows:

$$\frac{d}{dx}\left(h^3 \frac{dp}{dx}\right) = 6\eta U \frac{dh}{dx} \quad (2.100)$$

The solution of the above equation can be expressed in the following form:

$$p = \frac{\eta R U}{c^2} C_p(\theta) + p_0 \quad (2.101)$$

where, pressure coefficient C_p is a function of θ and is given by the following equation:

$$C_p = \frac{6\varepsilon(2 + \varepsilon\cos\theta)\sin\theta}{(2+\varepsilon^2)(1+\varepsilon\cos\theta)^2} \quad (2.102)$$

The pressure coefficient for $\varepsilon = 0.8$ is shown in Fig. 2.23 with a legend of 'Sommerfeld'.

Negative oil pressure exists in the range of $\theta = \pi$ to 2π in a form completely symmetrical to positive oil pressure that generates in the range of $\theta = 0$ to π. In the case, the shaft center always tend to deviate in the direction 90° to the load W working on the shaft and the shaft load W is expressed by the following equation:

$$W = \frac{\eta U R^2 L}{c^2} \cdot C_w$$

$$\text{where} \quad C_w = \frac{12\pi\varepsilon}{(2+\varepsilon^2)\sqrt{1-\varepsilon^2}} \quad (2.103)$$

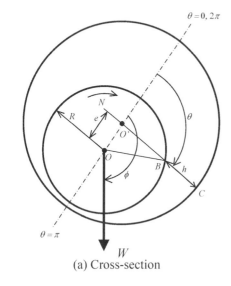

(a) Cross-section

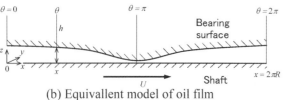

(b) Equivalent model of oil film

Fig. 2.22 Journal bearing with eccentric shaft center

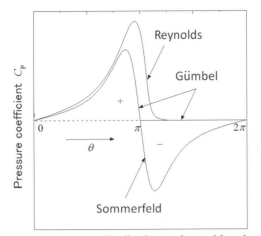

Fig. 2.23 Pressure distribution on journal bearing

C_w is load coefficient. Combined with Sommerfeld number S, the above equation can be re-written as follows:

$$S = \frac{1}{\pi C_w} \quad (2.104)$$

With Sommerfeld number S as a bearing operation status given, shaft eccentricity ε can be determined according to the above equation. The relationship is described in Fig. 2.24 by a dashed-dotted line and ε increases with decrease of S.

Frictional moment M_s working on the shaft is obtained as the sum of all oil film shear stresses over the shaft surface and is expressed by the following equation:

$$M_s = \frac{\eta U R^2 L}{c} \cdot C_{fs}$$
$$\text{where} \quad C_{fs} = \frac{4\pi(1+2\varepsilon^2)}{(2+\varepsilon^2)\sqrt{1-\varepsilon^2}} \quad (2.105)$$

And, friction moment M_b working on the bearing is:

$$M_b = \frac{\eta U R^2 L}{c} \cdot C_{fb}$$
$$\text{where} \quad C_{fb} = \frac{4\pi\sqrt{1-\varepsilon^2}}{2+\varepsilon^2} \quad (2.106)$$

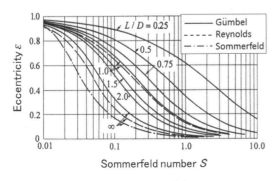

Fig. 2.24 Eccentricity

Here, C_{fs} and C_{fb} are friction coefficients. Figure 2.25 shows characteristics of C_w, C_{fs} and C_{fb} against eccentricity ε.

As a result, frictional coefficient μ of a journal bearing can be determined by the following equation:

$$\frac{R}{c}\mu = \frac{C_{fs}}{C_w} \quad (2.107)$$

Both load coefficient C_w and frictional coefficient C_{fs} are functions of eccentricity ε, which is, as shown in Fig. 2.24, a function of Sommerfeld number S. Figure 2.26 is a graphic presentation of frictional coefficient rate $(R/c)\mu$ with the horizontal axis representing S.

In order to improve the fundamental defect of the solution derived under the pressure condition of Sommerfeld shown in Fig. 2.23, the pressure condition is modified by Gümbel and Reynolds. The modified pressure distributions are shown in Fig. 2.23. The eccentricity solutions obtained under the Gümbel's and Reynolds'

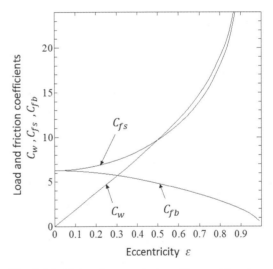

Fig. 2.25 Major characteristics of journal bearing

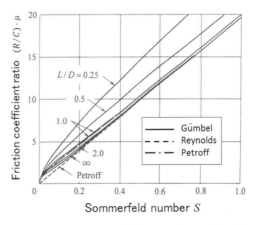

Fig. 2.26 Friction coefficient on journal bearing

conditions are shown in Fig. 2.24, by solid lines and dashed lines respectively. Figure 2.27 shows shaft center paths, describing the relationship between attitude angle ϕ and eccentricity ratio ε.

Regardless of whether or not to apply Gümbel's or Reynolds' conditions, the frictional coefficient rate $(R/c)\mu$ becomes as follows:

$$\left(\frac{\gamma}{C_\gamma}\right)\mu = \frac{2\pi^2 S}{\sqrt{1-\varepsilon^2}} \pm \left(\frac{\varepsilon}{2}\right)\sin\phi \qquad (2.108)$$

Here, + and − signs correspond to the shaft side and the bearing side respectively. Characteristics of the frictional coefficient rate are shown in Fig. 2.26.

(3) Infinitely short and finite-width bearing theories

Where the bearing's axial length L is finite in the range of $L/2R<1/4$, Reynolds Eq. (2.88) is expressed as follows:

$$\frac{d}{dy}\left(h^3 \frac{\partial p}{\partial y}\right) = 6\eta U \frac{\partial h}{\partial x} \qquad (2.109)$$

Under the Gümbel's boundary condition, an approximate solution of DuBois & Ocvirk and a theoretical solution of infinitely short bearing are obtained. Characteristics of infinitely short and finite-width bearing solutions are shown in Figs. 2.24 and 2.27 by solid lines.

2.3.5 Thrust bearing

A thrust bearing is a type of bearing that supports axial force working on the rotating body inside the compressor. Thrust bearing can be mainly classified into two types; one is rolling bearing and the other is sliding bearing, and each type should be applied properly. In addition, the contact area between the fixed and orbiting scrolls in a scroll compressor unit is also classified into a thrust bearing system, where thrust load is supported by a special lubrication mechanism.

(1) Rolling thrust bearings

Rolling thrust bearings are often installed at the end of a shaft to restrict axial movement of the shaft. To support a relatively light load in a small compressor, a single row deep groove ball bearing, which can also support radial load, may be used. On the other hand, types such as thrust ball bearings, angular-contact thrust ball bearings and tapered thrust roller bearings are more commonly used for larger compressors, where heavier load has to be supported. Most of these rolling-element bearings are normally assembled under an appropriate degree of pressurization, but bearings that only need to support the types of load such as, for example, the own weight of a longitudinal shaft, do not need such pressurization for assembly.

(2) Sliding thrust bearing

Sliding thrust bearings, that support thrust load by way of static pressure given by pressurized lubricating oil or dynamic pressure generated by shaft rotation, may be used for restricting the axial movement of a rotating shaft. To minimize failure risks in the case of intrusion of foreign substance into bearing gaps, sliding thrust bearings are commonly built using a relatively soft material.

A sliding thrust bearing system utilizing dynamic pressure is shown in Fig. 2.28. Here, the horizontal length of a single inclined pad is B, the width L, and the average radius R. Assuming that the number of rotations of the shaft over the pad is N [rps; 1/s], the average shaft rotation speed U will be $U=2\pi RN$. A schematic model of such inclined-plane bearing is shown in Fig. 2.29. With the exit gap height represented as h_o and the entrance gap height as h_i, height h that is located at x into the gap from the entrance is obtained by the following equation:

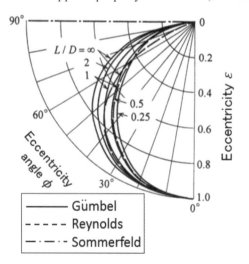

Fig. 2.27 Trajectory of shaft center

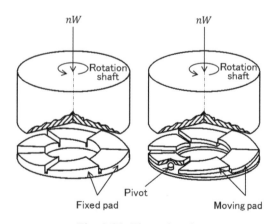

Fig. 2.28 Thrust bearing

Chapter 2 Basic Theory

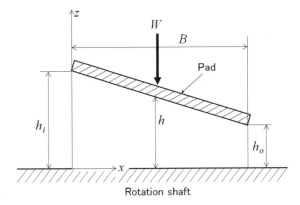

Fig. 2.29 Model of slide thrust bearing (inclined plain bearing)

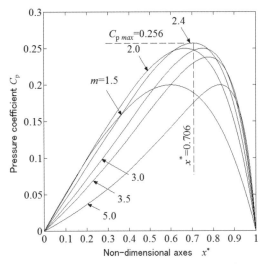

Fig. 2.30 Oil pressure distribution supporting inclined pad

$$h = h_i - \frac{x}{B}(h_i - h_o) \quad (2.110)$$

What governs the oil flow is, similarly to Eq. (2.100), the one-dimensional Reynolds equation. By solving that equation with the boundary conditions of $x = 0$, B and $P = P_0$, the following oil film pressure can be obtained:

$$p = \frac{\eta U B}{h_o^2} \cdot C_p(x^*) + p_0 \quad (2.111)$$

where pressure coefficient C_p is given by the following equation:

$$C_p(x^*) = \frac{6(m-1)(1-x^*)x^*}{(m+1)\{m-(m-1)x^*\}^2} \quad (2.112)$$

where x^* is the dimensionless coordinate x and m is the entrance-to-exit gap height ratio of:

$$x^* \equiv \frac{x}{B}, \; m \equiv \frac{h_i}{h_o} \quad (2.113)$$

Characteristics of the pressure coefficient C_p are shown in Fig. 2.30.

Thrust load W, expressed below, is equal to the oil film pressure integrated over the entire pad:

$$W = \frac{\eta U B^2 L}{h_o^2} C_w \quad (2.114)$$

where

$$C_w = \frac{6}{(m-1)^2}\left\{\ln m - \frac{2(m-1)}{m+1}\right\}$$

Sommerfeld number S^*, same as for Eq. (2.97), is defined as follows:

$$S^* = \frac{\eta N}{p_m}\left(\frac{B}{\Delta h}\right)^2 \quad (2.115)$$

where, $\Delta h \; (= h_i - h_o)$ is the height of the inclined pad, $\Delta h/B$ is the tilting angle and P_m the average surface pressure ($=W/BL$) caused by thrust load W. Combined with Sommerfeld number S^*, Eq. (2.114) is re-written as follows:

$$S^* = \frac{B/R}{2\pi(m-1)^2 C_w(m)} \quad (2.116)$$

The relationship between m and S^* is illustrated in Fig. 2.31. The frictional coefficient rate is expressed as follows:

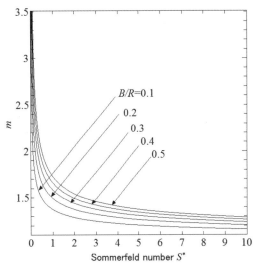

Fig. 2.31 Clearance ratio

25

$$\sqrt{\frac{WL}{\eta U}}\mu = \frac{C_{fs}}{\sqrt{C_w}} \qquad (2.117)$$

where

$$C_{fs} = \frac{1}{m-1}\left\{4\ln m - \frac{6(m-1)}{m+1}\right\}$$

Characteristics of the tilting pad bearing are illustrated in Fig. 2.32. With this, a thrust bearing design that is optimum in terms of both load capacity and frictional loss can be realized.

(3) Thrust bearing for special purpose (scroll compressors)

In a scroll compressor, a special thrust bearing is formed between the fixed and the orbiting scrolls to retain the orbiting motion specific to this type of compressor and also to support large thrust load. For this bearing application, sliding thrust bearings, where two flat plates are simply pressed against each other, and ball thrust bearings, composed of numerous balls placed between flat plates, can be used.

A specific example of a sliding thrust bearing for a scroll compressor is shown in Fig. 2.33. Here, a flat face on the outer edge of the orbiting scroll is pressed against a flat face of the fixed scroll. On the fixed scroll face, such devices as ring-shaped oil grooves are employed. A major difference from thrust bearings with an ordinary rotating shaft is that the sliding speed is very slow and the thrust load that needs to be supported is very large. It used to be believed that oil film formation cannot be easily achieved for such large-load, slow-sliding-speed application of sliding thrust bearings. However, it has been found that the elastic deformation of the orbiting scroll helps to form a wedge-shaped space along the circumference of the thrust bearing so as to generate dynamic pressure effect. Due to this, good lubricating condition is actually realized though the mechanism is composed of simple flat planes [23, 24].

The formation of such wedge-shaped space due to elastic deformation is illustrated in Fig. 2.34. Such elastic deformation observed in an orbiting scroll can be simulated relatively easily based on the finite element method (FEM). With that, the degree of wedge angle formed along the thrust bearing circumference can be determined, enabling theoretical analysis of the oil film pressure.

2.3.6 Rolling bearing

Rolling bearings can be built in various forms, such as deep-groove ball bearings, angular ball bearing, and cylindrical or tapered roller bearings, many of which are applied to compressors. The reason for such wide application is that these rolling bearings are easy to install and have high performance and relatively long service life, though they have some disadvantages such as rolling noises. In any of these bearings, the rolling elements (balls or rollers) and their track (bearing) are in a condition close to line or point contact. This is similar to engagement between gear teeth. In early years, it was unimaginable that a relatively thick oil film could be formed in and around such types of contact

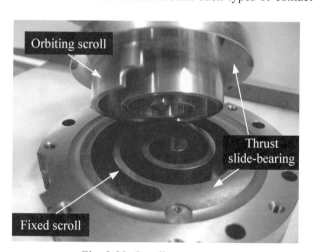

Fig. 2.33 Scroll compressor

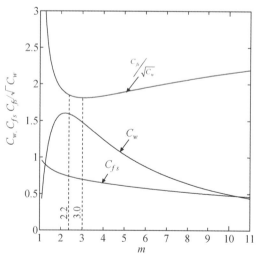

Fig. 2.32 Major characteristics of inclined plane bearing

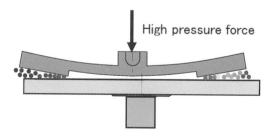

Fig. 2.34 Slide thrust bearing for scroll compressor

areas.

Martin's theory[25] of 1916 estimated that the oil film could only get as thick as 0.01 to 0.001 μm. However, this estimation had been questioned for years, based on such field experiences as scratches on gear tooth surface did not easily wear off after a considerable amount of gear operation, and a possibility of formation of much thicker oil film had been suspected. From around 1950s, many studies were conducted to swiftly clarify this question. Included in the results of such studies are the Weber-Saalfeld[26] solution (1954), the Ertel-Grubin[27] solution (1949) and the Dowson-Higginson[28] solution (1959), each of which are well-known. These solutions were obtained at around the same time when large-scale calculators started to come into use.

Where a line or point contact is formed, a significantly high Hertzian stress is caused by elastic deformation (Hertz deformation[29]). When lubricating oil is confined in such high-pressure space, its viscosity increases substantially, causing the oil film pressure to increase as well. The increased oil film pressure causes elastic deformation in the rolling elements and their track (bearing), resulting in change of oil film thickness and further change in the oil film pressure. The actual line or point contact between the rolling elements and their track is maintained through the mutually-affecting combination of high-pressure viscosity, oil film pressure, elastic deformation and oil film thickness. A lubrication analysis based on such approach is called "Elasto-Hydrodynamic Lubrication (EHL) Theory". However, EHL theory does not provide an analytical solution and therefore elaborate numerical computation is necessary to obtain a workable solution. This is why Dowson and Higginson were the first to successfully solve the problem, by combining a method of Grubin and Weber theories through the use of large-scale calculators that have just become available at the time. The Dowson and Higginson technique is now recognized as a monumental work and forms the basis of EHL analysis.

2.4 Compressor Efficiency

Losses occur in various parts of a compressor mechanism and decrease its efficiency. To clarify what type of loss reduces what aspect of compressor efficiency and by what degree, individual losses must be defined in terms of efficiency. Though these losses are not really independent of each other, this section discusses each type of loss separately for ease of efficiency definition. As the definition of compressor efficiency has not been standardized, it is important, when discussing compressor efficiencies, to clarify on what definition the efficiency concept being discussed is based.

Figure 2.35 illustrates the energy flow through a compressor. Compressors require to have some input from outside, such as electrical power or external motor drive. What the compression mechanism actually receives is this input minus the losses incurred, such as motor or inverter losses or belt transmission losses. The ratio of net compression mechanism input (L_a), which is the gross compressor input minus motor or transmission losses (ΔL_{mo}), to gross compressor input (L_i) is defined as motor efficiency η_{mo} or transmission efficiency, by the following equation:

$$\eta_{mo} = \frac{(L_i - \Delta L_{mo})}{L_i} = \frac{L_a}{L_i} \qquad (2.118)$$

Part of the power input to the compression mechanism is consumed in mechanical losses, and therefore the power input minus these mechanical losses is actually used for fluid compression. As mechanical losses include those by mechanical friction or fluid viscosity friction, fluid-stirring

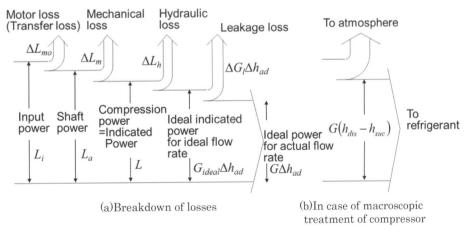

(a) Breakdown of losses (b) In case of macroscopic treatment of compressor

Fig. 2.35 Energy flow

power, which contributes to increase in shaft power but is not related to fluid compression, should also be treated as equivalent to mechanical loss. The ratio of the power actually used for fluid compression (L) to the power input to the compression mechanism (L_a) is defined as mechanical efficiency as follows:

$$\eta_m = \frac{(L_a - \Delta L_m)}{L_a} = \frac{L}{L_a} \quad (2.119)$$

where, ΔL_m represents mechanical loss. The power actually used for fluid compression (L) can be divided into the ideal compression power (L_s) and the hydraulic loss power (ΔL_h) which consists of the power consumed in flow losses like over-compression and the power used for the re-compression of refrigerant gas leaking back into the compressor chamber. The proportion of the ideal compression power (L_s) to the power actually used for fluid compression (L) is defined as hydraulic efficiency η_h as follows:

$$\eta_h = \frac{(L - \Delta L_h)}{L} = \frac{L_s}{L} \quad (2.120)$$

As shown in Fig. 2.36, flow and re-compression losses in a positive-displacement compressor mechanism lead to the PV diagram to expand, and then the hydraulic efficiency η_h is equal to the ratio of the ideal PV work (W_s) to the actual PV work (W), which is called indicated efficiency η_{ind}.

$$\eta_h = \frac{L_s}{L} = \frac{W_s}{W} = \eta_{ind} \quad (2.121)$$

Work indicated by an area surrounded by PV diagram is called indicated work, which gives the compression power by being multiplied by the number of compressions per unit time (N). The compression power corresponding to the ideal PV diagram expresses the ideal compression power for the ideal mass flow (G_{ideal}).

$$W_s N = G_{ideal} \Delta h_{ad} \quad (2.122)$$

Actual flow through the compressor is smaller than the ideal flow, due to such factors as leakage, pressure loss and suction gas heating. The ratio of actual flow (G) to ideal flow (G_{ideal}) is called volumetric efficiency η_v:

$$\eta_v = \frac{G}{G_{ideal}} \quad (2.123)$$

The volumetric efficiency is not merely the ratio of flow rates. It is also important in an energy viewpoint as the ratio of the ideal power for the actual flow rate to the ideal power for the ideal one expressed by Eq. (2.122):

$$\eta_v = \frac{(G_{ideal} - \Delta G_l)\Delta h_{ad}}{G_{ideal}\Delta h_{ad}} = \frac{G\Delta h_{ad}}{G_{ideal}\Delta h_{ad}} \quad (2.124)$$

The above equation indicates that the amount of compression power that is given to leaked fluid is wasted and therefore constitutes a loss in the energy viewpoint. Also, if the volumetric efficiency decreases due to heating loss in the suction process, it can be said that the amount of compression work per uit mass increases as much as the specific volume of the refrigerant gas increases due to heating.

Total efficiency η_t of compressor is given as the ratio of the ideal power for the actual flow rate to the power actually input to the compressor unit, which is equal to the product of the individual efficiencies explained in the above and defined by Eqs. (2.118) to (2.124).

$$\eta_t = \frac{G\Delta h_{ad}}{L_i} = \frac{G}{G_{ideal}} \frac{G_{ideal}\Delta h_{ad}}{WN} \frac{WN}{L_a} \frac{L_a}{L_i}$$
$$= \frac{G}{G_{ideal}} \frac{L_s}{L} \frac{L}{L_a} \frac{L_a}{L_i} = \eta_v \eta_{ind} \eta_m \eta_{mo} \quad (2.125)$$

In addition, the ratio of the ideal power for the actual flow to the compressor shaft power ($G\Delta h_{ad}/L_a$) is sometimes called "overall adiabatic efficiency".

If a significant leakage is involved, the area of work indicated by the PV diagram may become smaller, where as a result the indicated efficiency will increase. However, it should be noted that the total efficiency, depending on the multiplication of the indicated efficiency and the volumetric

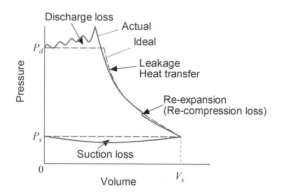

Fig. 2.36 Indicated (PV) work

efficiency, will always decrease. The above paragraphs have discussed individual losses independently so as to explain, in an easy-to-understand manner, how each type of loss contributes to compressor efficiency reduction. However, all losses and efficiencies are actually closely related to each other. To correctly evaluate the efficiency of a compressor, it is important to consider both the total efficiency and individual types of efficiency and not to focus on single type of efficiency only.

When a compressor system is viewed from macroscopic perspective, power input to the compressor can be classified into the part given to the refrigerant and the remaining part radiated as heat to the outside as shown in Fig. 2.35 (b). If the compressor is perfectly thermal-insulated, all the power input to the compressor is given to the refrigerant[*], in which case the total efficiency can be expressed by the following equation:

$$\eta_t = \frac{G\Delta h_{ad}}{L_i} = \frac{G\Delta h_{ad}}{G(h_{dis} - h_{suc})} \qquad (2.126)$$

It can be understood that, in the above case, all the types of losses indicated in the energy flow of Fig. 2.35 (a) have been ultimately transformed into heat and brought into the refrigerant. Assuming a constant specific heat at constant pressure, Eq. (2.126) can be rewritten as the following one, which is adopted in many fluid machinery textbooks as the definition of adiabatic efficiency.

$$\eta_{ad} = \frac{T_{ad} - T_{suc}}{T_{dis} - T_{suc}} \qquad (2.127)$$

where T_{ad} is the temperature after adiabatic compression, and T_{suc} and T_{dis} are suction and discharge temperatures, respectively. The adiabatic efficiency definition expressed by the above equation states that the compressor discharge temperature increases for the amount of losses incurred. As discussed in the preceding sections, this equation is only strictly true when heat radiation from the compressor to the outside is negligible.

2.5 Refrigerants

Refrigerant is "blood" of refrigeration and air conditioning systems. It is the working medium that carries heat, in the form of latent or sensible heat, through a refrigeration cycle. To serve this purpose effectively, a refrigerant needs to have various characteristics, such as having a proper pressure range in the temperature range of system operation.

The evolution of refrigerant technology has been central to the development of refrigeration and air conditioning systems. Due to increased awareness of environmental issues in recent years, such as ozone layer depletion and global warming, refrigerant selection now cannot be based on operating efficiency only. Stronger focus is being placed upon the environmental impacts of the refrigerant over its whole service life.

2.5.1 History of refrigerants

History of refrigerant evolution started with men's efforts for better use of naturally available refrigerants. When fluorine-based synthetic refrigerants (CFCs) were developed by advanced organic chemistry, it was lauded as a dream refrigerant. However, studies conducted in 1970s revealed that chlorine contained in CFCs contributes to the depletion of the ozone layer. Furthermore, it became known that not only the chlorine content but also the refrigerants themselves significantly contribute to global warning. Due to these findings, a shift to more environment-friendly refrigerants has been accelerated since around 2,000. Now refrigerant selection is being done with more focus on comprehensive environmental performance over the whole product life cycle, instead of being based solely on mechanical performance and personal safety. Table 2.1 summaries the history of refrigeration technology evolution. As shown in the table, hydrocarbon and CO_2 refrigerants are coming back to use, and other new low-environmental-impact refrigerants are also being developed.

2.5.2 Required characteristics of refrigerant

In general, a refrigerant must have the following characteristics:
(1) Chemical stability
(2) Low or zero toxicity and flammability for personal protection and safety consideration
(3) Least impact on global environment; zero ODP[**] and low GWP[***]
(4) Appropriate normal boiling point and critical temperature for the system's operating temperature range, and high latent heat of vaporization, small specific heat, low coefficient of viscosity and high thermal conductivity
(5) Low manufacturing cost

---

[*]Transmission losses that occur outside the compressor unit are excluded.

[**]Abbreviation for "Ozone Depletion Potential" The Ozone Depletion Potential of CFC-12 is 1.
[***]Abbreviation for "Global Warming Potential". The Global Warning Potential of CO_2 is 1.

(6) Oil-compatibility, low freezing point and high electrical insulation capability

Requirements such as chemical stability, low toxicity and flammability often conflict with ODP and GWP. In many cases, refrigerants that offer high chemical stability and non-flammability remain intact in the atmosphere for a long period of time after being released into the atmosphere and therefore has more environmental impacts. Therefore, the prevailing approach for refrigerant selection is to evaluate the available candidates from a broader perspective and

Table 2.1 The progress of refrigeration technology[30, 31]

Year	Event
1774	Priestley discovered NH_3
1834	Perkins filed a patent for a vapor compression refrigeration equipment
1850	Twining filed a patent for a NH_3 refrigeratior
1859	Karre developed an absorption NH_3 water solution refrigeration system
1861	Discovery of Joule Thomson effect
1872	Boyle filed a patent for a NH_3 compressor
1875	Pictet proposed SO_2 as a refrigerant
1882	Linde developed a CO_2 refrigeration system
1890	CO_2 refrigeration system widely adopted in maritime applications
1900	Starting of air-conditioning system with CO_2 refrigeration system
1913	Wolf developed and launched SO_2 household refrigerator "DOMELRE"
1930	Midgley invented Fluorocarbons[32]
1931	Commercialization of R11, R12 in USA
1935	Osaka Kinzoku(Now: Daikin Industries, Lted) developed Fluorocarbon in Japan
1974	Molina and Rowland presented a paper about the Ozone depletion by Fluorocarbons, in Nature[33]
1987	Adoption of Montreal Protocol*
1996	CFC phase-out in developed countries
2001	Commercialization of CO_2 heatpump hot water heater Eco-Cute in Japan
2002	Commercialization of Hydrocarbon household refrigerator in Japan
2006	EU announced F-gas regulation
2010	CFC phase-out in developing countries
2020	HCFC phase-out in developed countries
2040	HCFC phase-out in developing countries

*Montreal Protocol on Substances that Deplete the Ozone Layer is a protocol to the Vienna convention for the protection of the ozone layer. It was adopted in 1987 at Montreal, Canada, in order to designate ozone depletion substances and phase out these substance from production, to consumption and trade. According to Montreal Protocol, developed countries must phase out CFCs, halons, and carbon tetrachloride by 1996 and other HCFCs by 2020, for developing countries by 2015 and 2040, respectively.

to select the one that offers the best possible performance and the smallest possible detrimental effects for the specific use case. For example, household refrigerators are practically safe as they are all manufactured on the factory production lines and do not require any field installation work. Therefore, hydrocarbon refrigerants, which are flammable but is low in GWP is mainly selected based on environmental consideration. On the other hand, industrial refrigeration and air conditioning systems have higher leak-hazard risks as they require field installation work and contain a large amount of refrigerant. Therefore, based on safety consideration, R410A, which is a synthetic refrigerant that has a relatively high GWP but is non-flammable and non-toxic, is commonly selected.

2.5.3 Types of refrigerant

Refrigerants can be classified into natural and synthetic ones. Natural refrigerants are naturally available materials that are used as the working fluid for refrigeration and air conditioning systems. Commonly used natural refrigerants include air, water, ammonia, carbon dioxide and hydrocarbons such as propane and isobutane. For example, air is used as refrigerant for aircraft, carbon dioxide for heat pump-based water heaters and hydrocarbons for household refrigerators.

Synthetic refrigerants include CFCs (chloro fluoro carbons), HCFCs (hydro chloro fluoro carbons) and HFCs (hydro fluoro carbons). However, CFCs and HCFCs contain chlorine which is responsible to ozone layer depletion and are being phased out for eventual elimination based on the Montreal Protocol. As a result, HFCs are now the prevailing type of synthetic refrigerants. For example, most of the current automotive air conditioning systems use R134a, one of HFC refrigerants. In addition, development of HFOs (hydro fluoro olefins) and other new synthetic refrigerants that offer lower GWP is underway due to global warming concerns.

As it often happens that the requirements for a specific use case cannot be fully satisfied with a single- component refrigerant, mixed refrigerants are created by combining more than one single-component refrigerants. Mixed refrigerants can be classified into two types according to their composition; the azeotropic type where the boiling point and the condensation point are the same and the zeotropic type where the bubble point and the dew point are the different. For example, R410A is a quasi-azeotropic mixture** of R32 and R125, and R470C is a zeotropic mixture of R32,

**A mixed refrigerant of which difference between the boiling point and the condensation point is very small and which therefore can be practically classified as an azeotropic refrigerant.

R125 and R134a.

2.5.4 Designation of refrigerants

Designation of refrigerants is established by ASHRAE (American Society of Heating, Refrigerating and Air-conditioning Engineers, Inc.) and published in the form of ASHRAE Standard 34[34], "Designation and Safety Classification of Refrigerants". ASHRAE Standard 34 has been recognized as the de facto global standard for refrigerants and is adopted by the Japanese industry as well.

The ASHRAE designation for a refrigerant starts with an R, which is short for "refrigerant" and is followed by a combination of numbers (e.g. R410A). In addition to the numeric ASHRAE designation, refrigerant manufactures are allowed to give a commercial name to their products. Refrigerants is sometimes referred to with the combination of B, C and F, which respectively stand for bromine, chlorine and fluorine, followed by a combination of numbers, such as BCFC12B1, HCFC22, HFC134a. This type of designation indicates molecular structure of the refrigerant in a self-explanatory manner. "H" for hydrogen, if contained in that refrigerant, is inserted before B, C and F, and ether-based refrigerants use "E" for ether instead of "C" for carbon.

The numbering format in the refrigerant designation is defined as follows: The one's digit of the numeral represents the number of fluorine (F) atoms, the ten's digit the number of hydrogen (H) atoms plus 1, the hundred's digit the number of carbon (C) atoms minus 1, and the thousand's digit the number of unsaturated bonds of carbon atoms. The chlorine (Cl) number is calculated by deducting the number of atoms such as fluorine and hydrogen from the total number of carbon-bondable atoms. Isomers, that emerge due to differences in the atom-bonding arrangement, are suffixed, in the order of most symmetrical to least symmetrical molecular structures, first with no character and then lowercase alphabets of a, b, c and so forth.

For example, the R134a refrigerant has the following molecular structure:

$$CF_3 - CH_2F$$

Based on this molecular formula, R134a can also be called "HFC134a". According to the previously described designation rules, four fluorine atoms are represented as "4" in the one's digit position, two hydrogen atoms = 2 + 1 ="3" in the ten's digit position, two carbon atoms = 2 - 1 = "1" in the hundred's digit position. Due to the absence of unsaturated bonds, there is no number in the thousand's digit

position. Finally, as this molecular structure is second most symmetrical one next to "CHF_2 - CHF_2", a lowercase "a" is given as the suffix.

Azeotropic mixed refrigerants are designated with R 500-series numbers. As most of these refrigerants are CFC- or HCFC-based and are therefore restricted substances, R 500-series refrigerants have been almost completely removed from use since the end of 1990s. Zeotropic mixed refrigerants are designated with R 400-series numbers. Most refrigerants that are currently used for refrigeration and air conditioning systems are of the R 400-series. Individual refrigerants in the series are numbered in the order for which registration was applied. Multiple mixed refrigerants that are composed of the same combination of single-component refrigerants but with different percentage of compositions are distinguished with uppercase A, B and C. For example, a mixture of R32 and R125 with 50%-50% composition is called HFC410A or R410A, while the same mixture with 45%-55% composition is called HFC410B.

Refrigerants composed of non-organic molecules are designated with R 700-series numbers, with the lower two digits representing the refrigerant's molecular weight. For example, the carbon dioxide refrigerant, with a molecular weight of 44, is called R744.

The refrigerant number is designed to represent the molecular structure of each refrigerant and thus gives an indication of its characteristics. As described in the above, the last alphabetical lowercase and uppercase characters carry completely different meanings and therefore must not be confused.

2.5.5 Comparison of commonly used refrigerants

There are more than 50 types of single-component refrigerants and almost 60 types of mixed refrigerants (containing two to four different refrigerants) that are registered to ASHRAE Standard 34 and are globally recognized. Other than these, there are also numerous non-registered refrigerants.

This textbook hereinafter focuses on those refrigerants that have been put into actual use and are commonly used. Table 2.2 summarizes thermodynamic properties of such most commonly used refrigerants. Refrigerants in the table are listed in the order of the normal boiling point, from lowest to highest, which is more relevant to practical use than the order of numerical designations. Columns in the table represent, from left to right, the refrigerant number, the molecular formula or mass composition, the molecular weight, the critical temperature, the critical pressure, ODP,

Compressors for Air Conditioning and Refrigeration

Table 2.2 Refrigerant data and safety classifications (sorted by normal boiling point)

Refrigerant		Thermodynamic Properties				Environmental/Safety Data				
Number	Formula or composition	Molecular mass	NBP ℃	T_c ℃	P_c MPa	ODP	GWP 100yr	Safety Group	LFL %	OEL ppm
744	CO_2 Carbon dioxide	44.01	-78.4	31.0	7.38	0	1	A1	-	5000
32	CH_2F_2	52.02	-51.7	78.1	5.78	0	675	A2L	14.4	1000
410A	R 32/125 (50.0/50.0)	72.58	-51.4	70.5	4.81	0	2100	A1	-	1000
125	CHF_2CF_3	120.02	-48.1	66.0	3.62	0	3500	A1	-	1000
404A	R125/143a/134a (40.0/52.0/4.0)	97.60	-46.2	72.0	3.72	0	3900	A1	-	1000
407C	R 32/125/134a (23.0/25.0/52.0)	86.20	-43.6	85.8	4.60	0	1800	A1	-	1000
290	$CH_3CH_2CH_3$ Propane	44.10	-42.1	96.7	4.25	0	20	A3	2.1	2500
22	$CHClF_2$	86.47	-40.8	96.1	4.99	0.05	1810	A1	-	1000
717	NH_3 Ammonia	17.03	-33.3	132.3	11.33	0	<1	B2L	15.0	25
134a	CH_2FCF_3	102.03	-26.1	101.1	4.06	0	1430	A1	-	1000
152a	CH_3CHF_2	66.05	-24.0	113.3	4.52	0	124	A2	4.8	1000
600a	$CH(CH_3)_2$-CH_3 Isobutane	58.12	-11.7	134.7	3.63	0	20	A3	1.7	800
123	$CHCl_2CF_3$	152.93	27.8	183.7	3.66	0.02	77	B1	-	50
718	H_2O, Water	18.02	100.0	373.9	22.06	0	<1	A1	-	

GWP, ASHRAE 34 safety classification (refer to Table 2.3 for details), LFL (= lower flammability limit) and OEL (= occupational exposure limit). Natural refrigerants are shown shaded in Table 2.2. R22 and R123 are being phased out for eventual elimination. CFC refrigerants are already completely eliminated and therefore are not included in the table.

2.5.6 Applications of various refrigerants

Refrigeration systems are used for a wide variety of applications, each requiring a specific refrigerant appropriate for its operating conditions. In general, internal pressure of a refrigeration system pressure needs to be maintained at a level above the atmospheric pressure so that non-condensable gases outside the system will not leak into the system. To enable that, the normal boiling point of the refrigerant must be sufficiently lower than the working temperature of the system. Also, the refrigerant must have a critical temperature that is sufficiently higher than the system operating temperature so that phase transitions is most effectively utilized for optimum cycle efficiency. These and a number of other factors form the basis of refrigerant selection criteria.

As a refrigerant for household refrigerators, R12 has been replaced with R134a with zero ODP and nowadays isobutane, a type of hydrocarbon, is prevailing. The early refrigerant for household and commercial air conditioners was R22, which has now been mostly replaced with R410A. As a refrigerant for automotive air conditioners, after R134a has replaced R12, development of refrigerants with

even lower GWP (below 150) was called for due to the European "F-gas Regulation".[*] To cope with this requirement, a lot of engineering efforts have been put into utilization of CO_2 as a natural refrigerant, but actual implementation of CO_2 was hesitated due to its low energy efficiency and high pressure tendency. It was under such circumstances that some US refrigerant manufacturers proposed HFO1234yf, a new olefin-based low-GWP refrigerant. The automotive air-conditioner industry positively responded to this proposal. As of 2010, HFO 1234yf is in the final stage of preparation preceding actual implementation.

However, use of CO_2, which has a low critical temperature, in a trans-critical cycle[**] offers a specific advantage that high-efficiency heating on the high-temperature side is possible by employing a full counterflow heat exchanger with water. Such application has been realized as heat pump-type water heaters, bringing about a major commercial category that previously did not exist.

2.5.7 Refrigerant safety

Safety is an extremely important factor in refrigerant selection. Some types of refrigerant have toxicity, such as acute inhalation toxicity or chronic toxicity/carcinogenicity through exposure for a long period of time. Also, hydrocarbons and some other refrigerants are flammable and/or

*European F-gas Regulation: Regulation aimed at restricting the emission of F gases [HFCs (partly including gases that are not subject to Montreal Protocol restriction), PFCs, SF6] into the environment.
**A cycle of which high-pressure side is run above the critical pressure and the lower side is run below that pressure.

Chapter 2 Basic Theory

explosive. Use of these types of refrigerant requires sufficient safety measures to be taken in compliance with the applicable safety standards.

Until problems such as ozone layer depletion and global warning became an international issue, all that was required in refrigerant selection was to try to choose the safest product available. Now with stronger focus on environmental standards, there is much less choice of refrigerants that are acceptable for use. As a result, products with slight toxicity or flammability, those were once removed from the selection scope, are being considered for use again. Now more of these products are starting to be used on the condition that safety measures are taken as necessary and appropriate for the specific use case.

As a practical guideline, ASHRAE Standard 34[34] provides grouping of refrigerants according to their toxicity and flammability risk levels. Table 2.3 shows this risk grouping information. Toxicity grouping has two levels, those are Group A with "lower toxicity" and Group B with "higher toxicity". Flammability classification has three levels, those are Class 1 for "no flame propagation", Class 2 for "lower flammability" and Class 3 for "higher flammability". Increased environmental awareness leads to an urgent need for refrigerants with smaller global warming effect, while it often happens that refrigerants with less global warming effect are more flammable. To cope with these conflicting needs, a subclass 2L has been recently added to the existing "lower flammability" Class 2. The newly added Class 2L includes refrigerants with a burning velocity of less than or equal to 10 cm/s, for which sufficient fire safety can be practically assured in actual use cases. Newly classified into this Class 2L are NH_3, R32, R1234yf, R1234ze and so on.

2.5.8 Safety measures

The most basic safety measure in a refrigeration system against refrigerant hazard risks is to build a leak-proof system, while various rules and regulations are in place to make sure that no serious accident would result from unexpected refrigerant leaks. Regulations state typically that the system is only charged with a certain limited amount of refrigerant so that any refrigerant leak into the room cannot possibly result in a lower concentration limit condition*. For refrigerant with flammability risks, implementation of two or more mutually independent safety measures is required, for example installing a ventilation system in addition to refrigerant charge volume restriction.

Compliance with laws and regulations is an essential part of refrigeration and air conditioning system safety assurance. It is strongly recommended to actually check what laws and regulations are applicable to your operation.

References

1) M. Fukuta, T. Yanagisawa, T. Watanabe, Y. Ogi, S. Sakimichi, R. Radermacher: "Operating Characteristics of Expander for Carbon Dioxide Refrigeration Cycle", Proc of the 36th Japanese Joint Conference on Air-conditioning and Refrigeration, Tokyo, pp. 27-40 (2003). (in Japanese)

2) T. Ichikawa: "Hydraulics and Hydrodynamics", Asakura Publishing Co., Ltd., Tokyo, p. 50 (1981). (in Japanese)

3) T. Ikui, K. Matsuo: "Dynamics of Compressible fluid", Rikogakusha Publishing Co., Ltd., Tokyo, p. 54 (1986). (in Japanese)

4) Japan Society of Mechanical Engineers: "Flow Resistance in Pipes and Ducts", Japan Society of Mechanical Engineers, Tokyo, p. 39 (1979). (in Japanese)

5) K. Akagawa: "Gas-Liquid Two-Phase Flow", Corona Publishing Co., Ltd., Tokyo, p. 218 (1974). (in Japanese)

6) A. Cicchitti, C. Lombardi, M. Silvestri, G. Solddaini, R. Zavalluilli: "Two-phase cooling experiments-pressure drop, heat transfer and burnout measurements", Energia Nuclear., 7 (6), pp. 407-425 (1960).

7) A. E. Dukler, M. Wicks III, R. G. Cleveland: "Frictional pressure drop in two-phase flow: A. A comparison of existing correlations for pressure loss and holdup", AIChE J., 10 (1), pp. 38-43 (1964).

8) W. H. McAdams, W. K. Woods, R. L. Bryan: "Vaporization inside horizontal tubes -ii- benzene-oil mixtures", Trans. ASME, 64, pp. 193-200 (1942).

9) D. R. H. Beattie, P. B. Whalley: "A Simple Two-Phase

Table 2.3 Safety classification[34]

Higher flammability	A3	B3
Lower flammability	A2	B2
	A2L	B2L
No flame propagation	A1	B1
	Lower toxicity	Higher toxicity

* The respiration concentration limit where detrimental oxygen depletion does not occur and emergency leak control operations can be carried out, or the flammability concentration limit, whichever is lower. For more details, refer to applicable safety guidelines.

Frictional Pressure Drop Calculation Method", Int. J. Multiphase Flow, 8 (1), pp. 83-87 (1982).

10) S. Lin, C. C. K. Kwok, R. Y. Li, Z. H. Chen, Z. Y. Chen: "Local Frictional Pressure Drop during Vaporization of R12 through Capillary Tube", Int. J. Multiphase Flow, 17 (1), pp. 95-102 (1991).

11) R. W. Lockhart, R. C. Martinelli: "Proposed correlation data for isothermal two-phase two-component flow in pipes", Chemical Engineering Progress Symposium Series, 45 (1), pp. 39-48 (1949).

12) D. Chisholm: "A theoretical basis for the Lockhart-Martinelli correlation for two-phase flow", Int. J. Heat Mass Transfer, 10 (12), pp. 1767-1778 (1967).

13) H. Fujita, T. Ohara, M. Hirota, H. Furuta, H. Sugiyama: "Gas-Liquid Flows in Narrow Flat Channels (Influences of Liquid Properties)", Trans. JSME (Series B), 61 (592), pp. 4412-4419 (1995). (in Japanese)

14) S. M. Zivi: "Estimation of Steady State Steam Void Fraction by means of the Principle of Minimum Entropy Production", Trans. ASME, J. of Heat Transfer, Series C, 86, pp. 247-252 (1964).

15) D. Chisholm: "Pressure gradients due to friction during the flow of evaporating two-phase mixtures in smooth tubes and channels", Int. J. Heat Mass Transfer, 16 (2), pp. 347-358 (1973).

16) S. L. Smith: "Void fractions in two-phase flow: A correlation based upon an equal velocity head model", Proc. Inst. Mech. Engrs., London, 184 Pt.1 (36), pp. 647-664 (1969).

17) T. Yoshikawa, G. Matsui, N. Ishii: "Dynamics of machinery", Corona Publishing Co., Ltd., Tokyo, p.12 (1987). (in Japanese)

18) K. Imaichi, N. Ishii, N. Kagoroku: "Mechanical Vibration in Small Reciprocating Compressors", Trans. JSME, 41 (348), pp. 2333-2346 (1975). (in Japanese)

19) H. Lamb: "Hydrodynamics", Cambridge, pp. 576-584 (1879).

20) N. Patir, H. S. Cheng: "An Average Flow Model for Determining Effects of Three –Dimensional Roughness on Partial Hydrodynamic Lubrication", Trans. ASME, 100, pp. 12-17 (1978).

21) N. Patir, H. S. Cheng: "Application of Average Flow Model to Lubrication between Rough Sliding Surfaces", Trans ASME, 101, pp. 220-230 (1979).

22) J. A. Greenwood, J. B. P. Williamson: "Contact of nominally flat surfaces", Burndy Corporation Research Division, Norwalk, Connecticut, USA, pp. 300-319 (1966).

23) T. Oku, N. Ishii, K. Anami, C. W. Knisely, K. Sawai, T. Morimoto, A. Hiwata: "Theoretical Model of Lubrication Mechanism in the Thrust Slide-Bearing of Scroll Compressors", HVAC & R Research, ASHRAE, 14 (2), pp. 239-258 (2008).

24) N. Ishii, T. Oku, K. Anami, C. W. Knisely, K. Sawai, T. Morimoto, N. Iida: "Experimental Study of the Lubrication Mechanism for Thrust Slide Bearing in Scroll Compressors", HVAC & R Research, ASHRAE, 14 (3), pp. 453-465 (2008).

25) H. M. Martin: "The lubrication of gear teeth", Engineering, 102, p. 119 (1916).

26) C. Weber, K. Saalfeld: "Schmierfilm bei Walzen mit Verformung" Z. Angew. Math. Mech., 34, pp. 54-64 (1954). (in German)

27) A. N. Grubin, I. E. Vinogradova: "Contact stresses in toothed gears and worm gears", Central Scientific Research Institute for Technology and Mechanical Engineering, Book No. 30, Moscow (1949). (in Russian)

28) D. Dowson, G. R. Higginson: "A numerical solution to the elastohydrodynamic problem", J. Mech. Eng. Sci., 1 (1), p. 6 (1959).

29) H. Hertz: "Uber die Berührung fester elastisher Körper", Journal for die Reine und Angewandte Mathematik, 92, pp. 156-171 (1881). (in German)

30) O. B. Tsvetkov: "Thermophysical Aspects of Environmental Problems of Modern Refrigerating Engineering", Chemistry and Computational Simulation, 3 (10), pp. 74-78, (2002).

31) B. A. Nagengast: "A History of Refrigerants", ASHRAE Journal, pp. 3-14 (1988).

32) T. Midgley: "Organic Fluorides as Refrigerants", Industrial and Engineering Chemistry, 22 (5), pp. 542-545 (1930).

33) M. J. Molina: "Stratospheric sink for chlorofluoromethanes: chlorine atomc-atalysed destruction of ozone", Nature, 249, pp. 810-812 (1974).

34) ANSI/ASHRAE Standard 34-2010: ASHRAE, (2010).

Chapter 3 Reciprocating Compressors

3.1 History of Reciprocating Compressors

Reciprocating compressors have long been used for cooling and cold storage purposes, often as part of refrigeration related products such as refrigerators. It is difficult to accurately trace back when the technology was first developed or when the production of reciprocating compressors actually started. The use of reciprocating compressors became widespread in those days when fluorocarbons were created in 1926 and General Electric (GE) started the mass production of refrigerators using SO_2 as refrigerant. Before those days, there were refrigerators using air, ammonia or ether as refrigerant. It is generally said that a diaphragm-based compressor used by Perkins in 1834 was the first reciprocating compressor put to actual use. Concerning the first major Japanese manufacturer, in 1930 Tokyo Denki (now Toshiba) initiated the sales of electric refrigerators, which were designed after GE products and equipped with reciprocating compressor. Soon after, other Japanese manufacturers also started selling refrigerators. Figure 3.1 shows an early Japanese small-size hermetic reciprocating compressor.

For more details about the design and development history of those early products, refer to the corporate histories of the respective manufacturers. "Proceedings of JSRAE Annual Assembly 2007" published by the Japan Society of Refrigerating and Air Conditioning Engineers also provides a concise industry-wide chronicle of how reciprocating compressors for refrigerators were developed[1].

3.2 Operating Principle and Characteristics

3.2.1 Compression action

A reciprocating compressor has a mechanism where a piston moves in a reciprocating manner inside a cylinder to change the compression chamber volume thus compressing the gas inside. The main components of a reciprocating compressor are cylinder, piston, suction valve, discharge valve, crankshaft and connecting rod ("con rod") which converts a rotational motion to a reciprocating one.

Figure 3.2 shows a cross-section of belt-driven two-cylinder reciprocating compressor. Each cylinder has a pair of suction and discharge valves that are connected to suction and discharge lines respectively. Both the suction and the discharge valves are so-called "automatic valve" which, e.g. in the case of the suction valve, opens automatically when the cylinder pressure is lower than the suction line pressure and closes automatically when the cylinder pressure is higher than the suction line pressure. Compared with the suction and discharge valves of an internal combustion engine that are geared to the crank angle, the automatic valve system is simpler in structure and can maintain better efficiency even at different compression ratios.

Though the variation of stroke volume in a cylinder, which is essential to reciprocating compressors, can be intuitively understood and explanation about that may not be

Fig. 3.1 Small-size hermetic reciprocating compressor

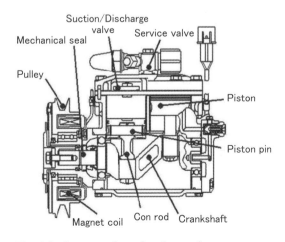

Fig. 3.2 Cross section of reciprocating compressor

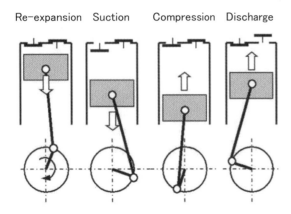

Fig. 3.3 Compression cycle

needed, the working principle of one cycle in a compressor cylinder is shown in Fig. 3.3. A compression cycle comprises of the following processes:

1) Re-expansion process: When the piston starts to move down from the top dead center, the gas confined in the small space left at the top of the cylinder and under the valves ("top clearance volume"), whose initial pressure is almost equal to the discharge pressure, begins to expand. Note that the piston may not actually "move down" depending on the compressor configuration, horizontal or vertical, but this textbook describes the piston movement as "up" and "down" for the sake of convenience.

2) Suction process: The volume of the space enclosed by the cylinder and the piston (referred as "cylinder volume") further expands so that the cylinder pressure drops below the suction line pressure, triggering the suction valve to open and thereby introducing the suction gas into the cylinder. The suction process is completed shortly after the piston reaches the bottom dead center, causing the suction valve to close.

3) Compression process: The piston starts to move up, causing the cylinder volume to decrease and the pressure to increase.

4) Discharge process: The discharge valve opens when the cylinder pressure rises above the discharge pressure. The gas in the cylinder is discharged to the discharge line.

Afterwards, the cycle repeats the above four processes. As both suction and discharge need to be completed in one rotation, it is critical to complete the discharge process in short time especially in high-pressure-ratio operation. The maximum pressure ratio attainable by a given cylinder is decided by the ratio of its stroke volume to the top clearance volume. If a compressor is expected to operate over the maximum pressure ratio, the compressor cannot discharge any flow.

Main factors that characterize the design of a reciprocating compressor include the number of cylinders, cylinder configuration and the arrangement of the crankshaft that converts the rotational motion of the shaft into the reciprocating motion of the piston. The upper two pictures in Fig. 3.4 show the possible configurations of the crankshaft and its offset directions. A crankshaft which has two eccentric pins in two opposed directions is called "two-throw" type, and one which has an eccentric pin in one direction is called "one-throw" type. When a crankshaft is offset in three directions, the crankshaft is called "three-throw" type.

When a compressor with multiple cylinders is applied for single-stage compressor, a multi-throw crankshaft is favorably used for the compressor to reduce the torque fluctuation of the shaft. However, it may be possible, in the case of a two-stage compressor whose crankcase chamber is filled with gas of an intermediate pressure, that having same-phase compression using a one-throw crankshaft actually reduces the fluctuation of shaft torque better than having different-phase compression using a multi-throw crankshaft. The lower three pictures in Fig. 3.4 show the typical cylinder configurations. Other than the most common in-line configuration, there are V-, W- and radial-type configurations. In addition, "opposed" configuration where the cylinder bank angle is 180° is a variation of the V-type configuration, and "cross-opposed" configuration where four cylinders are radially positioned 90° apart is a variation of the Radial configuration.

Commonly used rotational-to-reciprocating motion con-

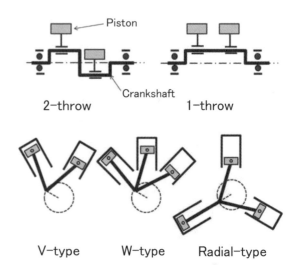

Fig. 3.4 Cylinder configuration

version mechanisms like the Scotch yoke type (Fig. 3.5), the crank-connecting rod ("con rod") type (Fig. 3.6) and the swash plate type (Fig. 3.7) will be explained later in this chapter. The swash plate type is widely used in automotive reciprocating compressors. The linear-motor-based compressors where the reciprocating motion is directly generated by a solenoid coil and a magnet have been recently employed in some compact refrigerators.

3.2.2 Theoretical stroke volume

The stroke volume V_{suc} of a reciprocating compressor including the top clearance volume V_{top} is given as follows:

$$V_{suc} = \frac{\pi}{4} D^2 \times 2 \times R + V_{top}$$
$$= \frac{\pi}{2} D^2 R + V_{top} \qquad (3.1)$$

Cylinder volume $V(\theta)$ including the top clearance volume at a given rotational angle θ, based on 0-degree top dead center, can be calculated by the next equation:

$$V(\theta) = \frac{\pi}{4} D^2 [R + L - \{\sqrt{(L^2 - R^2 sin^2\theta)} + Rcos\theta\}] + V_{top} \qquad (3.2)$$

where, D: cylinder bore (= piston diameter), R: crank radius, L: connecting rod length. On the other hand, the pressure $P(\theta)$ at a given rotational angle θ is obtained by the following equation with the suction-completion pressure denoted as P_{suc}:

$$P(\theta) = P_{suc}(V_{suc}/V(\theta))^n \qquad (3.3)$$

where, n is the polytropic index.

Load imposed on various parts of the mechanism by this pressure is explained in Section 3.4. It should be noted that, where the gas pressure load on the piston is supported by the connecting rod, lateral piston load may cause wear and scuffing (solid phase adhesion) of the cylinder and the piston. In the case, parameters such as the piston diameter-to-stroke (= $2R$) ratio and the connecting rod length-to-stroke ratio must be carefully selected when designing a reciprocating compressor. A commonly accepted practice

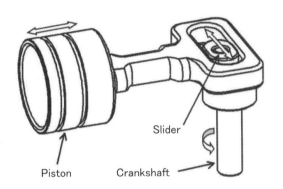

Fig. 3.5 Scotch yoke type

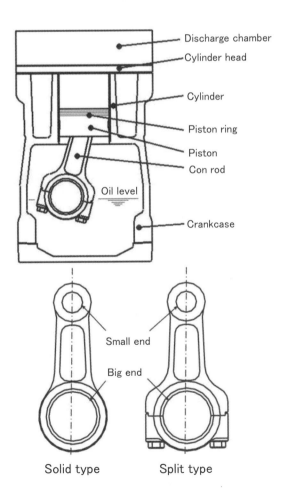

Fig. 3.6 Crank-con rod type

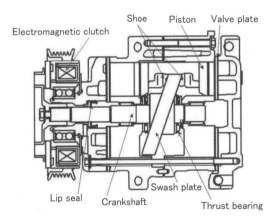

Fig. 3.7 Swash plate type

to realize a small-sized compressor is to make the piston diameter relatively larger with a shorter stroke, but this approach is sometimes associated with leakage risks which could result in performance deterioration and therefore must be used with great care. On the other hand, an excessively long stroke could cause piston overspeed which possibly leads to seizure or wear. Therefore, a stroke-to-piston diameter ratio in the range of 0.4 to 0.8 is normally used. In large capacity compressors, the ratio about 0.8 is generally adopted, but ratios higher 0.8 may be occasionally selected.

In addition, a commonly accepted ratio of the connecting rod length to the stroke is in the range of 1.7 to 3.5. A higher ratio will increase the compressor's mechanical efficiency though it may lead to a taller compressor.

3.3 Structure of Compressors

3.3.1 Motion conversion mechanism

This section explains the motion conversion mechanisms that are commonly used in reciprocating compressors, such as the Scotch yoke type, the connecting rod type and the swash plate type. Among a wide variety of mechanisms that can convert rotational motion into a reciprocating motion, hermetic compressors for refrigeration and air conditioning systems typically use either the Scotch yoke type or the connecting rod type mechanism due to their size, weight or complexity. On the other hand, the swash plate type mechanism is widely used in automotive open-type compressors.

(1) Scotch yoke mechanism

Figure 3.5 shows a cross section of the Scotch yoke compression mechanism. The mechanism comprises a slider ("yoke") with a slot that extends perpendicularly to the direction of piston motion and an eccentric pin that is fixed to a driving shaft to induce reciprocating motion of a piston while sliding back and forth in the slot.

Since the piston and the Scotch yoke are assembled in a body, lubrication of the mechanism is needed only at two spots, one is the pivot of the eccentric pin where rotational motion occurs and the other is the slot of the Scotch yoke where sliding motion occurs. These two lubrication spots are located closely together at the central portion of the mechanism and therefore can be easily lubricated simultaneously. However, the Scotch yoke mechanism has also drawbacks that the Scotch yoke as the core of motion conversion tends to be big and heavy and that mechanical loss becomes large since the eccentric pin slides back and forth in the slot while supporting the gas load working on the piston.

If the Scotch yoke is supported by some linear motion

mechanism in the piston moving direction, the lateral load between the piston and the cylinder would be fully supported by such mechanism and therefore no lateral load will act on the piston. In the case of the absence of such support structure, however, the discrepancy between the load acting position and the driving position will induce a moment which consequently arise a lateral piston load, similar to the situation of the normal connecting rod mechanism.

Currently, the Scotch yoke mechanism is mostly used in small-size compressors only.

(2) Connecting rod mechanism

Figure 3.6 shows a connecting rod mechanism and shapes of the connecting rod. This structure is similar to that of an automobile engine and is most popularly applied to reciprocating compressors. A portion of the connecting rod combined with the crankshaft is called the "big end", and one combined with the piston is called the "small end". Action of the big end is a combination of rotational motion by the shaft rotation and swing motion by the reciprocating motion of the piston. On the other hand, action of the small end is swinging motion only.

The longer the connecting rod is, the smaller the swinging angle will be, thus reducing the amount of mechanical loss at the small end. However, such approach would make the compressor unit longer in the piston direction and therefore heavier as a whole.

The connecting rod mechanism is used in a wide variety of compressors, from small to large ones. Many engineering studies have been conducted about the connecting rod mechanism, including analyses and measurements of the lubrication characteristics in the sliding area and various kinematic analyses taking account of the force of inertia[2].

(3) Swash plate mechanism

Figure 3.7 shows a swash plate mechanism. The mechanism comprises a shaft fitted up with an inclined plate ("swash plate") and pistons with shoes. As the shaft rotates, the pistons move back and forth with the help of the shoes sliding on the swash plate. If the shoes and the swash plate are simply slid against each other, it will result to a significant sliding loss. However, the sliding loss can be reduced by either providing a fluid lubrication or decreasing a contact load by introducing high-pressure compressed gas under the shoes. Due to its simple and compact structure, the swash plate mechanism is widely used in automotive compressors where weight reduction is critical.

3.3.2 Suspension mechanism

Some reciprocating compressors with a built-in compact motor have an internal suspension mechanism for vibro-isolating support which secures the compression mechanism to the compressor housing. Figure 3.8 shows a cross section of a compressor equipped with a spring-fitted internal suspension mechanism and a shock loop that reduces discharge pipe stress. Figure 3.9 shows details of an internal suspension mechanism.

In some cases, the internal suspension mechanism includes a displacement-restricting stopper. This will prevent contact between the compression mechanism and the housing in the event that the compression mechanism shifts significantly during the startup or shut-down operation of the compressor. If the spring's natural frequency is expected to be reached before the steady-state rotational speed of the compressor is attained, a rubber damper or other damping apparatus may be used in order to restrict the maximum amplitude.

Often used together with such suspension mechanism is a discharge shock loop. The shock loop connects the discharge muffler on the compressor mechanism side to the discharge pipe on the housing side. The shock loop is typically designed in a looped configuration in order to absorb potential displacement of the compression mechanism. The diameter and the winding number of shock loop must be carefully selected so that all the axial, radial and rotational stresses of the loop pipe will be kept below the allowable limit in all operating ranges of the compressor.

3.3.3 Lubrication system
(1) Centrifugal pump

In compressor technology, the biggest advantage of a centrifugal pump is that its lubrication system can be configured by combining a centrifugal pump with a smallest amount of additional parts. The centrifugal pump is commonly used in compact compressors that run at a constant speed and have a vertical shaft with one end submerged in an oil sump or reservoir. Figure 3.10 shows a shaft end structure containing a centrifugal pump mechanism. Lubrication oil entering from the shaft end develops pressure through the centrifugal force of the shaft rotating at an angular speed of ω. With the diameter of the lubricant outlet denoted as d, the lubricant supply pressure P can be obtained by the following equation:

$$P = \frac{\rho}{2} \left(\frac{d}{2} \omega\right)^2 \tag{3.4}$$

An important characteristic of this type of centrifugal pump is that it runs as a constant-pressure pump up to a certain flow rate. On the other hand, a drawback associated with this type of pump is that refrigerant dissolved in the oil becomes gasified due to the stirring action, which then possibly deteriorate the pump performance. To let the gasified refrigerant escape, many centrifugal pumps are designed with a vent hole on the side wall of the pump.

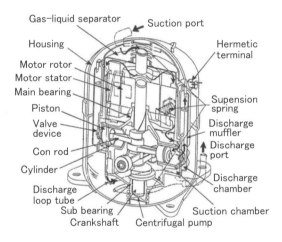

Fig. 3.8 Reciprocating compressor with internal suspension mechanism

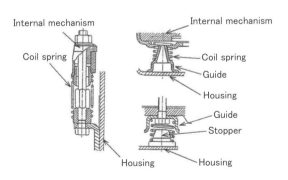

Fig. 3.9 Internal suspension mechanism

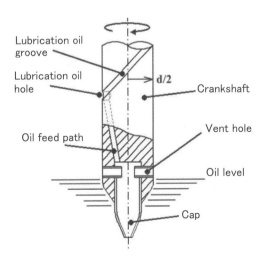

Fig. 3.10 Centrifugal pump

(2) Positive displacement pump

Figure 3.11 shows the structure of a trochoid pump. The trochoid pump comprises an outer rotor with a trochoidal profile, an inner rotor of which the number of teeth is smaller by one than that of the outer rotor, and a casing that houses these rotors. The inner rotor rotates by itself around the shaft axis, while the outer rotor set eccentrically from the shaft axis is driven by the inner rotor. As the outer rotor turns with the inner rotor, difference in the number of teeth causes volumes enclosed with the inner and outer teeth to expand and shrink, thus ejecting lubrication oil.

Aside from the trochoidal pump, another type of a positive displacement is the vane pump. Figure 3.12 shows an example of the vane pump. A vane pump consists of a rotor and vanes put in vane slots of the rotor. As the rotor turns, spaces enclosed with the vanes expand and shrink, thus ejecting lubrication oil.

(3) Mist lubrication and splash lubrication

Some automotive compressors do not have a lubrication oil sump or reservoir, but lubrication is provided in the form of oil mist migrating with refrigerant. In the case of such mist lubrication, the oil mist content in refrigerant must be kept above a certain level in order to assure reliable sliding actions. For this purpose, the oil mist lubrication system is equipped with an oil separator which separates the lubrication oil from compressed refrigerant at the discharge side of the compressor and returns the separated oil to the suction side.

A splash lubrication system has an oil reservoir at the bottom of the compressor. When the big end of the connecting rod hits oil in the reservoir, oil is splashed on components of the compressor and lubricates them.

(4) Oil feed path

Lubrication oil that has been pumped up typically flows through oil holes in the crankshaft to be channeled to mechanical seal areas and to big ends of the cylinder connecting rods. Part of the channeled oil lubricates the bearing in the big end and the remaining oil flows through an oil hole in the connecting rod to reach the small end, lubricating the bearing there. For the lubrication oil to be evenly distributed to all cylinder sets, diameters of the oil holes in the crankshaft must be carefully designed.

3.3.4 Valve system

Figure 3.13 shows an example of suction and discharge valves for refrigerant compressors. The suction and discharge valves shown are a type of reed valve, which is often used in relatively small compressors.

Figure 3.14 shows the structure of an annular valve, another type of commonly used compressor valve. The

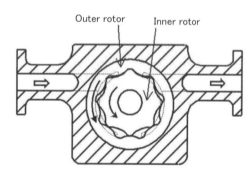

Fig. 3.11 Trochoid pump

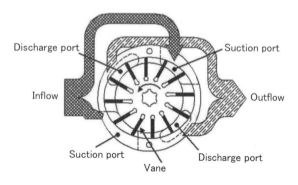

Fig. 3.12 Vane pump

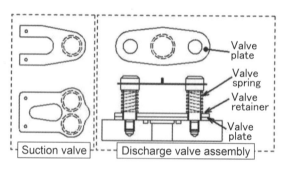

Fig. 3.13 Reed valves

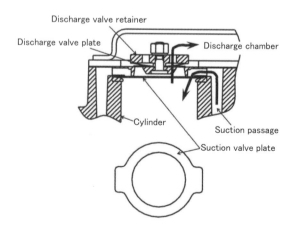

Fig. 3.14 Annular valve

discharge valve is ring-shaped and riveted or screwed at the center of a valve plate. A plate of the discharge valve covers and serves a number of discharge holes. This type of discharge valve can provide a larger discharge opening area than the reed valve, and therefore often used in large-capacity compressors. On the other hand, the suction valve, also ring-shaped, has protruding tabs on both ends which are loosely held between the cylinder and the valve plate body, thus securing the valve plate in place. During the suction process of compressor, the suction valve bends toward the cylinder to open and let gas flow into the cylinder as shown by an arrow in the figure. During the compression process, the suction valve is pressed against the cylinder head due to pressure difference, thus blocking any reverse flow into the suction path. Similar to the discharge valve, the suction valve can provide relatively large opening with a small amount of valve lift.

Since the compressor valve must complete both opening and closing in one rotation, the valve should have good high-speed response. This requirement can be easily satisfied if a thin material is used, as in the case of a reed or annular valves. However, the valve must also withstand deformation stress that develops due to its lifting action and fluid pressure difference as well as the impact stress that develops when the valve hits the retainer or the opening edge.

To summarize, the valve material must have high fatigue strength and must be wear- and impact-resistant. The finished valve sheet must also be defect-free, uniform in thickness and completely flat and have a smooth surface.

Materials that can satisfy such requirements generally have a high notch sensitivity. In order to increase the durability of the finished valve sheet, it is necessary to provide barrel polishing to round off sharp edges after the sheet has been machined into shape. Swedish steel (Sandvik) was the dominant material that satisfies the previously-mentioned valve sheet requirements in the past, but Japanese "Tatara" iron and SUS steel have been also commonly used in recent years.

3.3.5 Piston ring

Piston rings serve to block high-pressure gas in the cylinder from leaking out through the gap between the piston and the cylinder and also to prevent excessive migration of lubrication oil into the cylinder volume. Figure 3.15 shows the cross section of a large-size piston, which can be equipped with two seal rings at the upper part and one oil ring, serving to scrape off excessive oil to the bottom of the compressor, at the lower part. Some of small-size compressors, however, have only one or no seal ring on the piston.

Figure 3.16 shows shapes of the periphery shape and the closed gap of piston rings and also cross section shapes of oil rings. The piston rings must have a closed gap that is a cut on the ring, so that the ring can expand to the cylinder wall to seal the gap between the piston and the cylinder. The joint of the closed gap may be variously shaped, for example cut at an angle or stepped, to minimize leakage through the joint.

The periphery of the piston ring is also variously shaped in the cross section, some tapered and some barreled, so that an appropriate oil film can be formed on the cylinder wall. As the piston ring material, grey cast iron, which has superior impact- and wear-resistance, is most commonly used. Plastic rings made of carbon-fiber- or glass-fiber-reinforced PTFE may also be used.

One specific feature of the oil ring is that it has oil channels that help to collect excess oil on the cylinder wall and return it to the crank chamber.

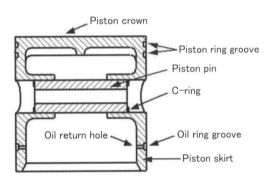

Fig. 3.15 Piston structure

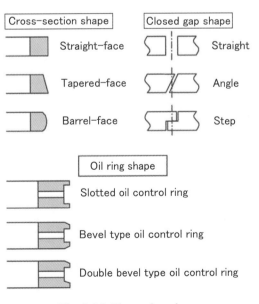

Fig. 3.16 Piston ring shape

3.3.6 Discharge muffler

The inside of the housing of a hermetic reciprocating compressor is typically at suction pressure. To control pressure pulsation of the gas being discharged, a discharge muffler is used in the compressor housing. A discharge muffler comprises a volume chamber and a pipe inserted in the volume. In the case of a multiple-cylinder compressor, the muffler is required to accommodate gas discharged from multiple cylinders working at different rotation phases. When designing the discharge muffler, its volume and length of its connecting pipe are the critical parameters that determine the muffler performance. Relatively low-level pulsation such as primary rotational pulses can be alleviated by the damping effect of the muffler volume itself. The length of pipe connected to the muffler must be carefully designed so that good enough damping effect can be obtained in the target frequency range where damping is most required.

In recent years, it has become possible to estimate the acoustic characteristics of the muffler and the manifold pipe through Finite Element Method (FEM). Estimation of the effect of pulsation reduction in the design process is often conducted using such techniques.

On the other hand, some muffler designs actively use the effect of discharge pulsation to reduce over-compression inside the cylinders. Figure 3.17 shows an example of the relationship between the cylinder pressure and the discharge cavity pressure, where multiple cylinders of different discharge timings are connected to a single muffler. There exist a response delay in the discharge cavity pressure because the discharge cavity is located downstream of the cylinder. The discharge cavity pressure increases during discharge phase of the cylinder and reaches its maximum higher than the discharge nominal pressure at about the time of discharge completion. After that, it decreases and becomes lower than the nominal discharge pressure at the time, designated at point A in Fig. 3.17, when discharge from the cylinder starts. The degree of over-compression inside the cylinder is controlled.

Some other compressor designs use a suction muffler to reduce suction-side pulsation. Studies have been conducted to visualize such flow field characteristics[3]. Mufflers may also be used to provide inertial supercharging with the help of a pipe connecting to the cylinder suction port, which can improve volumetric efficiency at a specific number of revolutions of the compressor.

3.3.7 Shaft seal mechanism

In the case of open-type compressors, a driving shaft have to penetrate the pressure housing of the compressor to supply energy from an external power source. Shaft seals are used for preventing leakage at the section of the shaft penetration. Most commonly used shaft seals are mechanical seals and lip seals.

(1) Mechanical seal

Many crank-piston type reciprocating compressors use a mechanical seal for the shaft sealing purpose. Mechanical seals are favorably used taking account of such advantages as structure, load, vibration, assemblicity, component shape, machinability, material selection and lubrication of the compressor. One notable advantage of using a mechanical seal is that it has higher tolerance in inclination and eccentricity of the crankshaft as compared with another type seal (i.e. lip seal which will be described later) where the seal contact force works in the radial direction.

Figure 3.18 shows the structure of a mechanical seal. Figure 3.19 shows an example of surface grooves which are formed on a sealing face with depth of several micrometers. The grooves work effectively only in one rotational direction but the pumping effect of the grooves helps to improve its sealing performance.

A mechanical seal comprises a fixed ring, a rotating ring, a setup that presses the rotating ring against the fixed ring

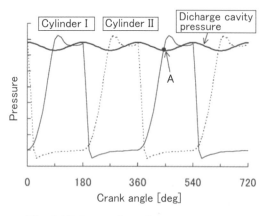

Fig. 3.17 Prevention of over-compression

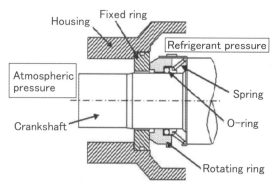

Fig. 3.18 Mechanical seal

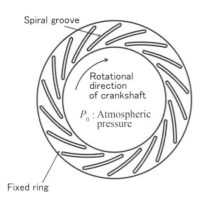

Fig. 3.19   Fixed ring

and a retainer to secure these components in position.

The first thing to consider in the selection of mechanical seal is the method to press the sealing faces each other. The sealing force is primarily given by a mechanical spring, and is assisted by a force based on the pressure difference across the mechanical seal. There are two methods to supply the pressure differential force; one is the balance type where the pressing force is controlled by a diameter and area of the sealing face to reduce a load by pressure difference, and the other is the unbalance type where the sealing force is simply added by the pressure differential force. The balance type has smaller risk of wear and seizure due to its lower surface pressure, but is likely to suffer from larger leakage than the unbalance type. On the other hand, the unbalance type causes little leakage owing to the sufficient sealing force by the pressure difference, but it requires lubrication and cooling to prevent wear and seizure of the sealing faces. The fixed ring is generally made of hard and wear-resistant material such as cast iron or ceramic. The rotating ring is commonly made of sintered carbon, which is less hard than the fixed-ring material.

Air conditioning and refrigeration compressors typically use an inner-type mechanical seal which positions the rotating ring on the refrigerant side. The structure is advantageous in a point that the sealing faces can be cooled and lubricated by refrigerant and oil inside the compressor. The inner type also provides better sealing performance than the outer type because any potential leakage must flow in an inward direction from the outer edge of the ring to the inner edge, against the centrifugal force by rotation.

As a structure to press the rotating ring against the fixed ring, a mechanical bellows or a spring pressing can be used. The bellows are made of metallic material or PTFE. The spring pressing is possible in two types; a single-spring type where only a spring presses the sliding face and a multiple-spring type where more than one springs press the sliding face. Small reciprocating compressors with a stroke volume of up to approximately 500 ml typically use a single-spring type, and larger compressors sometimes use a multiple-spring type depending on the situation. The multiple-spring type has a feature that the rotating ring is pressed against the fixed ring more compliantly under the case that the shaft axis is tilting unfavorably.

When using a mechanical seal, it should be noted that tilt and run-out of the shaft are within a level allowable for the mechanical seal, that the sealing face is effectively lubricated and cooled and that get-in of solid materials foreign to the compressor between sealing faces is structurally prevented. The mechanical seal shown in Fig. 3.19 should not be used in a reciprocating compressor without controlling the rotational direction because it can work effectively only in one rotational direction. When a sealing problem is reported from the market field, the direction of shaft rotation should firstly be checked in addition to wear, seizure by inadequate lubrication and poor alignment by excessive run-out of the shaft.

(2) Lip seal

Figure 3.20 shows the structure of a lip seal which is used in a swash-plate type reciprocating compressor. The lip seal has a fewer number of components since the shaft itself serves as a seal surface on the rotating side while the rubber or PTFE lip is in direct contact with the shaft, making it more suitable for small-diameter shaft designs where the circumferential sliding speed is lower.

In an air conditioning and refrigeration system, a rubber lip may deform excessively due to large pressure difference, which will lead to deterioration of the sealing performance. A backup lip made of PTFE is commonly used for double lip sealing to support the rubber lip, and furthermore a dust lip is added for triple rip sealing to block external foreign objects.

Crisscross pattern or thread pattern, caused by feeding of a lathe, left on the shaft surface may cause a leakage as

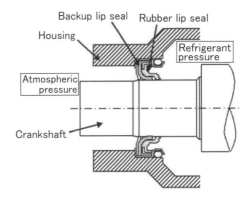

Fig. 3.20 Lip seal

refrigerant can travel along such surface imperfections. Therefore, it is important to maintain the surface finish without any machining defects, and the surface roughness level of approximately $R_a = 0.1$ must be maintained. Since a high surface hardness is required, the shaft must be heat-treated. Note that metallic materials which are prone to carbon segregation on the surface should be avoided since pits left after decarburization may aggravate leakage or lip wear.

After a prolonged use of a lip seal, the shaft gets worn as well as the rubber or PTFE lip. Therefore if any leakage from the lip seal area is found or suspected, both the shaft and the lip seal should be replaced. Due to this drawback, it is difficult to use a lip seal in large reciprocating compressors. Lip seals can only be practically used in small reciprocating compressors where it is more common to replace the whole compressor in the event of a failure.

3.3.8 Capacity control mechanism

Capacity control generally refers to a feature or mechanism that adjusts the compressor capacity according to the operating condition of air conditioning and refrigeration systems. Compressor capacity Q can be obtained by the following equation:

$$\left. \begin{array}{l} Q = G \times \Delta h \\ G = n\rho\eta_v V_{th} m \end{array} \right\} \quad (3.5)$$

where G is mass flow rate, Δh is the specific enthalpy variation of the refrigerant, n is number of revolutions, η_v is volumetric efficiency, V_{th} is the stroke volume of a single cylinder, m is the number of cylinders, and ρ is suction gas density. Capacity control is possible by changing any of the parameters that determine capacity Q. The following are examples of capacity control techniques:

(1) Changing the number of revolution n
- In the case of motor-driven applications: Speed variation with the use of an inverter, changing the number of motor poles.
- In the case of belt-driven applications: Use of a planetary drive or CVT drive.

(2) Changing the stroke volume
- Early closure of the suction valve: Have the suction valve closed before the piston reaches the bottom dead center, that is suction completion.
- Delay in suction valve closure: Have the suction valve open even after the piston has passed the bottom dead center. This technique may also be called "compression stroke bypath" and may be classified as a volumetric efficiency reduction feature through intentional leakage. An actual compression stroke bypath technique provides an additional port on the cylinder side wall for connection to the suction line, so that the opening and closing of the valve causes stroke volume variation.
- Changing the stroke length: In a swash-plate type compressor, the piston stroke can be increased or decreased by changing the swash plate angle. In a crank-piston type unit, the stroke length can be increased or decreased by changing the link geometry.

(3) Changing the volumetric efficiency η_v:
- Varying the top clearance volume: The volumetric efficiency can be reduced by increasing the top clearance volume by changing the linkage configuration. (Refer to Fig. 3.21)
- Discharge gas bypath: Fully compressed gas is returned to the suction side.

(4) Changing the number of cylinders:
- Have the suction valve of a particular cylinder continuously open or closed so that no actual work can be done by that cylinder. Both the "continuously open" and "continuously closed" techniques have been actually implemented in multiple-cylinder reciprocating compressors.

(5) Changing other operating conditions:
- Changing density ρ: Intentional restriction of the suction inlet of the compressor so as to lower pressure on the low-pressure side will cause the gas density to decrease, resulting in mass flow rate reduction.

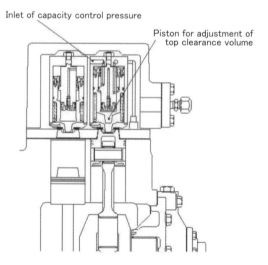

Fig. 3.21 Structure of variable top clearance volume

The above-described capacity control techniques have all been actually implemented in reciprocating compressors. The same principles can be applied to rotary and scroll compressors for similar capacity control effects.

3.4 Analysis of Kinematics

3.4.1 Kinematic analysis

The most typical and most commonly used mechanism for reciprocating compressors is the connecting rod mechanism, or commonly called "piston-crank mechanism". This section provides analytical method of mechanical dynamics of the piston-crank mechanism.

In this analysis, the piston-crank mechanism is represented by a "lumped mass system" shown in Fig. 3.22. This figure is based on a Cartesian coordinate of x and y axes, where the crankshaft center is the origin. Here the mass of the connecting rod is denoted by m_c and its center of gravity locates at (x_c, y_c). Its moment of inertia around the center of gravity is denoted by I_c. The piston mass m_p is gathered around the piston pin center $(x_p, 0)$. The mass of crank arm and crank pin is denoted by m_e and its center of gravity locates at (x_e, y_e). The moment of inertia of the entire crankshaft system including the motor rotor around the crankshaft axis is denoted by I_0.

With the counterclockwise rotational angle θ of the crankshaft from the x axis and the clockwise rotational angle ϕ of the connecting rod, the relationship between θ and ϕ can be expressed by the following equation:

$$r \sin \theta = l \sin \phi \tag{3.6}$$

where r is the rotational radius of the crank pin center and l is the distance between the centers of big end and small end of the connecting rod. The piston pin center coordinate x_p can be obtained by the following equation:

$$x_p = r \cos \theta + l \cos \phi \tag{3.7}$$

Eliminating ϕ from the above equation by using Eq. (3.6), the following equation is obtained:

$$x_p = r \cos \theta + l \sqrt{1 - (r/l)^2 \sin^2 \theta} \tag{3.8}$$

Thus, the piston displacement x_p can be determined by the crank rotational angle θ. Figure 3.23 expresses the forces and torques acting on the individual components of the piston-crank mechanism. Motor torque M_D works on the crankshaft as the drive power, causing the gas inside the cylinder to be compressed via piston-crank mechanism. The force of gas working on the piston is expressed as P. This force is supported by forces of constraint developed at the linking points between each component in mutually opposing directions. These constraint forces are denoted as Q_x, Q_y, S_x, S_y, T_x and T_y. At the same time, frictional torques M_Q, M_S and M_T develops at the turning pairs and frictional force F_f develops at the sliding pair.

Force of inertia caused by linear motion of the piston is denoted by $(-m_p \ddot{x}_p)$, forces of inertia caused by linear motion of the connecting rod $(-m_c \ddot{x}_c)$ and $(-m_c \ddot{y}_c)$, torque of inertia caused by rotational motion of the connecting rod $(-I_c \ddot{\phi})$, forces of inertia caused by linear action of the crankshaft system $(-m_e \ddot{x}_e)$ and $(-m_e \ddot{y}_e)$, torque of inertia caused by rotational motion of the crankshaft system $(-I_0 \ddot{\theta})$. Each force works in the direction defined on the coordinate plane. Applying the d'Alembert's principle with consideration of the above factors, equations of motion for all the movable links can be derived as follows:

Piston

$$-m_p \ddot{x}_p - P + F_f - T_x = 0 \tag{3.9}$$

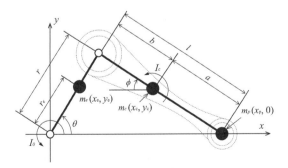

Fig. 3.22 Lumped mass system of con rod type compressor

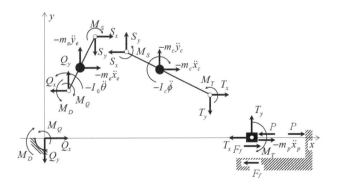

Fig. 3.23 Forces and torques acting on links

Connecting rod

$$-m_c \ddot{x}_c + T_x - S_x = 0 \qquad (3.10)$$

$$-m_c \ddot{y}_c - T_y + S_y = 0 \qquad (3.11)$$

$$\begin{aligned} &-I_c \ddot{\phi} - T_x a \sin\phi + T_y a \cos\phi \\ &-S_x b \sin\phi + S_y b \cos\phi \\ &-M_S - M_T = 0 \end{aligned} \qquad (3.12)$$

Crankshaft system

$$-m_e \ddot{x}_e + S_x - Q_x = 0 \qquad (3.13)$$

$$-m_e \ddot{y}_e - S_y + Q_y = 0 \qquad (3.14)$$

$$\begin{aligned} &-I_0 \ddot{\theta} + M_D - S_x r \sin\theta \\ &- S_y r \cos\theta + N - M_Q - M_S = 0 \end{aligned} \qquad (3.15)$$

Putting Eqs. (3.9) to (3.14) together, the six constraint forces that develop at the linking points between the components can be obtained as follows:

$$T_x = -P + F_f + \left(-m_p \ddot{x}_p\right)$$

$$\begin{aligned} T_y &= \left\{-P + F_f + \left(-m'_p \ddot{x}_p\right)\right\}\tan\phi \\ &+ \frac{1}{l\cos\phi}\left\{I'_c \ddot{\phi} + M_S + M_T\right\} \end{aligned}$$

$$\begin{aligned} Q_x &= -P + F_f + \left(-m'_p \ddot{x}_p\right) \\ &+ (m_{ro}r + m_e r_e)\frac{d}{dt}(\sin\theta\,\dot{\theta}) \end{aligned}$$

$$\begin{aligned} Q_y &= \left\{-P + F_f + \left(-m'_p \ddot{x}_p\right)\right\}\tan\phi \\ &+ \frac{1}{l\cos\phi}\left\{I'_c \ddot{\phi} + M_S + M_T\right\} \\ &+ (m_{ro}r + m_e r_e)\frac{d}{dt}(\cos\theta\,\dot{\theta}) \end{aligned} \qquad (3.16)$$

$$\begin{aligned} S_x &= -P + F_f + \left(-m'_p \ddot{x}_p\right) \\ &+ m_{ro}r\frac{d}{dt}(\sin\theta\,\dot{\theta}) \end{aligned}$$

$$\begin{aligned} S_y &= \left\{-P + F_f + \left(-m'_p \ddot{x}_p\right)\right\}\tan\phi \\ &+ \frac{1}{l\cos\phi}\left\{I'_c \ddot{\phi} + M_S + M_T\right\} \\ &+ m_{ro}r\frac{d}{dt}(\cos\theta\,\ddot{\theta}) \end{aligned}$$

where m'_p is modified mass of the piston:

$$m'_p \equiv m_p + m_{re} \qquad (3.17)$$

and I'_c is modified moment of inertia of the connecting rod:

$$I'_c \equiv I_c - m_c\, a \times b \qquad (3.18)$$

where a and b are distances between the center of gravity and the centers of the small and big end of the connecting rod respectively, as shown in Fig. 3.22. m_{re} and m_{ro} are reciprocating mass and rotational mass of the connecting rod respectively, and can be expressed by the following equation:

$$m_{re} \equiv \frac{b}{l}m_c \qquad m_{ro} \equiv \frac{a}{l}m_c \qquad (3.19)$$

Equation (3.16) indicates that the constraint force is determined not only by the force of gas working on the piston but also by the forces of inertia, inertia toques, the frictional forces and the frictional torques.

By substituting the constraint forces S_x and S_y of Eq. (3.16) into Eq. (3.15), the equation of motion can be obtained as follows:

$$\begin{aligned} &I'_0 \ddot{\theta} + \left(-m'_p \ddot{x}_p\right) x_p \tan\phi + \alpha\frac{\cos\theta}{\cos\phi}I'_c \ddot{\phi} \\ &= M_D - (-P x_p \tan\phi) \\ &- \left[F_f x_p \tan\phi + \alpha\frac{\cos\theta}{\cos\phi}M_T + \alpha\frac{x_p/r}{\cos\phi}M_S + M_Q\right] \end{aligned} \qquad (3.20)$$

where I'_0 is the modified inertia moment of the crankshaft system and can be given by the following equation:

$$I'_0 \equiv I_0 + m_{ro}\, r^2 \qquad (3.21)$$

Concerning Eq. (3.20), the left hand side expresses inertial terms and the first, the second and the third terms of the right hand side correspond to the motor drive torque, the gas compression load torque and the load torque induced by mechanical friction respectively. When the values for the motor torque and the gas pressure characteristics are given, the solution to this equation of motion, i.e. rotational behavior of the crankshaft, can be obtained by using a numerical computation method such as the Runge-Kutta-Gill method. Note that iterative calculation is needed to obtain the numerical solution of Eq. (3.20) because the frictional forces and torques are influenced by the constraint forces of Eq. (3.16). The solution will converge fairly quickly. After converging the solution, the constraint forces at the connections between components of the mechanism are finally decided using Eq. (3.16).

3.4.2 Equation of energy and mechanical efficiency

The equation of energy can be obtained by multiplying the equation of motion, Eq. (3.20), by angular displacement and then integrating the multiplied equation over one rotation of the crankshaft. Since the integral of the inertia term is zero in a steady-state operation, the equation can be derived as:

$$E_{motor} = E_{gas} + E_{piston} + E_{p-pin} \\ + E_{c-pin} + E_{c-shaft} \tag{3.22}$$

where

$$E_{motor} = \int_0^{2\pi} M_D \, d\theta$$

$$E_{gas} = \int_0^{2\pi} \left(-P x_p \tan\phi \right) d\theta$$

$$E_{piston} = \int_0^{2\pi} F_f x_p \tan\phi \, d\theta$$

$$E_{p-pin} = \int_0^{2\pi} \alpha \frac{\cos\theta}{\cos\phi} M_T \, d\theta \tag{3.23}$$

$$E_{c-pin} = \int_0^{2\pi} \alpha \frac{x_p/r}{\cos\phi} M_S \, d\theta$$

$$E_{c-shaft} = \int_0^{2\pi} M_Q d\theta$$

Equation (3.22) means that the motor-supplied energy E_{motor} is consumed by the gas compression energy E_{gas} and such frictional energies as E_{piston} on the piston side wall, E_{p-pin} at the piston pin, E_{c-pin} crank pin and $E_{c-shaft}$ on the crankshaft. Mechanical efficiency η_m can be calculated by the following equation:

$$\eta_m \\ = 1 - \frac{\left(E_{piston} + E_{p-pin} + E_{c-pin} + E_{c-shaft} \right)}{E_{motor}} \tag{3.24}$$

3.4.3 Excitation forces and vibration of compressor body

As shown in Fig. 3.23, constraint forces Q_x and Q_y and frictional torque M_Q act on the crank bearing of the compressor, while constraint force T_y, frictional force F_f and frictional torque M_T act on the cylinder in the respective

directions in the figure. In addition, torque M_D acts on the motor stator. By integrating all these forces and torques into x-axis force F_x, y-axis force F_y and counterclockwise torque M_z around the crankshaft axis, the following equations can be obtained:

$$F_x = Q_x + P - F_f$$

$$F_y = -Q_y + T_y \tag{3.25}$$

$$M_z = T_y \cdot x_p - M_D + M_Q - M_T$$

By substituting the constraint forces Q_x, Q_y, T_y of Eq. (3.16) and M_D obtained from Eq. (3.20) of motion in the above equation, the following equation can be obtained:

$$F_x = \left(-m'_p \ddot{x}_p \right) \\ + (m_{ro}r + m_e r_e) \frac{d}{dt} \left(\sin\theta \dot{\theta} \right)$$

$$F_y = -(m_{ro}r + m_e r_e) \frac{d}{dt} \left(\cos\theta \dot{\theta} \right) \tag{3.26}$$

$$M_z = -I'_0 \ddot{\theta} + I'_c \ddot{\phi}$$

In this equation, all the terms concerning internal forces and internal torques such as gas force, frictional forces and frictional torques are not included, and only inertia forces and inertia torques remain. Vibration of the compressor body is caused as a result of excitation by these unbalanced inertia forces and inertia torques.

To minimize the excitation forces F_x and F_y, careful static and dynamic balancing of the compressor body is critical. Note that only the primary component of the piston's reciprocating inertia force ($- m'_p \ddot{x}_p$) needs to be considered in such balancing process. Excitation forces F_x and F_y are expressed by the following equations:

$$F_x = \left(m'_p r + m_{ro}r + m_e r_e \right) \dot{\theta}^2 \cos\theta$$

$$F_y = (m_{ro}r + m_e r_e) \dot{\theta}^2 \sin\theta \tag{3.27}$$

In general, a compressor is fitted with balancers in two opposed locations along the circumference of the motor rotor, one at the top of the rotor and the other at the bottom. Figure 3.24 shows an example of such balancer installation. Masses of the top and bottom balancers are expressed as m_{bU} and m_{bL} respectively, and their rotational radius as r_b. Here only the centrifugal force is shown, assuming that the crankshaft is rotating at constant speed. On the principle of partial balancing that only a half of the piston's reciprocating inertia force needs to be balanced, the unbalanced centrifugal force is shown at the center of the

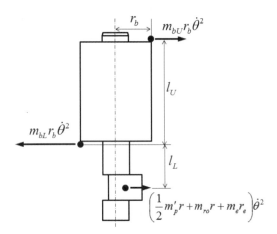

Fig. 3.24 Static and dynamic balance

crank pin.

The height from the crank pin center to the two balancers are expressed as $l_U + l_L$ and l_L respectively. Planar balancing of centrifugal forces, or static balancing, can be given as follows:

$$-m_{bL}r_b\dot{\theta}^2 + m_{bU}r_b\dot{\theta}^2 + \left(\frac{1}{2}m'_p r + m_{ro}r + m_e r_e\right)\dot{\theta}^2 = 0 \tag{3.28}$$

Dynamic balancing is given as follows:

$$m_{bU}r_b\dot{\theta}^2 l_U - \left(\frac{1}{2}m'_p r + m_{ro}r + m_e r_e\right)\dot{\theta}^2 l_L = 0 \tag{3.29}$$

Based on the above two equations, specifications of the balancers are determined as follows:

$$m_{bU}r_b = \left(\frac{1}{2}m'_p r + m_{ro}r + m_e r_e\right)\frac{l_L}{l_U}$$

$$m_{bL}r_b = \left(\frac{1}{2}m'_p r + m_{ro}r + m_e r_e\right)\frac{l_L + l_U}{l_U} \tag{3.30}$$

In a balancing design approach like the one described above, excitation forces in x- and y-directions can be obtained by the following equations. These equations mean that unbalanced excitation forces of x- and y-directions are cosine and sine components of a half of the centrifugal force about the piston modified mass.

$$F_x = \left(m'_p r + m_{ro}r + m_e r_e - m_{bL}r_b + m_{bU}r_b\right)\dot{\theta}^2 \cos\theta$$

$$= \frac{1}{2}m'_p r \cdot \dot{\theta}^2 \cos\theta$$

$$F_y = \left(m_{ro}r + m_e r_e - m_{bL}r_b + m_{bU}r_b\right)\dot{\theta}^2 \sin\theta \tag{3.31}$$

$$= -\frac{1}{2}m'_p r \cdot \dot{\theta}^2 \sin\theta$$

The principle of partial balancing is to evenly disperse and thereby minimize the effect of excitation forces, as described above. This approach is applied to the balancing design of all single-cylinder machines. The excitation forces F_x and F_y given by Eq. (3.26) can be finally expressed as follows:

$$F_x = -m'_p \ddot{x}_p$$
$$+ (m_{ro}r + m_e r_e - m_{bL}r_b + m_{bU}r_b)\frac{d}{dt}(\sin\theta \cdot \dot{\theta})$$
$$= -m'_p \ddot{x}_p - \frac{1}{2}m'_p r \frac{d}{dt}(\sin\theta \cdot \dot{\theta})$$
$$\tag{3.32}$$
$$F_y = -(m_{ro}r + m_e r_e - m_{bL}r_b + m_{bU}r_b)\frac{d}{dt}(\cos\theta \cdot \dot{\theta})$$
$$= \frac{1}{2}m'_p r \frac{d}{dt}(\sin\theta \cdot \dot{\theta})$$

Meanwhile, excitation moment M_z of Eq. (3.26) does not change.

Figure 3.25 shows a typical vibration system associated with a compressor. This figure is based on an absolute coordinate system of X, Y and Z axes, where the center of gravity G of the compressor is the origin. It is assumed that the three coordinate axes are identical to the principal axes of inertia of the compressor. Rotational displacements around the respective axes are denoted as $(\Theta_X, \Theta_Y, \Theta_Z)$. The equation of motion to determine the vibration of the compressor can be expressed as the following determinant:

$$[M]\left[\ddot{X}\right] + [\gamma]\left[\dot{X}\right] + [K][X] = [F] \tag{3.33}$$

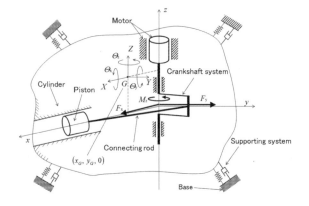

Fig. 3.25 Vibration system of compressor

where $[X]$ is matrix of the vibration displacement, $[M]$ is matrix of the mass and inertia moment, $[\gamma]$ is matrix of the viscous damping coefficient and $[K]$ is matrix of the stiffness of supporting spring system, each of which can be expressed as follows:

$$[X] = \begin{bmatrix} X \\ Y \\ Z \\ \Theta_x \\ \Theta_y \\ \Theta_z \end{bmatrix}, [M] = \begin{bmatrix} M & \cdots\cdots & 0 \\ & M & & \vdots \\ & & M & & \vdots \\ & & & I_x & \vdots \\ Sym. & & & I_y & \vdots \\ & & & & & I_z \end{bmatrix} \quad (3.34)$$

$$[\gamma] = \begin{bmatrix} \gamma_{11} & \cdots\cdots\cdots & \gamma_{16} \\ & \gamma_{22} & & \vdots \\ & & \gamma_{33} & & \vdots \\ & & & \gamma_{44} & \vdots \\ Sym. & & & \gamma_{55} & \vdots \\ & & & & & \gamma_{66} \end{bmatrix} \quad (3.35)$$

$$[K] = \begin{bmatrix} K_{11} & \cdots\cdots\cdots & K_{16} \\ & K_{22} & & \vdots \\ & & K_{33} & & \vdots \\ & & & K_{44} & \vdots \\ Sym. & & & K_{55} & \vdots \\ & & & & & K_{66} \end{bmatrix} \quad (3.36)$$

On the x, y and z coordinate system where the crankshaft is the origin and with the center of gravity G expressed as $(x_G, 0, z_G)$ shown in Fig. 3.25, matrix $[F]$ of the excitation force of compressor, is basically given as follows:

$$[F] = \begin{bmatrix} F_x \\ F_y \\ 0 \\ z_G F_y \\ -z_G F_x \\ M_z - x_G F_y \end{bmatrix} \quad (3.37)$$

The excitation force matrix can be expressed as a Fourier series like Eq. (3.38). By using Laplace transform, the periodic solution of Eq. (3.33) is given by Eq. (3.39) as follows:

$$[F] = \sum_{i=0}^{\infty} \begin{bmatrix} F_i^S & F_i^C \end{bmatrix} \begin{bmatrix} \sin i\omega t \\ \cos i\omega t \end{bmatrix} \quad (3.38)$$

$$[X] = \sum_{i=0}^{\infty} \frac{1}{[A(ji\omega)]} \begin{bmatrix} |\Delta^S(ji\omega)| & |\Delta^C(ji\omega)| \end{bmatrix} \\ \times \begin{bmatrix} \sin i\omega t \\ \cos i\omega t \end{bmatrix} \quad (j = \sqrt{-1}) \quad (3.39)$$

$$[A(p)] = \begin{bmatrix} Mp^2 + \gamma_x p + K_{11} & \cdots\cdots & K_{16} \\ & Mp^2 + \gamma_y p + K_{22} & & \vdots \\ & & Mp^2 + \gamma_z p + K_{33} & & \vdots \\ & & & I_x p^2 + \gamma_{mx} p + K_{44} & \vdots \\ Sym. & & & I_y p^2 + \gamma_{my} p + K_{55} & \vdots \\ & & & & I_z p^2 + \gamma_{mz} p + K_{66} \end{bmatrix} \quad (3.40)$$

(p: Laplace transform)

where $[A(p)]$ is a frequency response function. $[\Delta^S(p)]$ and $[\Delta^C(p)]$ are matrices where each line of the $[A(p)]$ matrix is replaced by the excitation force amplitude F_i^S and F_i^C respectively

3.5 Application and Performance

The reciprocating compressor was put to practical use for the first time in refrigeration and air conditioning systems due to its simple compression principle. It is being used in a wide variety of applications not only for general refrigeration/cold storage and air conditioning but also for cryogenic refrigeration systems[4]. The reciprocating compressor can cover almost all the technically attainable temperature ranges. It is also available in a wide variety of sizes, from smaller ones for electric refrigerators with a stroke volume of several milliliters to larger ones with a stroke volume of several liters that can cool an entire refrigeration warehouse. It is also being used with various refrigerants such as fluorocarbons as well as ammonia, carbon dioxide, helium and so forth. The reason for such a diverse and widespread use is that load control is possible even at high pressure ranges by adjusting the piston diameter without adversely affecting the mechanical reliability. The total stroke volume can also be increased or decreased easily by changing the number of cylinders. With these advantages, the reciprocating compressor is capable of serving a wider capacity range than other types of positive placement compressors such as the rolling piston type, the scroll type or the screw type.

However, the production quantity of reciprocating compressors has been declining in the Japanese household and commercial air conditioner industry due to the increasing use of rolling piston and scroll compressors which have better adaptability for inverter-based variable speed operation. In the refrigerator industry, the reciprocating compressors had been replaced for a period of time with inverter-driven rolling piston compressors. At around 2000, flammable carbon-hydrate refrigerants such as propane and butane started to be used in household refrigerators from a viewpoint of environmental issues such as ozone layer depletion and global warning. In order to use the flammable refrigerants safely, it is necessary to develop a refrigeration system that works with the smallest possible amount of refrigerant by reducing the system volume with high pressure. For this requirement, using the rolling piston compressor having a large space on the high-pressure side was refrained from. In consequence, compact reciprocating compressors began to be increasingly used again. In recent years, higher-efficiency reciprocating compressors that provide variable-speed operation in combination with a brushless permanent-magnet motor and an inverter have been put into practical use.

3.6 Other Mechanisms of Reciprocating Compressors

3.6.1 Ball-joint mechanism

Figure 3.26 shows a ball-joint type piston mechanism. This type mechanism can also be classified as a variation of the connecting rod mechanism. A ball-joint is used instead of a piston pin to connect the piston and the small end of the connecting rod. The ball-joint mechanism is self-aligning like the Scotch yoke mechanism and is therefore easier to manufacture than the standard connecting rod mechanism in terms of perpendicularity and parallelism tolerances. Performance of the ball-joint mechanism is equivalent to that of the connecting rod mechanism. It offers advantages of both the Scotch yoke type and the connecting rod type.

3.6.2 Two-cylinder mechanism

Figure 3.27 shows a 300 W class two-cylinder compressor that is used for cold-chain storage systems. The compressor uses the previously described ball-joint mechanism and a double-layered big end. This two-cylinder mechanism realizes a compact design by having two cylinders in horizontally opposed positions and one driving crankshaft. This structure enables the two-cylinder compressor to have a larger stroke volume than that of a one-cylinder compressor with an equivalent outer diameter. A variation of the above two-cylinder mechanism is an inverter-controlled compressor where one of the cylinders works on the low-pressure side and the other on the high-pressure side, while the remaining space inside the compressor housing is filled with an intermediate-pressure refrigerant. This type of compressor has been actually installed to household refrigerators[5].

3.6.3 Two-stage compression

In refrigeration applications, it is not rare that the ratio of the saturation pressure at the condensing temperature to the one at the evaporating temperature exceeds 15. Some of such applications use a two-stage compression system, where a gas that has been compressed once is introduced into another cylinder for re-compression.

Figure 3.28 shows the cross section of a two-stage reciprocating compressor. This example is a three-cylinder unit, where the suction gas is first channeled from the crankcase chamber into one of two-lower-stage cylinders, for compression to the intermediate pressure. The gas is then channeled out of the lower-stage cylinders through the cylinder head and into a one-higher-stage cylinder. The gas fully compressed in the higher-stage cylinder is discharged through a discharge pipe to the refrigeration system.

In this example, the crankcase chamber is maintained at suction pressure. The higher-stage cylinder is subjected to a differential pressure between the higher-stage cylinder

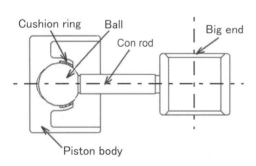

Fig. 3.26 Ball-joint type piston

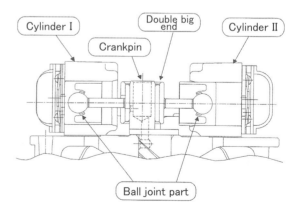

Fig. 3.27 Two-cylinder compression mechanism

Chapter 3 Reciprocating Compressors

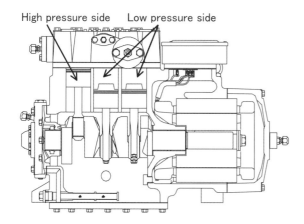

Fig. 3.28 Two-stage compressor

pressure and the suction pressure, which causes a larger leakage. In addition, since back pressure of the high-stage piston is equal to the suction pressure, a larger torque is required to drive the higher-stage piston than what would be normally necessary under the higher-stage operating conditions.

In some other two-stage compression units, the crank chamber is maintained at intermediate pressure. In the lower-stage cylinder of such unit, the piston's back pressure during the compression phase is at intermediate pressure equal to the discharge pressure from the lower stage cylinder, which means that the lower-stage piston receives additional work for power recovery during the compression phase. On the other hand, it has to work more during the suction phase because it needs to overcome the intermediate pressure in order to expand the stroke volume. Pressure relationship on the higher-stage piston is the same as that on a piston of a normally operating single-stage compressor.

It is common in many reciprocating compressors to use a crankshaft with multi throws for different cylinders (for example 180° opposed for two cylinders) in order to control torque fluctuation of the crankshaft. In the case of the above mentioned two-stage compressor having intermediate-pressure crank chamber, however, "one-throw" crankshaft which has throws for all of the three cylinders in the same direction is typically used.

3.6.4 Linear motor-driven mechanism

Newer type of reciprocating compressors uses a linear motor-driven system, where a linear motor directly generates the linear action without any mechanism of rotational-to-reciprocating motion conversion. The elimination of motion conversion mechanism significantly contributes to size and weight reduction and also makes the unit totally free from motion-conversion-induced mechanical losses.

Though linear motors are inherently lower in efficiency than rotary motors, it is believed that the improvement is achieved by the elimination of motion conversion that compensates the inferior motor efficiency. Such linear-motor-driven compressors are now used in small refrigerators. Another advantage of a linear-motor-driven compressor is that it is free from lateral piston load. One concerned drawback is inertia-induced vibration due to the large mass of the motor moving core that generates the driving power. However, improvement will be possible through such techniques as damper design and opposed multiple-cylinder configuration.

References

1) I. Umeoka: "History of Reciprocating Compressor Technology for Refrigerator", Proc. of 2007 JSRAE Annual Conference, Tokyo, pp. 37-40 (2007). (in Japanese)
2) T. Yoshimura, T. Shimizu, T. Yanagisawa, T. Nagao: "A study of the Swing Journal Bearing Characteristics at the Small End of a Connecting Rod in Reciprocating Compressors", Trans. of the JAR, 12 (3), pp. 313-324 (1995). (in Japanese)
3) I. Lee, A. Kaga, K. Yamaguchi: "An Analysis of Reciprocating Compressor Suction System Using Digital Image Processing", Trans. of the JSRAE, 15 (1), pp. 53-61 (1998). (in Japanese)
4) K. Murakami, T. Otaka, M. Sakamoto, H. Yamaguchi, M. Ota: "A Study of a Regenerator for a Personal Stirling Refrigerator", Trans. of the JSRAE, 17 (3), pp. 381-390 (2000). (in Japanese)
5) T. Okamoto: "Refrigeration Cycle with 2-Stage Compressor", Toshiba Review, 59 (5), pp. 60-61 (2004). (in Japanese)

Chapter 4 Rotary Compressors

4.1 History of Rotary Compressors

Unlike reciprocating compressors, rotary compressors have the working fluid compressed directly by the rotational motion of the electric motor without conversion into a reciprocating motion. Rotary compressors belong to the family of rotational compressors. Most rotary compressors use one of the following two types of compression mechanism; the rolling piston type that is often used in the hermetic compressors for residential air conditioners, and the rotary vane (sliding vane) type that is commonly used in the semi-hermetic compressors for automotive air conditioners[1] (see Figs. 4.1 and 4.2). This chapter mostly focuses on the rolling piston-type compressor.

The rotary compression mechanism has a long history. It is said that the basic principle of rotary compression has been considered for use in liquid pumping as early as 1857. From around 1930 onward, numerous engineering and commercialization efforts have been made, mostly by US companies, to use the rotary compression mechanism for refrigerators and air conditioners. However, rotary compression did not become as common as reciprocating compression as it required high levels of component machining and assembly accuracy to work properly.

With the significant improvement in manufacturing technologies in 1960s, it became possible to control the machining accuracy and the assembling clearance on the order of micrometer, which led to a wider use of rotary compressors, mostly for residential air conditioners for their superior efficiency and reliability. Since the oil crisis in 1970s, rotary compressors became even more favored over reciprocating compressors due to their compactness, light weight and high efficiency. In the Japanese market, rotary compressors began to rapidly replace reciprocating compressors in this period. It was also during this period that the basic engineering and manufacturing technologies of rotary compressor have become firmly established.

In 1980s, newly developed large-capacity rotary compressors started to be used, first in commercial packaged air conditioners and then in refrigerators, freezers and cold food showcases as well as dehumidifiers and other medium-temperature applications, thus significantly expanding the range of rotary compressor usage.

One particular turning point that led to the widespread use of rotary compressor technology in the Japanese market today has been the introduction of inverter drive into residential air conditioners. The inverter drive enables to decrease or increase the compressor operational speed (number of rotations per unit time), thus adjusting the cooling capacity as required to realize better comfort and superior power saving. Heat pump type air conditioners based on such inverter technology became extremely popular for residential cooling and heating.

Rotary compressors continued to be further improved for better adaptability to inverter-based variable speed operation and for more refined operating characteristics. Eventually, two-cylinder-type inverter rotary compressors with superior adaptability for variable speed operation were developed and commercialized.

A two-cylinder type rotary compressor has two compression chambers, each running a compression cycle with a 180-degree phase offset from each other. This design contributes to a superior rotational balance and thereby

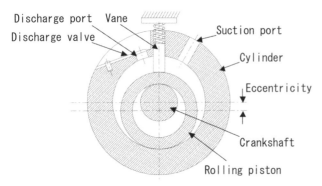

Fig. 4.1 Rolling piston type

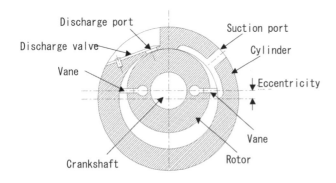

Fig. 4.2 Sliding vane type

provides a low-vibration, high-efficiency performance in a wide operational range, from low to high speeds. With the introduction of this two-cylinder rotary compressor technology, the variable speed operation range has been significantly extended, improving air conditioning comfort and power efficiency and also reducing noise and vibration[2].

As the amount of electric power consumption keeps rising with the increasing number of refrigeration and air conditioning systems being used around the world, greater focus is now placed on how to address global environmental issues to achieve sustainable development. Under the Montreal Protocol of 1987, CFC and HCFC refrigerants, which contain ozone-depleting chlorine, are now restricted and scheduled for gradual switchover to chlorine-free HFC refrigerants. More recently, heat pump-based water heaters using naturally available and environment-friendly CO_2 refrigerant have been commercialized.

However, switching to different types of refrigerant requires a comprehensive study of all the factors involved, including safety and legal requirements as well as energy efficiency and reliability. The entire system needs to be designed anew to be adaptable to the new refrigerant. In the field of rotary compressor technology, significant amounts of engineering development were undertaken to accommodate various alternative refrigerants. For example, engineering efforts on adequate pressure resistance, adaptation of refrigeration oils, compressor materials and surface treatments and introduction of new mechanical features have helped realize superior abrasion resistance in sliding components, thus achieving better reliability and compatibility with the new groups of refrigerants. For performance maximization, compression mechanism dimensions were closely analyzed and refined for stroke volume optimization as well as minimization of leakage, pressure and mechanical losses[3-6].

In Japan, the revised Energy Conservation Act (Partial Amendment of the Act on the Rational Use of Energy), a law that encourages more efficient use of energy resources, was put into effect in 1999, further accelerating the drive for power efficiency improvement. In various parts of the world including Europe, the United States, China and Thailand, similar EER- or COP-based energy conservation laws and regulations are in place.

In a commonly used residential wall-mounted or commercial packaged air conditioner, it is generally believed that compressor input accounts for more than 80% of electrical power consumption. This indicates that compressor efficiency is one of the most critical factors that determine the energy saving performance of the entire system. In the field of rotary compressor technology, many engineering efforts are being made for energy efficiency improvement. Techniques that provided successful efficiency improvement include the replacement of AC motors with more efficient brushless DC motors, introduction of concentrated winding technique, and use of strong neodymium magnets with higher magnetic flux density. The engineering race for better efficiency still continues today[7].

As described above, the evolution of rotary compressor technology has been an essential part of the global growth and environmental adaptation of refrigeration and air conditioning systems, to which the Japanese industry's long time efforts in basic technology development and production engineering advancement greatly contributed. While the majority of the country's rotary compressor production capacity has been relocated to China and Southeast Asia where demand keeps rising, rotary compressors are the most economical form of compression technology in the capacity range of 0.75 to 6 horsepower that are suitable for residential and small-scale commercial packaged air-conditioners. More than 90 million units of them are globally produced each year.

4.2 Operating Principle and Characteristics

4.2.1 Compression action

Figure 4.3 shows the compression cycle of a rotary compressor. A rolling piston (or roller) coupled to the rotating shaft turns inside the cylinder in an eccentric manner. A vane, pressed from behind by the force of a spring plus that of gas pressure, divides the space inside the cylinder into a

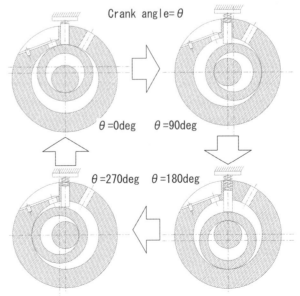

Fig. 4.3 Compression cycle

higher-pressure compression chamber and a lower-pressure suction chamber, where compression and suction processes are run concurrently. The discharge valve does not open until the pressure of the gas inside the compression chamber reaches the discharge pressure level, thus preventing already discharged gas from flowing back into the compression chamber.

Two major differences of the rotary compressor from the conventional reciprocating compressor are that the rotational motion of the motor is directly utilized for gas compression without conversion into a reciprocating motion, and that it does not require a suction valve. These characteristics lead to the following advantages:

(1) Smaller variation in the compression load due to the suction and compression processes being run continuously and concurrently in the two chambers. Also, the smaller reciprocating mass leads to better rotational balance and low vibration.
(2) Fewer number of parts due to having no rotational-to-reciprocating motion conversion mechanism. Also, due to low vibration characteristics the mechanism can be directly secured to the hermetic casing (housing), which helps size and weight reduction.
(3) Without a suction valve, the non-effective volume can be made very small, allowing suction pressure loss minimization and superior volumetric and pressure efficiency.

On the other hand, the following tradeoffs must be noted:
(1) As the compression and suction chambers are separated only by the vane and rolling piston without any mechanical sealing parts, significant leakage loss could result unless component clearances are tightly and uniformly controlled, which can only be achieved with a very high level of machining and assembly accuracy.
(2) For the refrigeration oil to effectively seal compression chamber clearances, the inside of the hermetic casing containing the compression mechanism and the drive motor must be maintained at high pressure, and as a result the motor and the compression mechanism could become overheated easily.

4.2.2 Displacement volume

A rotary compressor's theoretical volume (variously referred to as "stroke volume", "displacement", etc.), V_{th}, can be obtained by the following equation:

$$V_{th} = \pi(R^2 - r^2)H \quad (4.1)$$

where: H: Cylinder height
R: Cylinder radius
r: Rolling piston radius

Vane extension x, shown in Fig. 4.4, can be obtained as follows:

$$x = R(1 - \cos\theta) - r(\cos\phi - \cos\theta)$$
$$\text{where: } \phi = \sin^{-1}\left(e\sin\theta/r\right) \quad (4.2)$$
$$e = R - r$$

The compression chamber volume V_c is:

$$V_c = \frac{1}{2}\left[\{R^2\theta - r^2(\theta + \phi)\} - (R - x)e\sin\theta\right]H \quad (4.3)$$

Assuming that the compression stage is an isentropy compression, compression chamber pressure P_c is:

$$P_c = P_s\left(V_{th}/V_c\right)^\kappa \quad (4.4)$$

where, $P_c = P_d$ if $P_c > P_d$
κ: Adiabatic compression exponent
P_s: Suction pressure
P_d: Discharge pressure

Typical compression chamber volume curve V_c and pressure curve P_c are shown in Fig. 4.5, where the horizontal axis represents the crankshaft crank angle.

4.3 Structure and Design of Compressor Components

4.3.1 Basic structure

Figure 4.6 shows the basic structure of a vertical rotary compressor commonly used in air conditioners. The

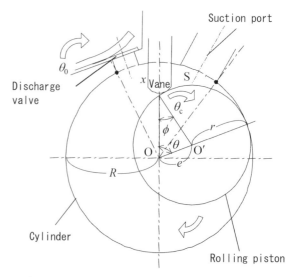

Fig. 4.4 Change of compression-room volume

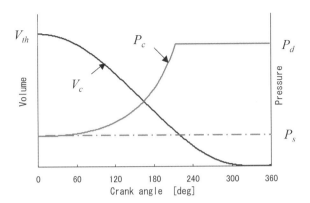

Fig. 4.5 Change of compression-room volume and pressure

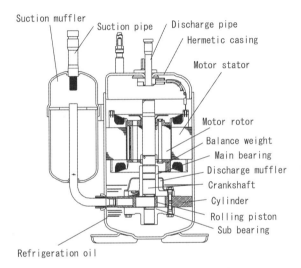

Fig. 4.6 Longitudinal section drawing of rotary compressor

previously described compression mechanism is located in the lower part of a hermetic casing and is coupled to a motor above it via a crankshaft. In these vertical rotary compressors, the inside of the hermetic casing is typically maintained at high pressure as it is filled with compressed discharge gas. The bottom of the hermetic casing serves as a refrigeration oil reservoir to lubricate the components.

Motor rotation is transmitted to the compression mechanism via the crankshaft, which is supported by main and sub bearings across the cylinder. The refrigerant gas flows in through the suction muffler adjacent to the hermetic casing and then through the suction line and into the compression cylinder.

Inside the cylinder, the refrigerant is compressed by the eccentric motion of the rolling piston. The fully compressed refrigerant flows out of the cylinder through the discharge valve and discharge muffler to enter into the space inside the hermetic casing. There, the compressed refrigerant flows upward through motor gaps as well as gas holes in the motor core, thus cooling the motor along the way. Finally, the refrigerant exits from the hermetic casing through the discharge pipe at the top, to be discharged into the refrigeration cycle. Before the compressed refrigerant gas exits from the hermetic casing, refrigeration oil mixed with the gas is separated by the centrifugal force of motor rotation and returned to the oil reservoir at the bottom of the casing.

The suction muffler serves as an accumulator to separate and temporarily store the liquid portion of the refrigerant, which might otherwise flow into the cylinder through the suction pipe and get compressed there.

Normally, balance weights are fitted at the top and the bottom of the motor rotor so as to maintain a rotational balance against the eccentric motion of the crank pin and the rolling piston.

4.3.2 Crankshaft

Figure 4.7 shows the components of a rotary compression mechanism. The crankshaft of a rotary compressor is coupled to the motor rotor to transmit motor rotation to the compression mechanism. To perform this function, the crankshaft must have sufficient torsional rigidity to withstand the compression load torque and the maximum motor torque, especially during various transition periods such as the liquid compression phase immediately after start-up. The crankshaft must also have enough bending rigidity to resist the centrifugal force of the motor rotor, particularly during high speed operation under inverter control. If excessive deformation is caused by the bending moment, the journal bearing's load capacity decreases, possibly leading to a boundary lubrication state or metal-to-metal contact.

The crankshaft will be subject to axial thrust generated by motor magnetic thrust force as well as the motor rotor's own weight. These thrust forces must be supported at the bottom of the crank pin or on the thrust face at the crankshaft bottom.

The crankshaft comprises the crank pin over which the rolling piston is fitted, the main (long) shaft section, the sub (short) shaft section and the motor-rotor-mounting section to which the motor rotor is shrunk-fit. The shaft center has oil supply holes to lubricate the sliding areas. The shaft also contains a centrifugal pump mechanism to feed oil, by utilizing rotation-induced centrifugal force as well as the force of pressure difference, to have the bearings and other compression components lubricated and oil-sealed.

The compression volume is determined by the amount of eccentricity in the crank pin. The greater the eccentricity

Compressors for Air Conditioning and Refrigeration

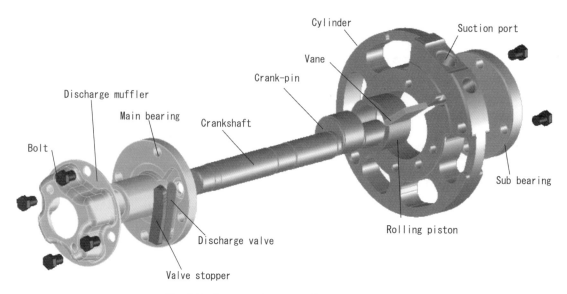

Fig. 4.7 Compression part of rotary compressor

is, the larger the compression volume will be.

Typical crankshaft materials include spheroidal graphite cast iron, eutectic graphite cast iron and hardened carbon steel. For better initial conformability and boundary lubricity, surface treatment with manganese phosphates or molybdenum disulfide may be employed.

4.3.3 Bearings

The crankshaft of a rotary compressor is supported by the main bearing, which is located on the motor side of the compression chamber, and the sub bearing, which is located on the opposite side. The two bearings each constitute part of the compression chamber enclosure by providing a closure seal at the end of the cylinder. Therefore, the bearings must be truly circular on their inner surface and also must be perfectly parallel and perpendicular on the end faces. During the assembly process, a very tight coaxiality must be satisfied between the main and the sub bearings.

The opposite-to-load phase of the bearing inner surface has spiraling oil grooves for oil feed purpose. The bearing end face also has a discharge port through which to eject compressed refrigerant gas out of the compression chamber. The discharge port is fitted with a discharge valve and a valve stopper to retain the valve.

In general, bearings used in a rotary compressor are journal bearings that subject to variable load. Requirements for bearing load capacity and minimum oil film thickness, which are determined based on the oil film characteristics obtained by the Reynolds equation for incompressible fluids, must be satisfied[8].

Due to recent demand for further size and weight reduction and more advanced inverter-regulated variable speed functionality, rotary compressors are required to operate under very severe lubrication conditions and in a wider operational speed range, from very low speed to very high speed levels. Therefore, bearing specifications such as structure, rigidity and journal clearances must be carefully determined, taking into consideration how the bearing oil film characteristics will be affected by the degree of crankshaft and bearing deformation caused by gas compression load and the motor rotor's centrifugal force. Improved crankshaft designs, for example one where the bearing root is intentionally made to have the flexible structure for locally reduced rigidity so that it can flexibly comply with crankshaft deformation to increase the bearing load capacity, are being used.

The main and the sub bearings are usually built with either grey cast iron or steam-oxide-film-sealed, oil-impregnated sintered iron alloy. Materials for the crankshaft and the bearings must provide good wear resistance, which should be verified through exhaustive in-situ reliability testing.

4.3.4 Rolling piston

The rolling piston is coupled with the crank pin with the vane pressed on the rolling piston outer circumference, the rolling piston forms compression chambers inside the cylinder where suction and compression actions take place. The peripheral surface of the rolling piston and the tip of the vane form a line contact with a composite rolling-and-sliding action.

The rolling piston's rotational speed is determined by the combination of the vane-pressing force, the amount of load working on the crank pin bearing and the lubrication condition, and it typically rotates at a speed a several tenth of the crankshaft speed. With this, the rolling piston will not slide very fast against the vane tip, thereby reducing sliding loss and minimizing abrasion and seizure risks.

Considering various factors such as the rolling piston's thermal expansion potential due to material difference with the cylinder and the gas compression load direction, the compressor is designed so that clearances at the ends of the rolling piston and one between the rolling piston outer circumference and the cylinder inner circumference will all be in the range of several micrometers, so that they can be dependably sealed by refrigeration oil film during compression cycle to prevent refrigerant leakage.

The rolling piston is typically made of wear-resistant eutectic graphite cast iron or other special cast steels such as Ni-Cr-Mo alloy steel.

4.3.5 Vane

The vane serves as a partition to divide the space inside the cylinder into compression and suction chambers. It is fitted into a slot formed in the cylinder wall and is triggered into action by spring force upon start-up. After start-up, the vane moves in a reciprocating motion by being pushed from behind by the spring force plus gas pressure inside the hermetic casing while also being pushed back from the front by the rolling piston. The lubrication condition between the rounded tip of the vane and the rolling piston outer circumference and the one between the sides of the vane and the cylinder slot wall will all be boundary lubrication. To assure compressor reliability, it is critical to reduce abrasion in these areas. Most critically, the machining accuracy (including both surface roughness and geometric accuracy such as straightness) of the rounded vane tip contour, which forms a line contact with the rolling piston outer circumference, affects significantly the sliding characteristics between the vane and the rolling piston.

As the vane divides the space inside the cylinder into compression and suction chambers, clearances between the vane ends and the bearing end faces and between the vane sides and the cylinder slot wall would allow the refrigerant gas to leak from the compression chamber into the suction chamber. To improve and maintain the compressor efficiency, it is critical to control these clearances to an appropriate magnitude so that they can be effectively sealed by oil film to minimize leakage loss.

The vane also have a function of pressure relief for when the compression chamber pressure gets too high as a result of liquid or oil compression condition that might occur, for example, when the compressor is started up after a long time stopping or when incompletely gasified refrigerant is sucked into the cylinder. When such situation happens, the pressure relief function moves the tip of the vane temporarily away from the rolling piston surface to let the built up pressure escape, thus preventing component damage due to excessive stress.

The rotary compressor vane is usually made of heat-hardened high speed tool steel or bearing steel. For wear resistance improvement under high-temperature, high-pressure and high-speed operating conditions, surface hardening treatment such as nitriding may be employed.

4.3.6 Cylinder

The cylinder has its top and bottom ends closed off by the main and the sub bearings to form a compression chamber enclosure. The cylinder has a suction port to take the refrigerant inside, a vane slot to hold the vane in place, and a vane spring hole. The rolling piston turns eccentrically inside the cylinder to compress the refrigerant. The compressor is designed so that some amount of clearance always exist between the rolling piston outer circumference and the cylinder inner circumference to avoid contact between them.

As the outer wall of the cylinder is directly welded to the hermetic casing in general, the cylinder must be designed to have a certain level of rigidity. In some compressors, areas of the cylinder outer wall that are welded to the hermetic casing are ribbed so that the geometric accuracy of the cylinder inner circumference and the vane slot will be least affected by welding stress.

Inner diameter D and height H of the cylinder are one of the most important compression chamber specifications as they determine the approximate stroke volume and aspect ratio of the compression chamber. The D/H ratio also determines the bearing-load-to-vane-load ratio as well as the radial-to-heightwise leakage channel area ratio and therefore can significantly affect compression efficiency of the compressor. These cylinder dimensions and other compression chamber specifications should be determined with compression efficiency optimization taken into consideration.

The compressor cylinder is generally made of grey cast iron for larger compressors and sintered iron alloys for smaller products.

4.3.7 Discharge valve

The discharge valve of a compressor is typically a reed type valve, which opens and closes by the force of pressure difference between the inside and the outside of the cylinder. The valve lift amount is restricted by the valve stopper.

Pressure loss through the discharge valve, often referred to as "over-compression loss" or "overshoot loss", is one of the major elements of compressor loss. Discharge valve pressure loss will be greater at higher operating speeds where the refrigerant flow speed is higher. To assure compressor efficiency, it is critical to optimally design the valve lift amount, valve plate thickness, the valve port diameter and the gas channel geometry to minimize the pressure loss.

Discharge valve pressure loss can be reduced by increasing the amount of valve lift, but this in turn could increase the collision speed against the valve stopper and the valve seat and could detrimentally affect the valve's fatigue and impact strength. The degree of valve impact stress may be decreased or increased by such factors as the geometry and thickness of the valve plate as well as valve seat and valve stopper shapes. Valve closure delay could also have serious consequences as it could lead to significant impact stress upon valve closure due to pressure difference between the discharge pressure in the back of the valve and the suction pressure inside the compression chamber. Such situation potentially leads to component damages.

Discharge valves are commonly provided on the main and sub bearings, either on one of them or on both of them. Or, as shown in Fig. 4.8, discharge valves may be provided in a cylindrical form and adjacent to the inner cylinder surface[9]. The discharge valve is typically made of Swedish steel, which is a high grade steel with good fatigue and impact strength.

4.4 Dynamic-mechanical Analysis

Rolling piston compressors are the most commonly used type of rotary compressors. With rolling friction between the vane and the rolling piston and with fluid film lubrication between the rolling piston and the crank pin, rolling piston compressors are inherently capable of attaining much better mechanical efficiency than reciprocating compressors[10, 11]. The following sections provide a dynamic-mechanical analysis of the rolling piston type rotary compressor.[12]

4.4.1 Mechanism and movements

The rolling piston type rotary compressor is structurally similar to reciprocating compressors. The main difference is that a rolling piston compressor has a larger-diameter crank pin so that a direct circumscribed contact can be made with the "piston pin", which actually is called a "vane" in a rolling piston compressor, eliminating the need for a connecting rod. Figure 4.9 shows the mechanism of a rolling piston compressor. Unlike the piston pin in a reciprocating compressor, the vane is a flat thin plate without shaped circle. It moves in a reciprocating action inside a slot in the cylinder block. If the tip of the vane directly touches the crank pin, a very high-speed sliding state would occur at the line of contact between the vane and the crank pin, resulting in significant friction loss. To avoid such situation, a ring is fitted to the outer circumference of the crank pin, which is "rolling piston". The rolling piston is so called because it moves in a rolling motion and is directly involved in gas compression. The invention of the rolling piston has drastically improved the mechanical efficiency characteristics of the compression mechanism.

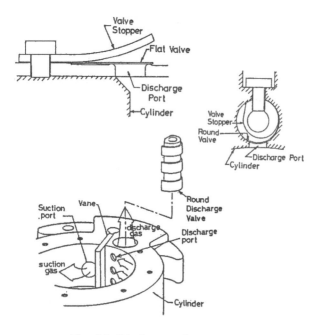

Fig. 4.8 Discharge valve strucure

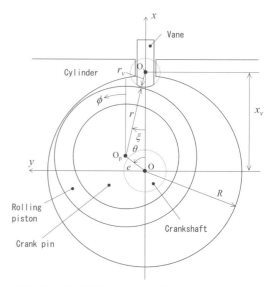

Fig. 4.9 Definition of coordinate and variables

As described above, the rolling piston compressor can be basically considered a reciprocating mechanism only without a connecting rod, and therefore its movements are fundamentally the same as those of a reciprocating compressor. As shown in Fig. 4.9, a triangle formed by three points of the vane tip O_v, the crank pin center O_p and the crankshaft center O represents the same piston-crank contrivance as that of a reciprocating compressor. The figure shown is based on a Cartesian coordinate plane of x and y axes, whose origin is the crankshaft center O and whose x axis matches the vane centerline.

With the crankshaft rotational angle from the x axis denoted as θ, the distance x_v between O and O_v, or vane displacement, can be given as follows:

$$x_v = e\cos\theta + (r_v + r)\cos\xi \tag{4.5}$$

Here, e is the crank pin's rotational radius, r is the rolling piston radius and r_v is the radius of the arc described by the vane tip. ξ represents the rotational angle of line segment $\overline{O_v O_p}$ and it has the following relationship with θ:

$$e\sin\theta = (r_v + r)\sin\xi \tag{4.6}$$

4.4.2 Constraint forces and equations of motion

(1) Vane: As the rolling piston compression mechanism does not have a connecting rod, the vane needs to be constantly pressed against the rolling piston to maintain contact, or the compressor cannot be function at all. This is achieved by the force of discharge gas pressure P_d working on the rear end of the vane. Under the condition, when the crankshaft is turned counterclockwise, the right-hand chamber becomes a compression chamber of pressure P_c and the left-hand chamber becomes a suction chamber of pressure P_s. The vane, being pushed by predominant compression pressure P_c, tilts in the cylinder slot. Therefore, as shown in Fig. 4.10, the vane will be pressed against the cylinder slot at points G_1 and G_2. The gas forces working on the vane in x and y directions, or F_{qx} and F_{qy} respectively, and the gas moment M_q around O_v can be calculated by the following equations:

$$F_{qx} = \{-2aP_d + (a + r_v\sin\xi)P_c + (a - r_v\sin\xi)P_s\} \times l$$

$$F_{qy} = \{-bP_d + (R + b - x_v + r_v\cos\xi)P_c$$

$$M_q = \left[-b\left(R - x_v + \frac{b}{2}\right)P_d \right.$$
$$+ \frac{1}{2}\{(R + b - x_v)^2 + a^2 - r_v^2\}P_c$$
$$\left. - \frac{1}{2}\{(R - x_v)^2 + a^2 - r_v^2\}P_s\right] \times l \tag{4.7}$$

where a is half the vane thickness, l is the vane height (= cylinder height) and b is the length of the cylinder slot. It is defined that a positive gas moment runs counterclockwise.

Also, constraint forces at points G_1 and G_2 are denoted as F_{gn1} and F_{gn2}, respectively. Friction forces at the same points are denoted as F_{gt1} and F_{gt2}. F_{vn} and F_{vt} represent the reactive and frictional forces generated by the rolling piston. F_d represents oil-viscous frictional force on the vane ends, and F_s represents the pressing force of the spring fitted to the vane rear end. With the vane's (mass m_v) inertia force ($-m_v\ddot{x}_v$) taken into consideration, the balance equations of all the forces working on the vane can be derived at as follows:

$$-m_v\ddot{x}_v - F_s + F_{qx} + F_{gt1} + F_{gt2} + F_{vn}\cos\xi + F_{vt}\sin\xi + F_d = 0 \tag{4.8}$$

$$F_{qy} + F_{vt}\cos\xi - F_{vn}\sin\xi + F_{gn1} - F_{gn2} = 0 \tag{4.9}$$

And the moment balance equation around O_v is:

$$(R + b - x_v)F_{gn1} + aF_{gt1} - (R - x_v)F_{gn2} - aF_{gt2} + M_q - r_v F_{vt} = 0 \tag{4.10}$$

From the x-axis balance Eq. (4.8), the constraint force F_{vn} between the vane and the rolling piston can be determined, as follows:

$$F_{vn}\cos\xi = m_v\ddot{x}_v + F_s - F_{qx} - F_d - F_{gt1} - F_{gt2} - F_{vt}\sin\xi \tag{4.11}$$

From Eqs. (4.9) and (4.10), constraint forces F_{gn1} and F_{gn2} can be determined as follows:

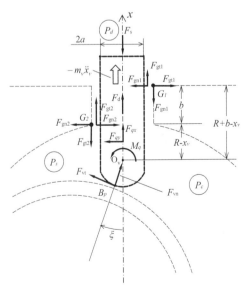

Fig. 4.10 Force and moment on vane

$$bF_{gn1} = (R - x_v)(F_{qy} - F_{vn}\sin\xi)$$
$$- a(F_{gt1} - F_{gt2}) - M_q \qquad (4.12)$$
$$+ \{(R - x_v)\cos\xi + r_v\}F_{vt}$$

$$bF_{gn2} = (R + b - x_v)(F_{qy} - F_{vn}\sin\xi)$$
$$- a(F_{gt1} - F_{gt2}) - M_q \qquad (4.13)$$
$$+ \{(R + b - x_v)\cos\xi + r_v\}F_{vt}$$

(2) Rolling piston: Figure 4.11 illustrates the forces and moments working on the rolling piston. The compression chamber and the suction chamber are partitioned by a minimum clearance point A_p, which exists between the rolling piston circumference and the cylinder inner wall, and the vane. The rolling-piston-pressing force F_p, which is a resultant of forces by compression chamber pressure P_c and suction chamber pressure P_s, acts in the direction normal to the line connecting points A_p and B_p and is through the rolling piston center O_p. The magnitude of the force F_p can be expressed by the following equation:

$$F_p = 2r\sin\{(\theta + \xi)/2\} \times l(P_c - P_s) \qquad (4.14)$$

At the vane contact point B_p, the vane-reaction force F_{vn} determined by Eq. (4.11) and its frictional force F_{vt}, work in the respective directions shown in Fig. 4.11. The area between the rolling piston inside and the crank pin outside is force-fed by a lubrication oil pump, thus constituting a high-speed sliding bearing (relative sliding speed $v_{pc} = r_c(\dot\theta - \dot\phi)$; r_c: crank pin radius). With this, this area is generally in a hydrodynamic lubricating condition. An oil-film viscous torque M_p that is proportional to the relative sliding speed, will work on the rolling piston. The magnitude of this torque can be determined based on the journal lubrication theory. Also, a resultant force F_{en} by oil-film pressure force works on the inner surface of the rolling piston. The direction of this force, which actually is determined by the balance of all the forces involved, is assumed to be an angle of η from the x axis.

In the vicinity of the minimum clearance point A_p on the rolling piston outer wall, a normal pressing force F_{cn} by refrigerant pressure and a viscous frictional force F_{ct} by shearing effect of refrigerant gas will work. The magnitudes of these forces can be obtained based on the plane bearing theory. Furthermore, oil-viscous frictional force F_a and frictional moment M_a will work on both end of the rolling piston, in the respective directions shown in Fig. 4.11. These forces can also be obtained from the plane bearing theory. The direction of F_a is normal to moving radius $\overline{OO_p}$.

With all the forces working on the rolling piston considered, equations of motion in x- and y-direction can be derived at as follows:

$$-m_p\ddot{x}_{op} + F_{en}\cos\eta - F_{vn}\cos\xi - F_{vt}\sin\xi$$
$$- F_{cn}\cos\theta + F_{ct}\sin\theta + F_p\cos(\theta - \xi)/2 \qquad (4.15)$$
$$+ F_a\sin\theta = 0$$

$$-m_p\ddot{y}_{op} + F_{en}\sin\eta + F_{vn}\sin\xi - F_{vt}\cos\xi$$
$$- F_{cn}\cos\theta - F_{ct}\cos\theta + F_p\cos(\theta - \xi)/2 \qquad (4.16)$$
$$- F_a\sin\theta = 0$$

Here, (x_{op}, y_{op}) represents the coordinates for point O_p, which can be given by $x_{op} = e\cos\theta$ and $y_{op} = e\sin\theta$. With the balance of moments around point O_p considered, an equation of rotational motion of the rolling piston can be obtained as follows:

$$-I_p\ddot\phi - r(F_{vt} + F_{ct}) + M_p - M_a = 0 \qquad (4.17)$$

(3) Crankshaft system: As shown in Fig. 4.12, the crankshaft is driven by motor drive torque M_D to turn counterclockwise. At the same time, oil-film reaction force F_{en} from the rolling piston works on the crank pin in the direction toward point O_p. Furthermore, oil-film frictional moment M_p works in the clockwise direction. The crankshaft receives constraint force due to connection with the main and sub bearings. The x- and y-axis direction components of such force are denoted as F_{sx} and F_{sy}. The

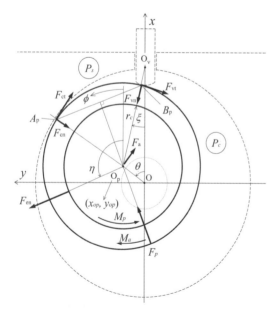

Fig. 4.11 Force and moment on rolling piston

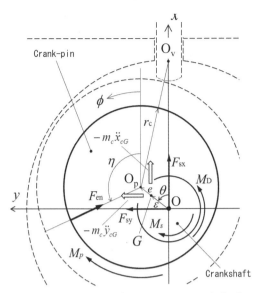

Fig. 4.12 Force and moment on crankshaft

crankshaft also receives friction torque M_s against the bearings, in clockwise direction. With the combined mass of the crank pin and the crank arm denoted as m_c and the coordinates for its center G of gravity as (x_{cG}, y_{cG}), inertia forces are expressed as $(-m_c \ddot{x}_{cG})$ and $(-m_c \ddot{y}_{cG})$. From the x-axis and y-axis direction balance equations of all the forces involved, crankshaft constraint forces F_{sx} and F_{sy} can be determined.

$$F_{sx} = m_c \ddot{x}_{cG} + F_{en} \cos\eta \\ F_{sy} = m_c \ddot{y}_{cG} + F_{en} \sin\eta \quad (4.18)$$

With a rotational radius of the gravity center G denoted as ε, x_{cG} and y_{cG} can be given as $x_{cG} = \varepsilon \cos\theta$ and $y_{cG} = \varepsilon \sin\theta$, respectively. The equation of motion that governs the rotational behavior of the crankshaft can be obtained, from the balance equation of all the moments involved, as follows:

$$-I_c \ddot{\theta} + M_D - e F_{en} \sin(\eta - \theta) - M_p - M_s = 0 \quad (4.19)$$

Here, I_c is the moment of inertia of the entire crankshaft system including the motor rotor around the crankshaft axis. The unknown oil film force F_{en} and its direction η represented in the third term can be eliminated using Eqs. (4.15) and (4.16). The result can be derived at as follows:

$$(I_c + m_p e^2)\ddot{\theta} - e\frac{\sin(\xi+\theta)}{\cos\xi}(m_v \ddot{x}_v - F_s) \\ = M_D - e[\sin\frac{\xi+\theta}{2} F_p + \frac{\sin(\xi+\theta)}{\cos\xi} F_{qx}] \\ - e[\{\sin(\xi+\theta)\tan\xi + \cos(\xi+\theta)\}F_{vt} \\ + \frac{\sin(\xi+\theta)}{\cos\xi}(F_{gt1} + F_{gt2} + F_d)] \\ - (eF_{ct} + eF_a + M_p) - M_s \quad (4.20)$$

The left side of the equation contains the inertia term and the spring term. The first term of the right side represents the motor drive torque, the second term the gas compression load torque, the fourth term the load torque by the frictional forces working on the tip, sides and ends of the vane, the fifth term the load torque by frictional forces working on the outer and inner circumferences and ends of the rolling piston, and the last term the load torque by frictional forces working on the crankshaft.

Basically, if values for the motor torque and the gas pressure characteristics are given, the solution to Eq. (4.20) of rotational motion can be obtained by using a numerical computation method such as the Runge-Kutta-Gill method to find rotational behavior θ of the crankshaft. Note that the frictional forces and torques are affected by constraint forces and sliding speeds, which must be included in the actual computation process. The solution will converge fairly quickly. After the crankshaft rotational behavior is obtained at the end, constraint forces at kinematic pairs of the compressor components can be determined.

4.4.3 Equation of energy and mechanical efficiency

By multiplying Eq. (4.20) of motion by angular displacement and then integrating the result over one crankshaft rotation, an equation of energy can be obtained. As the integral for the inertia term and the spring term will be zero in a steady-state operation, the equation of energy can be derived as:

$$E_{motor} = E_{gas} + E_{vane} + E_{piston} + E_{shaft} \quad (4.21)$$

where:

$$E_{motor} = \int_{1rev} M_D \dot{\theta} dt \\ E_{gas} = \int_{1rev} e\left(\sin\frac{\xi+\theta}{2} F_p\right)\dot{\theta} dt \\ E_{vane} = \int_{1rev} e[\{\sin(\xi+\theta)\tan\xi + \cos(\xi+\theta)\}F_{vt} \\ \qquad + \frac{\sin(\xi+\theta)}{\cos\xi}(F_{gt1}+F_{gt2}+F_d)]\dot{\theta} dt \\ E_{piston} = \int_{1rev}(eF_{ct} + eF_a + M_p)\dot{\theta} dt \\ E_{shaft} = \int_{1rev} M_s \dot{\theta} dt \quad (4.22)$$

The above equation indicates that motor supply energy E_{motor} will be consumed as gas compression energy E_{gas} and frictional energies of E_{vane}, E_{piston} and E_{shaft} at the vane, the rolling piston and the crankshaft respectively.

Mechanical efficiency η_m can be calculated by the following equation:

$$\eta_m = \frac{E_{motor} - (E_{vane} + E_{piston} + E_{shaft})}{E_{motor}} \quad (4.23)$$

4.4.4 Mechanical excitation forces and vibration of a compressor body

With all the forces and moments working on the cylinder block expressed as x- and y-axis direction forces F_x and F_y and the counterclockwise crankshaft torque M_z, the following equations can be obtained:

$$\begin{aligned}F_x &= (-m_v \ddot{x}_v) + (m_p e + m_c \varepsilon)\frac{d}{dt}(\sin\theta\dot{\theta}) \\ F_y &= -(m_p e + m_c \varepsilon)\frac{d}{dt}(\cos\theta\dot{\theta}) \\ M_z &= -(I_c + m_p e^2)\ddot{\theta} + I_p \ddot{\phi}\end{aligned} \quad (4.24)$$

Here, all the internal force terms such as the gas force, frictional forces and torques have been eliminated, leaving only inertia forces and torques. Vibration of the compressor unit is induced as a result of excitation by these unbalanced inertia forces and torques.

Static and dynamic balance design of a rolling piston compressor can be done in exactly the same steps as for reciprocating compressors. By designing the balancers described in Section 3.4.3 according to the Eq. (4.25), excitation forces F_x and F_y of Eq. (4.24) can be minimized, based on the principle of partial balancing, as Eq. (4.26).

$$\begin{aligned}m_{bU} r_b &= \left(\frac{1}{2}m_v r + m_{ro} r + m_e r_e\right)\frac{l_L}{l_U} \\ m_{bL} r_b &= \left(\frac{1}{2}m_v r + m_{ro} r + m_e r_e\right)\frac{l_L + l_U}{l_U}\end{aligned} \quad (4.25)$$

$$\begin{aligned}F_x &= -m'_v \ddot{x}_v - \frac{1}{2}m'_v r \frac{d}{dt}(\sin\theta \cdot \dot{\theta}) \\ F_y &= \frac{1}{2}m'_v r \frac{d}{dt}(\sin\theta \cdot \dot{\theta})\end{aligned} \quad (4.26)$$

Vibrations of a rolling-piston compressor can also be determined in exactly the same steps as for reciprocating compressors. The determinant [F] of excitation forces working on the compressor unit around the center of gravity can be given by Eq. (3.37), and the equation of motion that describes the vibration can be given by Eq. (3.33). The numerical solution for the motion equation can also be obtained in exactly the same way.

4.5 Applications and Performance

4.5.1 High-temperature application rotary compressors

Depending on the purpose and the equipment it is installed in, compressors may use different refrigerants or work under different temperature and pressure conditions. As a general rule, compressors can be classified according to the evaporation temperature range of the equipment they are used for, into high-temperature, medium-temperature and low-temperature application categories. Of those, the following paragraphs explain about the rotary compressors for high-temperature applications, such as those employed in residential air conditioners and small-capacity commercial packaged air conditioners and heat-pump-based water heater systems.

(1) Rotary compressors for air conditioners: Figure 4.13 shows a cross section of a rotary compressor for an air conditioner.

In general, rotary compressors for air conditioning purpose use R410A refrigerant and are operated, when in a steady state, in the evaporating temperature range of -10°C to +15°C. However, with wide use of air conditioners as heat pumps under low outside temperature conditions, rotary compressors, especially those intended for use in cold climate regions, are now being employed in lower evaporating temperature ranges.

Rotary compressors are typically used in the condensing temperature range is 28°C to 65°C, but some of rotary compressors, especially those installed in systems intended for use in hot climate regions such as the Middle East or South Asia, are designed to be compatible with higher condensing temperature ranges, as high as 70°C in some cases.

For compatibility with such lower evaporating or higher condensing temperature ranges, the unit may sometimes be

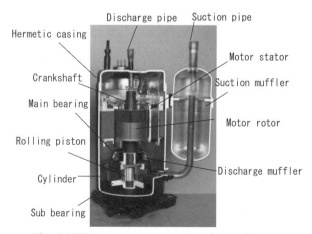

Fig. 4.13 Rotary compressor for air conditioner

designed with a built-in self-cooling feature such as liquid injection or use more heat tolerant materials in sliding parts or motor insulation, so that the discharge gas temperature or motor coil temperature will not fall out of the range where reliability can be assured.

The paragraphs below analyze the effective work and loss in a rotary compressor unit for air conditioners, and explain how each compressor structural component may influence the compressor performance.

The following equations define various efficiencies related to a rotary compressor.

Motor efficiency: $\eta_m = L_m / L_c$ (4.27)

Mechanical efficiency: $\eta_{me} = L_i / L_m$ (4.28)

Indicated efficiency: $\eta_C = L_{th} / L_i$ (4.29)

Compressor efficiency: $\eta_{comp} = \eta_m \cdot \eta_{me} \cdot \eta_C$ (4.30)

where, L_m: motor output, L_i: effective compression power, L_c: motor input power, L_{th}: theoretical compression power;

$L_{th} = \eta_V \cdot Gr_{th} \cdot \Delta h_{th}$ (4.31)

Volumetric efficiency: $\eta_V = Gr / Gr_{th}$ (4.32)

where, Gr_{th} = theoretical refrigerant mass flow rate, Δh_{th} = theoretical enthalpy variation, Gr = actual refrigerant mass flow rate.

The definition of compressor efficiency by Eq. (4.30) is slightly different from that by Eq. (2.125) in Chapter 2. This is due to the theoretical compression power L_{th} in this section being defined as a value including volumetric efficiency η_V.

Figure 4.14 shows an analysis of the effective work and loss in a rotary compressor.

(i) Motor loss

Motor loss is the difference between motor output power L_m and motor input power L_c, and is comprised of iron loss in the motor iron core, copper loss in motor windings and stray load loss. Motor characteristics will be explained in more details in Chapter 10.

(ii) Mechanical loss

Mechanical loss in an air conditioning rotary compressor includes sliding losses, or friction losses between such compressor components that slide against each other, as the vane tip and the rolling piston outer circumference, the vane side and the cylinder slot wall, the crankshaft pin and the rolling piston inner circumference, the crankshaft and the inner circumference of the main or sub bearing, and the crankshaft thrust area and the sub bearing end face. Loss generated by each sliding pair of components can be expressed as:

$L = \mu \cdot F \cdot V$ (4.33)

Here, μ is the friction coefficient, the value of which is determined by the lubrication condition. F is the load and V is the sliding speed.

(iii) Indicated loss

Indicated loss is the difference between the ideal adiabatic compression power (effective compression power) and the actual compression power used. Indicated loss can be divided into suction pressure loss, superheated loss, leakage loss, re-compression loss and discharge loss.

Suction pressure loss is the flow channel loss through a suction muffler and a suction pipe. As rotary compressors do not have a suction valve, the amount of suction pressure loss is relatively smaller than that of reciprocating compressors.

Superheated loss is a thermal loss that occurs as a result of the suction gas being superheated inside the cylinder before the compression stage starts, which is caused by

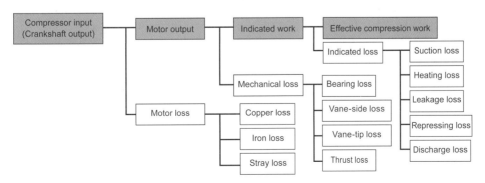

Fig. 4.14 Efficiency analysis of rotary compressor

heat transfer from the cylinder with high temperature in the high pressure casing to the suction gas. The amount of superheated loss is equal to refrigeration capacity reduction caused by the increase in the specific volume of suction gas. The loss can be calculated as the product of the incoming heat flux, the specific heat at constant pressure C_p and the ratio of change in the specific volume to change in the temperature $\partial v_s / \partial T_s$.

Leakage loss refers to the amount of compressed gas that leaks inside the compressor, from the high-pressure side into the low-pressure side through various clearances between compressor components, thus nullifying the compression work previously done on the leaking gas. The amount of leakage loss varies according to various factors including clearance size, pressure difference, compression ratio, sonic speed of refrigerant and oil-sealing effect. Figure 4.15 shows potential leakage paths in a rotary compressor.

Re-compression loss refers to loss to compress refrigerant gas remaining in the non-discharged space ("top clearance") including the discharge port in the compression chamber. The residual gas in the top clearance volume expands into the suction chamber after the discharge process finished, then the amount of compression work previously done on the re-expanding gas is nullified. The degree of re-compression loss is proportional to the ratio of the re-expanded gas mass to the suctioned gas mass. To minimize re-compression loss, it is critical to make the top clearance volume as small as possible.

Discharge loss includes over-compression loss of refrigerant gas through the discharge valve and also flow channel loss that occurs along the discharge path. Discharge loss may also be referred to as "overshoot loss".

During a low-speed operation of the compressor, leakage and superheated losses will account for the majority of indicated losses, while during high-speed operation the proportion of discharge pressure loss will be larger. When an air conditioner is operated with inverter control, the compressor installed in it is running more frequently under the low-speed operation. For reduction of the annual electric power consumption of the air conditioner, it is important to reduce such compressor indicated loss that will occur dominantly under the low-speed operation.

Indicated losses can be analyzed using the *P-V* diagram. Figure 4.16 is an example of actually measured *P-V* diagram of a rotary compressor. The figure specifies what parts of the diagram correspond to the suction, the compression and the discharge processes, respectively. The thin line represents the theoretical diagram, while the bold line describe the actually measured one. The line-surrounded area represents the amount of work done by the compression action. The areal difference between the theoretical and the actual diagrams indicates the amount of indicated loss.

In some parts of the suction process, the actually measured pressure is lower than the theoretical value due to pressure loss and other factors. The amount of difference indicates how much loss is incurred during the suction stage.

During the compression process, the actually measured pressure curve is even higher than the theoretical curve due to such factors as re-compression of residual gas in the top clearance volume, compression of refrigerant leaked into the compression chamber and heat transfer from high-temperature compressor components and leaked-in oil to the refrigerant gas under compression. These factors result in an extra amount of compression work, constituting a loss.

Lastly, in the discharge process, the pressure inside the cylinder will actually get higher than the required discharge pressure due to the time lag of discharge valve opening to release refrigerant gas and also due to pressure loss that oc-

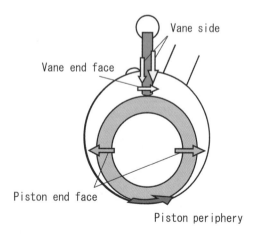

Fig. 4.15 Leak-path of rotary compressor

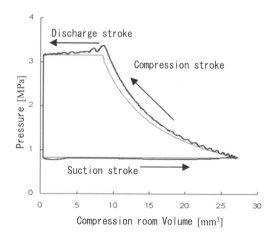

Fig. 4.16 *P-V* diagram of rotary compressor

curs along the discharge line. These lead to an extra amount of work, thereby a loss.

(2) Rotary compressors for water heater systems:
Heat pump-type residential water heater systems using naturally available CO_2 as refrigerant were first commercialized in 2001, and since then their use has rapidly spread. The advantages of using the CO_2 refrigerant include that it does not deplete the ozone layer and also that it has an extremely small global warming effect, only a several-thousandth of that of R410A and other HFC refrigerants that have been traditionally used in refrigeration and air conditioning equipment. CO_2 is also superior from safety and environmental viewpoints as it is nonflammable and is not strongly toxic. However, as shown in Fig. 4.17, CO_2 requires more than 10 MPa operating pressure to work properly as refrigerant, which is more than three times higher than that of the conventional R410A refrigerant. Therefore, any compressor that is used in a CO_2-based heat pump water heater system must provide sufficiently high performance and reliability under such high pressure conditions. The following paragraphs explain the engineering aspects of rotary compressors that are suitable for the above-described CO_2-based water heater applications.

Figure 4.18 shows the longitudinal section of a rotary compressor employed in a CO_2-based heat pump water heater system. The hermetic casing and the compression elements are highly pressure resistant so as to be compatible with high operating pressure requirement of CO_2 refrigerant. In a rotary type compressor, the vane is pressed against the rolling piston outer circumference by gas pressure from behind, and is also pressed against the cylinder slot wall by the pressure difference between the compression chamber and the suction chamber. As the CO_2 refrigerant requires high operating pressure, these pressing forces will also become much higher than those in a compressor running with

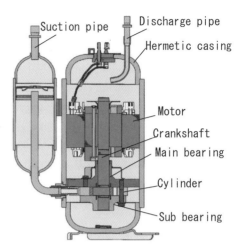

Fig. 4.18 CO_2 rotary compressor

the conventional R410A refrigerant. As a result, the sliding actions between the vane tip and the rolling piston outer circumference and also between the vane side and the vane slot wall have to be performed under very severe conditions.

To address this issue, CO_2-based rotary compressors such as the one shown in Fig. 4.18 typically employ high-hardness, high-strength surface treatment on the vane (DLC-Si coating). Also, as shown in Fig. 4.19, certain areas of the cylinder's vane slot may be designed to be less stiff so that they can minutely deflect under pressure from the vane, thus expanding the available sliding surface and thereby reducing the contact surface pressure per unit area. With this structural design and the surface treatment of the vane, a sufficiency pressure-tolerant sliding performance can be achieved even under the high operating pressure conditions required for the CO_2 refrigerant.

Another requirement to assure good efficiency under the high-pressure operating conditions specific to the CO_2 refrigerant is to minimize the amount of leakage through

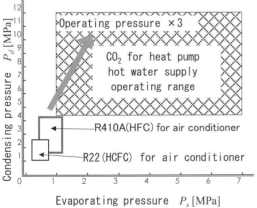

Fig. 4.17 Operating envelope of high-temperature compressor

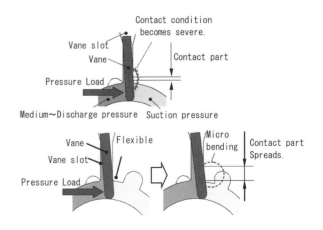

Fig. 4.19 Flexible structure of vane slot

65

clearances that are present in the compression mechanism. For this purpose, engineering efforts are made to reduce radial clearances and to optimize the amount of refrigeration oil to be fed into the compression chamber.

Figure 4.20 is a graph showing changes in the clearance between the cylinder inner circumference and the rolling piston outer circumference ("radial clearance") both on a R410A-based rotary compressor and on a CO_2-based rotary compressor, during one revolution of the crankshaft. It is indicated that the radial clearance on the CO_2-based compressor has been successfully reduced to half that on the R410A-based compressor.

With these improvements, the compressor efficiency of a CO_2-based rotary compressor has now been increased to almost the same level of that of a R410A-based unit[13].

4.5.2 Low-temperature application rotary compressors

Low-temperature application rotary compressors are commonly employed in refrigeration systems, household refrigerators, commercial refrigerator-freezers, and cold food showcases where the evaporation temperature is typically in the range of -35°C to -15°C. Refrigerants used in such systems include R410A, R134a and the naturally available R600a.

Figure 4.21 shows a cross section of a low-temperature application rotary compressor. The general characteristics of a low-temperature application rotary compressor include that it is operated at low evaporating temperature and that it is commonly designed in a horizontal configuration to save installation space inside the machinery room. Due to these factors, a low-temperature application rotary compressor must satisfy the following requirements:

(1) Capability to lower the discharge gas temperature: As this type of compressor is operated at low evaporating temperature for an extended period of time, the resultant high compression ratio leads to higher discharge gas temperatures. Also, the relatively small mass flow rate of refrigerant circulating in the system is not sufficient to dependably cool the motor. As a result, careful consideration must be given to wear resistance of the compression elements and also to heat resistance of the organic motor components. This requires designing an additional cooling feature into the system.

Commonly used cooling features include forced air cooling by a fan, gas cooling, oil cooling and liquid injection cooling. Figure 4.22 shows the schematic of a gas cooler-based refrigeration circuit in which a low-temperature application rotary compressor is installed, as well as a *P-h* diagram representing its refrigeration cycle.

Gas discharged from the compressor is first cooled as it passes through a gas cooler line that runs through a part of the condenser unit and then is returned to the compressor again, where it takes heat away from the compressor before flowing into the condenser unit again. Here, Δi_1 is the cooler's heat radiation amount per unit mass flow of refrigerant and Δi_2 is the amount of heat taken away from the compres-

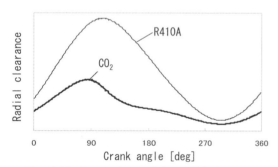

Fig. 4.20 Comparison of the radial clearance change

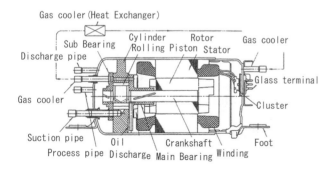

Fig. 4.21 Low-temperature rotary compressor

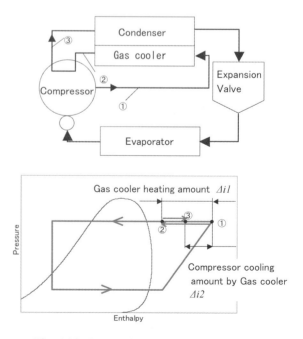

Fig. 4.22 Gas cooler type refrigeration cycle

sor by unit mass flow of refrigerant.

(2) Optimization of the lubrication system: A horizontal rotary compressor unit requires a lubrication system to distribute the refrigeration oil, which otherwise sits in the reservoir at the bottom of the hermetic casing, to its sliding components. Figure 4.23 shows the lubrication system employed in a low-temperature rotary compressor. This lubrication system utilizes the force of pressure difference. The oil feed pipe from the main bearing connects the oil reservoir under discharge pressure to the rolling piston inner circumference under the intermediate pressure of the compression stage. Difference between these two pressures causes the oil to be sucked up and fed to various components and areas such as the bearings and the rolling piston end faces for lubrication and oil-sealing purposes. This is a highly reliable system as it does not use mechanical moving parts.

(3) Assurance of oil return: When the evaporating temperature is as low as -40°C, refrigeration oil that has flowed out of the compressor casing together with the refrigerant gas will increase its viscosity, thus harder to circulate back to the compressor. If the temperature of the refrigeration oil drops below its two-layer separation temperature on the low-temperature side of the refrigeration system, the oil will no more be miscible with the refrigerant and will remarkably increase its viscosity, thus becoming less fluid. In worst cases, wax precipitation may occur. To prevent that, the refrigeration oil that is used in a low-temperature rotary compressor must have low viscosity, good fluidity even at low temperatures and high miscibility with the refrigerant used.

4.5.3 Medium-temperature application rotary compressors

Medium-temperature application rotary compressors are commonly employed in dehumidifiers and industrial equipment where the evaporating temperature is typically in the range of -20°C to 0°C. R134a is the dominantly used as refrigerant for such systems.

Figure 4.24 shows the longitudinal section of a medium-temperature application rotary compressor used in a residential dehumidifier. Compared to rotary compressors for air conditioning purpose, these medium-temperature application compressors have a smaller refrigeration capacity and operate in narrower evaporating and condensing temperature ranges. Also, as the amount of refrigerant charged in the refrigeration cycle is small, the suction muffler is designed with a relatively small capacity. It is common that exhaustive engineering efforts for size and weight reduction are made in the design of dehumidifier compressors.

4.6 Other Types of Rotary Compressor Mechanisms

4.6.1 Two-cylinder type

Figure 4.25 shows a cross section of a two-cylinder type rotary compressor, which contains two compression segments. As shown in Fig. 4.26, the crankshaft of this type of compressor typically has 180-degree opposed crank pins so that compression processes run by the two compression segments can be offset by 180 degrees in terms of shaft rotational angle. This contributes to smaller load on the shaft and thus higher reliability, leading to reduced torque variation (see Fig. 4.27). As a result, vibration in the circumferential direction will be significantly reduced (see Fig.4.28).

With the development of such two-cylinder rotary compressors, a much wider rotational speed range has become possible in inverter-based operations. Now rotary compressors that offer rotational speed ranges where the lowest region can be as low as 10 rps are being comercialized[15].

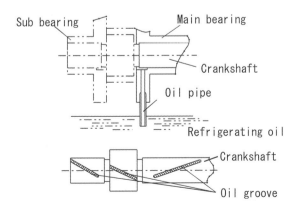

Fig. 4.23 Differential pressure lubrication

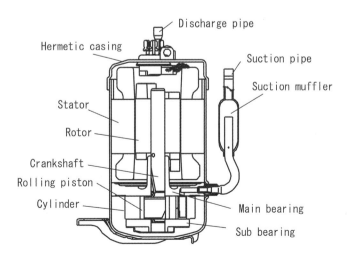

Fig. 4.24 Middle-temperature rotary compressor

Compressors for Air Conditioning and Refrigeration

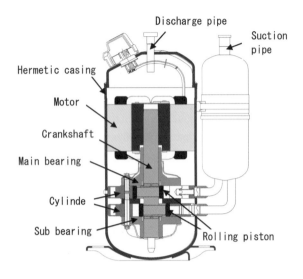

Fig. 4.25 Two-cylinder rotary compressor

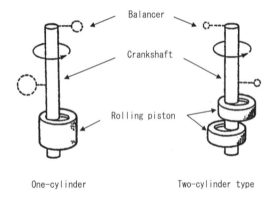

Fig. 4.26 Comparison between one-cylinder rotary and two-cylinder rotary conpressors

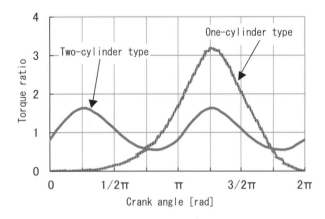

Fig. 4.27 Comparison of torque variation normalized by the average torque

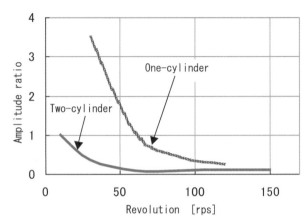

Fig. 4.28 Comparison of circumferential vibration amplitude normalized by that of one-cylinder type at 60 rps.

These two-cylinder compressors are also advantageous in that it has superior rotational balance and therefore is capable of high-efficiency, low-vibration performance from low to high rotational speed ranges. As a result, two-cylinder rotary compressors are now employed in large-scale equipment including industrial air conditioning systems.

More recently, rotary compressors with variable cylinder management have been put into the market, where one of the two cylinders can be deactivated in low capacity regions to offer an even wider variable range of capacity[17].

4.6.2 Swing compressors

(1) Introduction: Swing-type rotary compressors started to be mass-produced in 1995 as refrigerant compressor for air conditioners. Their use has since been widened to include CO_2-based systems. The swing compressor has been developed based on the Kinney pump designed in early 20th century. The Kinney design has been employed in industrial applications such as vacuum pumps[18].

(2) Structure and operating principle: Figure 4.29 shows the structure of a swing compressor. The most important characteristic of the structure of a swing compressor is that the vane and the rolling piston, which are separate in a standard rolling piston compressor, are integrated into one "piston". To drive this integrated piston in a swinging action, a pair of semi-cylindrical "swing bushes" are employed.

Figure 4.30 illustrates the compression actions that take place in a swing compressor. As the crankshaft rotates, the protruding part of the piston (equivalent to the "vane" in a rolling piston compressor) is held and moves back and forth between the pair of swing bushes inside the cylinder, thus the piston swings about the rotational center of

Chapter 4 Rotary Compressors

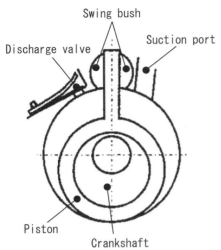

Fig. 4.29 Strcture of swing rotary compressor

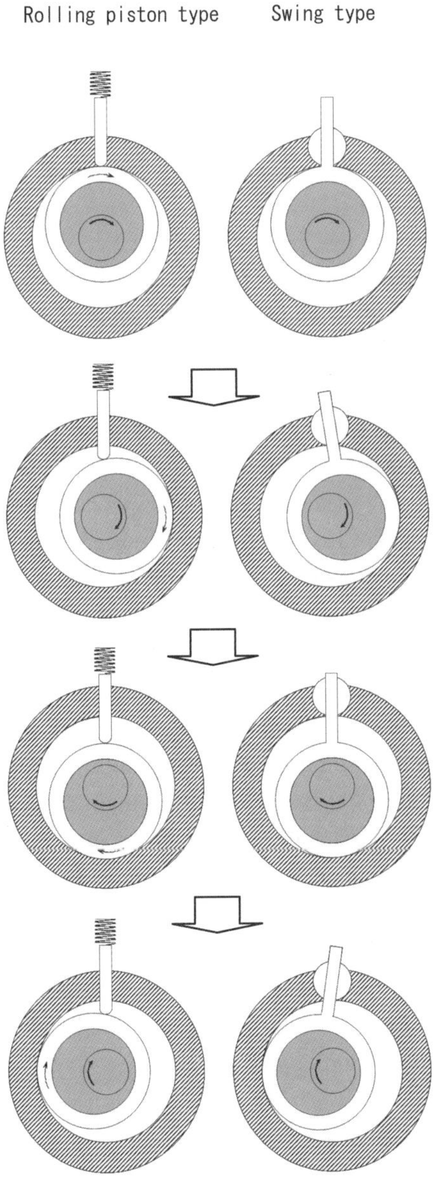

Fig. 4.30 Compression movement

the bushes like a clock pendulum. Similarly to the rolling piston compressor, the inside of the cylinder of a swing compressor is divided into a low-pressure suction chamber and a high-pressure compression chamber. As the volume of the compression chamber decreases, the refrigerant will be compressed to reach the discharge-level pressure, which triggers the discharge valve to open to let the compressed refrigerant escape into the discharge space. The theoretical volume can be obtained by the same calculation steps as for a rolling-piston compressor.

(3) Features: By integrating the vane and the rolling piston into a single component, the swing compressor successfully avoids having a severe mixed-lubrication condition along the line of contact that is experienced at the vane tip of a rolling piston compressor. The vane and rolling piston integration also contributes to the reduction of vane leakage and vane sliding loss[19]. With no demanding mixed-lubrication condition, the piston can be built of ordinary cast metal, which can be used even for CO_2-based systems that are subject to larger differential pressure[20].

However, tradeoffs for having no sliding loss at the vane tip include that, as the piston does not turn by itself, the relative speed between the pin shaft (= crank pin) and the piston will be higher than that in a rolling piston compressor, leading to a greater mechanical loss between the pin shaft and the piston. There are also additional regions of loss that do not exist in a rolling piston compressor, such as those on the bush outer surface. Table 4.1 provides an example of theoretical mechanical loss comparison between the rolling piston compressor and the swing compressor. According to this example, the swing compressor offers an overall mechanical loss reduction of several percent compared to the rolling piston compressor[21].

Table 4.1 Compression of mechanical loss

	Swing	Rotary
Main/Sub bearing	24.1%	24.0%
Pin bearing	34.5%	31.7%
Blade tip	----	12.5%
Blade side	24.0%	26.5%
Bush side	6.6%	----
Thrust etc.	6.0%	5.3%
Total loss	95.3%	100%
※ Comparison on the same design at 3/4HP		

4.6.3 Liquid injection

Figure 4.31 provides a refrigeration cycle using a liquid injection-based compressor and its P-h diagram. In the liquid injection-based refrigeration system, a portion of liquid refrigerant is diverged from the condenser exit into an injection line before the expansion valve. After flowing through the injection line so that the pressure drops to the intermediate level, the liquid refrigerant is injected into the compressor cylinder when the cylinder is halfway through the compression process, so that the inside of the cylinder is cooled by the evaporation latent heat of the liquid refrigerant, lowering the discharge gas and motor temperatures.

This technique is used, for example, in low-temperature application or cold-climate air conditioning compressors that need to run at higher compression ratio and in lower evaporating temperature range, or compressors intended for use in Middle East or South Asian countries that need to operate in high condensing temperature ranges. The purpose of utilizing such liquid injection feature is to provide additional cooling for the discharge gas temperature and the motor coil temperature that may otherwise exceed the allowable reliability range.

The following paragraphs further explain about the liquid injection technique using the P-h diagram in Fig. 4.31. Refrigerant mass flow rate G sucked in to the compressor is expressed by the product of compressor displacement, suction gas density and the compressor rotational speed.

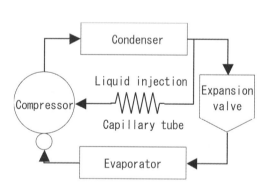

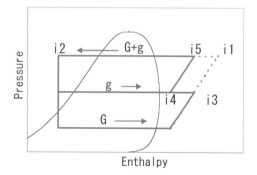

Fig. 4.31 Liquid injection circuit

In a refrigeration system without liquid injection feature, this flow rate G circulates through the entire system. Upon liquid injection, refrigerant that was being compressed (enthalpy i_3, refrigerant mass flow rate G) is mixed with the newly injected refrigerant (enthalpy i_2, refrigerant mass flow rate g). Enthalpy i_4 after mixing can be expressed by the following equation:

$$i_4 = \frac{G \cdot i_3 + g \cdot i_2}{G + g} \quad (4.34)$$

After the above mixing, the refrigerant flow rate becomes $(G + g)$. Then the refrigerant is further compressed to the discharge pressure so as to reach enthalpy i_5. As $i_5 < i_1$, the discharge gas temperature is lower than that without liquid injection.

Next, the heating capacity of a refrigeration system with and without liquid injection is compared. Heating capacities Q_0 without liquid injection and Q_1 with liquid injection can be expressed by the following equations:

$$Q_0 = G \cdot (i_1 - i_2) \quad (4.35)$$

$$Q_1 = (G + g) \cdot (i_5 - i_2) \quad (4.36)$$

In the above equations $(G + g) > G$ and $(i_5 - i_2) < (i_1 - i_2)$, which means that liquid injection increases the refrigerant mass flow rate while it decreases the heating enthalpy variation. This leads to $Q_0 \fallingdotseq Q_1$, where the final heating capacity almost does not change at all (minute increase or decrease is possible depending on the refrigerant properties).

The degree of discharge gas temperature drop depends on the amount of liquid refrigerant injected, which is dependent on the intermediate pressure that is determined by such factors as injection path restriction and the injection port location inside the cylinder compressor chamber. Therefore, the size and location of the injection port is critical in the design of a liquid injection-based compressor.

In a rotary compressor, the injection port is typically provided in the cylinder or in the main or sub bearing, opening into the compression chamber inside the cylinder. The timing of injection port opening in the compression stage, which influences the intermediate pressure, is dependent on the rolling piston's passage timing, which varies depending on where the injection port is located and how large it is. Figure 4.32 provides an example of an injection port provided in a rotary compressor[22].

4.6.4 Capacity control

Two types of compressor capacity control techniques may be used; one is rotational speed control where the re-

Chapter 4 Rotary Compressors

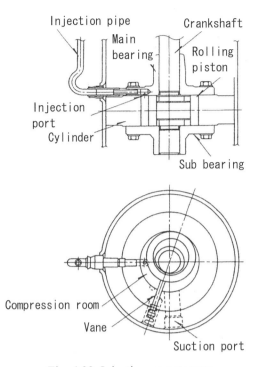

Fig. 4.32 Injection port strcutre

frigerant circulation volume per unit time is controlled by varying the compressor rotational speed using an inverter drive or changing the number of motor poles, and the other is mechanical capacity control where the refrigerant circulation volume per unit time is controlled by mechanically varying the cylinder's internal volume, or displacement, per compressor revolution.

With the progress of electronic control technology, inverter-based rotational speed control has become the dominant capacity control method. However, the following sections explain about mechanical capacity control techniques based on the rotary compressor mechanism.

(1) Suction gas bypass (release) method: Figure 4.33 shows the refrigeration cycle based on the suction gas bypass (release) technique. The technique uses a bypass channel that bypasses refrigerant gas from the compression chamber to the suction side after the suction process is completed, thus delaying the compression start timing. As a result, the effective compression chamber volume, and thereby the compressor capacity, are reduced.

Ideally, while the compression chamber is connected to the suction side by the gas bypass channel, pressure in the compression chamber remains at the same suction pressure even after the volume of compression chamber starts to decrease. At the time when the gas bypass connection is blocked, pressure in the compression chamber begins to rise.

In a rotary compressor, the location of the gas bypass port controls the timing when the rolling piston moves over the port to block the gas bypass connection, which in turn determines how long the bypass channel stays open and how much refrigerant flows through it.

In theory, both capacity and power input of the compressor with the gas bypass connection reduces in proportion to the effective reduction in the compression chamber volume under the gas bypass condition. However, in practice, the actual power input does not decrease to the ideal one due to never decreasing mechanical loss and additional pressure increase in the compression chamber, the latter of which is subject to the pressure loss through the bypass channel. As a result, the overall compressor efficiency tends to decrease.

(2) Gas injection method: As shown in Fig. 4.34, the refrigeration cycle with a gas injection-based compressor has two separate decompression steps (expansion valves),

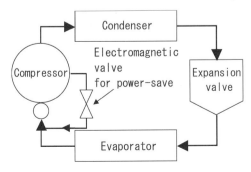

Fig. 4.33 Suciton gas bypass (release) type

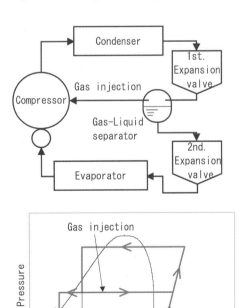

Fig. 4.34 Gas injection

71

with a liquid-gas separator provided in between. After the refrigerant is separated into gas and liquid portions through the separator, the gas portion only is directly routed into the compressor cylinder when it is halfway through the compression process. As shown in the P-h diagram, the above technique provides a larger enthalpy variation on the evaporator side, resulting in cooling capacity increase.

At the same time, heating capacity is also increased on the condenser side as the refrigerant circulation amount is increased by the mass of intermediate-pressure gas refrigerant injected into the compression chamber from the injection channel.

Furthermore, cooling and heating capacities can be made variable by opening and closing the gas injection channel.

The range of such capacity variation depends on the amount of gas refrigerant injected. It is critical, similarly to the liquid-refrigerant injection method, to optimally design the location and size of the injection port that opens into the compression chamber, with taking the system's operating conditions and the anticipated pressure losses into consideration. The injection port into the compression chamber and its associated piping will be a non-effective volume during the compression stage. Factors that are detrimental to the compression efficiency, such as increase in the re-compression work and decrease in volumetric efficiency, should be carefully considered[23].

(3) Variable cylinder management method: Under the variable cylinder management method, the vane for one of the cylinders in a two-cylinder rotary compressor can be disengaged from the rolling piston so that that cylinder runs idly, thus providing only one cylinder's worth of compression work[24].

Figure 4.35 shows the structure of a compressor with such variable cylinder management feature. In the compressor shown, the vane for the upper cylinder is always pressed against the outer circumference of the rolling piston by a spring, while the vane for the lower cylinder can be pressed against or moved away from the rolling piston according to the differential pressure working on the vane. In the two-cylinder operation mode, pressure inside the compressor casing increases as a result of the compression action taking place in the upper cylinder. This generates pressure difference between the inside and the outside of the lower cylinder. The vane for the lower cylinder is pressed by the differential pressure against the rolling piston to start compressing the gas in the lower compression chamber as well. For a one-cylinder operation, a switching valve is activated to introduce high-pressure discharge gas, instead of the suction gas that would otherwise be introduced, into the lower compression chamber. With no force of differential pressure pushing the vane from behind, the vane is attracted by a magnet set at the rear of the vane slot and stays back in the vane slot. As a result, function of the lower cylinder is nullified so that only the upper cylinder performs compression action.

With this method, the minimum capacity of the compressor can be easily reduced to only half that of the two-cylinder mode, leading to avoid frequent cycling operation of the compressor, which realizes energy saving and better comfort. Figure 4.36 is an example of energy saving effect of such variable cylinder management feature when installed to a residential air conditioner. The example shows that running the air conditioner continuously in a one-cylinder mode under the more frequent low-cooling-capacity

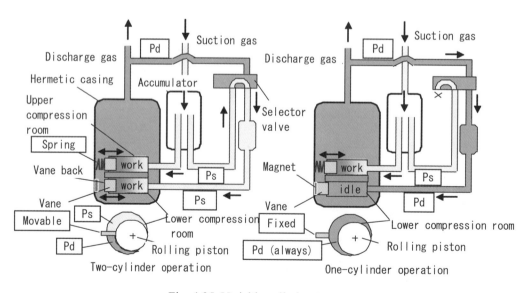

Fig. 4.35 Variable cylinder structure

Chapter 4 Rotary Compressors

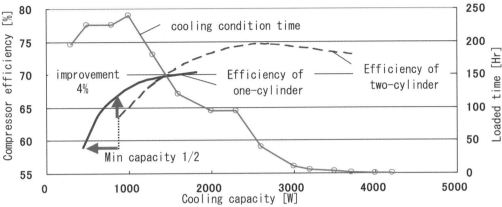

Fig. 4.36 Efficiency improvement of variable cylinder

ranges, where compressor efficiency at one-cylinder mode is much higher than that of two-cylinder mode, can reduce the cooling power consumption by 44%[17].

4.6.5 Two-stage compressors

A two-stage compressor has two serially-connected compression chambers, one lower pressure chamber and one higher pressure chamber, so as to compress the refrigerant gas in multiple stages. An example of such two-stage compressor used in a residential air conditioner is shown in Fig. 4.37. This compressor is realized by modifying the base model, a two-cylinder one-stage compressor housed in a high-pressure casing, so that it has two compression chambers serially connected through a linkage connecting pipe[25].

As the refrigerant is compressed in multiple stages, refrigerant leakage losses and gas load-related mechanical losses will theoretically decrease while hydrodynamic losses of the refrigerant increase. One of the main design factors is the volume ratio between the low-pressure and high-pressure compression chambers, which must be optimized according to the operating conditions such as the suction and discharge pressures and rotational speed of the compressor. To further reduce hydrodynamic losses, the geometry of the discharge ports and cylinders must also be optimized.

Figure 4.38 shows normalized overall adiabatic compressor efficiencies of a dimensionally optimized two-stage compressor and of a one-stage compressor respectively. This figure reveals that the efficiency of the two-stage compressor is much higher than that of the one-stage one in low rotational speed ranges where the effect of hydrodynamic loss is relatively small.

In a two-stage compressor, a space in the connecting pipe between the low-pressure and high-pressure compression chambers will be always at intermediate pressure. Some refrigeration systems utilize this characteristic for a "two-stage-compression and two-stage-expansion refrigera-

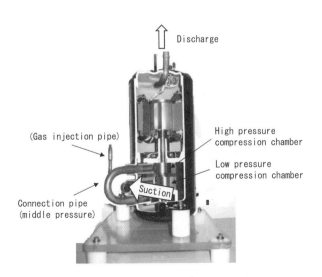

Fig. 4.37 Two-stage compression rotary compressor

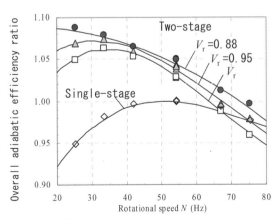

V_r: Low/High pressure compression volume ratio
$\eta_{ad,0}$: Single-stage compressor efficiency at 54 Hz

Fig. 4.38 Compressor effifiency of two-stage/one-stage

tion cycle" to achieve higher cycle efficiency. Figure 4.39 shows the schematics of the "two-stage-compression and two-stage-expansion cycle" for a residential air conditioner and Fig. 4.40 illustrates the operating principle of such cycle on *P-h* diagram.

This cycle will be capable of stable gas injection from the gas-liquid separator to the intermediate pressure region. With this, an ideal injection performance, which would not be attainable with a "one-stage-compression and two-stage-expansion cycle" where gas needs to be injected into a compression chamber with significant pressure variations, can be achieved.

4.6.6 Two-stage compressors with an intermediate pressure casing

Some CO_2-based refrigeration systems adopt a two-stage compression method in order to accommodate to high operating pressure and large pressure difference. One example of such two-stage compressor is shown in Fig. 4.41, where refrigerant that has been compressed to the intermediate pressure through the first compression mechanism is dis-

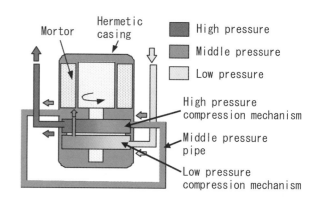

Fig. 4.41 Two-stage compressor mechanism (intermediate pressure casing type)

charged into the casing and then flows through an intermediate pressure piping that extends out of the casing before being introduced into the second compression mechanism and compressed to the final discharge pressure. In this design, the casing only needs to withstand the intermediate pressure and therefore can be made thinner than if the inside of the casing is filled with the final discharge pressure. This allows size and weight reduction of the compressor.

4.7 Lubrication and Oil Separation

4.7.1 Lubrication system for vertical rotary compressors

Due to the widespread use of inverter-driven variable-speed functionality, the compressor is now required to reliably operate under severe lubrication conditions in wider operational speed ranges, from very low to very high speeds. To realize that, it is critical to optimize the lubrication system in the compressor. A commonly used lubrication system in a vertical rotary compressor is shown in Fig. 4.42.

Refrigeration oil stored at the bottom of the hermetic casing is pressurized in a hollow space inside the crankshaft by the centrifugal pumping effect of the crankshaft rotation and distributed to various compressor components. To augment the centrifugal oil pumping effect, a twisted stirring vane and a cap with a reduced end diameter are fitted inside the crankshaft hollow. Also, oil feed holes are drilled radially on the crankshaft to connect the crankshaft hollow to the main- and sub-bearings and the rolling piston, supplying refrigeration oil to these sliding components.

At the main- and sub-bearings, refrigeration oil supplied through the oil feed holes flows through spiral oil grooves formed on the bearing bores to lubricate the bearing clearances, before returning to the bottom space of the hermetic

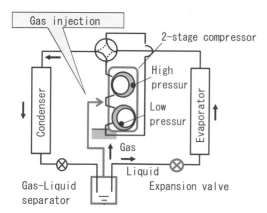

Fig. 4.39 Two-stage compressor and two-stage expansion cycle (at cooling operation)

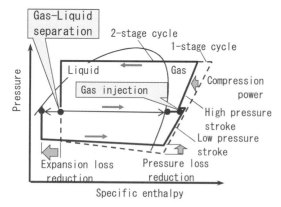

Fig. 4.40 Two-stage compressor and two-stage expansion cycle chart (at cooling operation)

Chapter 4 Rotary Compressors

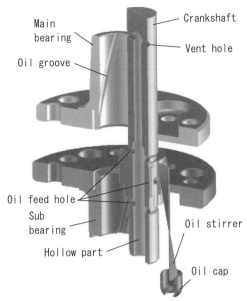

Fig. 4.42 Oil supply mechanism

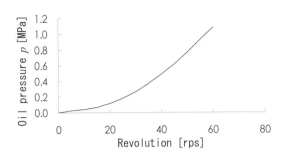

Fig. 4.43 Oil pressure of centrifugal lubrication

casing as an oil reservoir.

At the rolling piston, oil coming out from the oil feed hole flows through the lubrication groove formed on the crank pin surface to lubricate the rolling piston bearing clearance. After lubrication, the oil leaks into the suction and compression chambers in the cylinder through clearances on the end faces of the rolling piston by the pressure difference between the inside and outside of the rolling piston. The oil does not only lubricate sliding components in the cylinder but also serves to seal gaps and clearances around them, thus contributing to the reduction of refrigerant gas leakage loss in the compressor.

Oil pressure when the crankshaft rotates at angular speed ω can be obtained, based on the centrifugal force working on the fluid and also on the degree of pressure drop, by the following equation[26]:

$$p = \frac{\rho}{2} r_0^2 \omega^2 - \frac{\rho}{4} r_c^2 \omega^2 \qquad (4.37)$$

where: ρ: Density of oil
r_0: Radius of the oil feed hole exit (= outer radius of crankshaft at oil feed hole)
r_c: Radius of crankshaft lower end opening (= radius of crankshaft oil cap hole)

Figure 4.43 shows the calculation result. As this lubrication system utilizes the centrifugal force of the rotating shaft, the amount of oil feed is greater in higher speed operation ranges and smaller in lower speed operation ranges. During low speed operation of the compressor, risks of bearing damage due to insufficient lubrication and increase of internal leakage due to insufficient oil seal must be cared. During high speed operation, a greater amount of oil is supplied and introduced to the inside of the cylinder, and as a result more oil must escape from the compressor to the refrigeration circuit with discharge refrigerant gas, which increases "oil circulation ratio" in the refrigeration cycle. This may cause deterioration of heat exchange efficiency at the condenser and the evaporator and abrasion or seizure at compressor sliding parts due to oil shortage in the hermetic casing. These risks must be carefully addressed.

The inside of the hermetic casing of a rotary compressor is typically maintained at high discharge pressure, and as a result the refrigeration oil stored at the bottom of the casing is constantly exposed to high temperature and high pressure. As the solubility of refrigerant into the refrigeration oil increases with pressure increase, refrigeration oil under high pressure in the compressor bottom always has a certain amount of refrigerant dissolved into it. Such refrigerant-dissolved oil may foam in oil channels as a result of depressurization in the compressor or pressure loss in the oil feed path. Such foaming may block the flow of oil and prevent the sliding components from being lubricated.

Such foaming risk is greater in situations where massive amounts of liquid refrigerant are temporarily dissolved or mixed in the refrigeration oil, for example when the compressor is being started after an extended period of non-operation or when it is continuously operated under the condition that considerable amount of liquid refrigerant is contained in the suction refrigerant gas. Careful monitoring and evaluation of refrigerant foaming risk is important.

As a countermeasure against refrigerant foaming, the crankshaft of a rotary compressor commonly may have a degassing vent hole above the oil feed holes, as shown in Figure 4.42, that connects the crankshaft hollow and the inner space of the hermetic casing. With this, foaming of the refrigerant dissolved in the refrigeration oil will not obstruct oil supply to and through the oil feed holes.

4.7.2 Lubrication system for horizontal rotary compressors

One of the most difficult engineering challenges concerning a horizontal rotary compressor is that it requires a lubrication system to distribute the refrigeration oil, which otherwise sits in the reservoir at the bottom of the hermetic casing, to its sliding components. Various types of lubrication systems are used. Shown in Fig. 4.44 is a system that utilizes the force of pressure difference. The one shown in Fig. 4.45 is an ejector-type system utilizing the flow energy of the discharge gas to distribute refrigeration oil to sliding components[27].

Another one, shown in Fig. 4.46, is a helix centrifugal pump-based system that provide lubrication using a spiraled spring directly connected to the crankshaft. Figure 4.47 shows yet another one, a fluid diode system that utilizes the reciprocating action of the vane to feed refrigeration oil from one side to the other side.

4.7.3 Oil separation in rotary compressors

Refrigeration oil that flows out of the compressor together with the discharge gas and circulates in the refrigeration cycle could become a source of refrigeration performance deterioration, for example by causing heat exchange efficiency reduction at the condenser or evaporator or increasing pressure loss in the heat exchangers, connecting pipes and suction pipes. Meanwhile, decrease in the amount of oil at the bottom of the hermetic casing and the resulting lubrication deficiency may cause excessive abrasion or seizure of sliding components in the compressor. To maintain performance and reliability, it is critical to minimize the oil circulation ratio, which is the ratio of the oil amount escaping from the compressor into the system with discharge refrigerant gas.

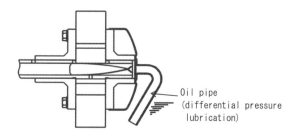

Fig. 4.44 Differential pressure lubrication mechanism

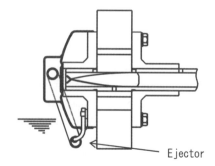

Fig. 4.45 Ejector lubrication mechanism

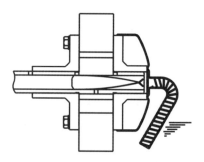

Fig. 4.46 Helix centrifugal pump mechanism

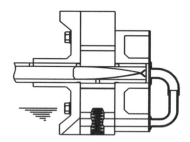

Fig. 4.47 Fluid diode type mechanism

The oil circulation ratio of a rotary compressor gets greater at elevated rotational speeds. Or, under the same rotational speed condition, oil circulation ratio will be greater as the evaporation temperature gets higher and the refrigerant circulation amount increases. These tendencies are due to the following factors: (1) As refrigeration oil is fed to compressor components using the centrifugal effect of the rotating shaft, more oil will be fed at higher rotational speeds. (2) As rotational speed and evaporation temperature get higher, refrigerant circulation amount will increase and discharge gas will flow at greater speeds. This will deteriorate the efficiency of oil separation inside the hermetic casing.

To reduce the amount of oil circulation in the refrigeration system, oil separation efficiency in the hermetic casing must be improved. To do that, the speed of discharge gas flow inside the hermetic casing must be reduced, by increasing the cross section area of flow channels, for example motor air gaps and air holes penetrating the motor core. Figure 4.48 shows the relationship between the cross section area ratio of gas flow channel and the oil circulation ratio.

Figure 4.49 shows another design intended for oil circulation minimization, a centrifugal separation mechanism

Chapter 4 Rotary Compressors

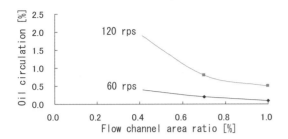

Fig. 4.48 Relationship between flow channel area ratio and oil circulation

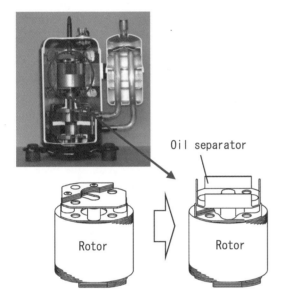

Fig. 4.49 Centrifugal separating structure

that uses an oil separator above the motor rotor. In this type of mechanism, the centrifugal effect accompanying the shaft rotation plus the specific gravity difference between the refrigerant gas and oil will help reduce the amount of oil escaping into the system through the discharge pipe located at the center of the hermetic casing. Figure 4.50 shows the effect of such centrifugal separation mechanism.

In transitional operation modes, oil circulation could present even more serious problems. For example, when the compressor is being started up after an extended period of non-operation time, a large amount of refrigerant that migrated and trapped inside the compressor during the non-operation period may bring with it a large quantity of refrigeration oil dissolved in it out of the compressor and into the refrigeration cycle. To address these risks, the refrigeration system must be designed to assure good oil returnability.

For example, an accumulator aimed to hold extra amount of refrigerant in the system is also prone to hold oil. Where such accumulator is used, an additional oil return feature should be provided. Or, an additional oil separator may be provided along the refrigeration system, especially if the system has long piping lengths.

Rotational speed control pattern at the compressor start-up may also influence the escape and return characteristics of the oil. Rapid acceleration immediately after startup or continuous operation at the maximum rotational speed will increase the amount of oil escape. Excessively short intervals of between the compressor on and off may also detrimentally affect the oil returnability. Figure 4.51 shows an example of continuous oil circulation ratio measurements during startup operation of an inverter driven rotary compressor[29].

4.8 Noise and Vibration of Rotary Compressors

4.8.1 Excitation forces and vibration reduction

In Japan, inverter-based refrigeration systems that offer superior comfort and energy saving effect are dominantly used in residential air conditioners as well as commercial packaged air conditioners. As increasingly greater focus is placed on energy conservation and environmental protection on a global scale, inverter-based systems will be used even more widely throughout the world in the future. To use an inverter drive effectively, it is desirable that the compressor

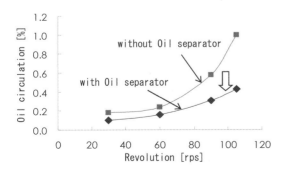

Fig. 4.50 Effect of oil separator

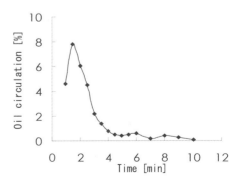

Fig. 4.51 Change of oil circulation during compressor starting

has a wide variable speed operation range. However, rotary compressors tend to have greater vibration amplitudes in lower speed ranges, where the allowable stress limit of the unit piping or the allowable vibration limit of the outer unit panels may potentially be exceeded. Reduction of vibration amplitude in the rotational direction is one of the most difficult engineering challenges when attempting to widen the operation range of an inverter-based compressor.

The following paragraphs explain about rotational-direction vibration that increases in low speed operation ranges, which presents the most serious operational issue in an inverter-driven rotary compressor, and what measures are taken to reduce it, using a simplified model to clarify the excitation forces involved and the vibro-isolating spring system used.

Figure 4.52 shows the excitation forces and the vibro-isolating features in a rotary compressor. The motor rotor is coupled to the rolling piston via the crankshaft, together constituting a rotational system. The motor stator, the cylinder and the bearings are fixed to the hermetic casing and together constitute a fixed system. Vibro-isolating rubber (a spring element) is used to support the entire compressor.

When the motor is powered to generate output torque T_m, the motor rotor drives the rotating shaft. The resulting eccentric rotational motion of the rolling piston changes the volume of the compression chamber to compress the refrigerant gas, so that compression load torque T_g is generated. The magnitude of torque T_g varies in repetitive cycles (one such cycle occurs per every shaft rotation).

If it is assumed that the rotational and the fixed systems of the compressor are fully rigid, that the rotational axis and the supporting spring system are fully coaxial, and that the mass of the rigid components are distributed in a fully

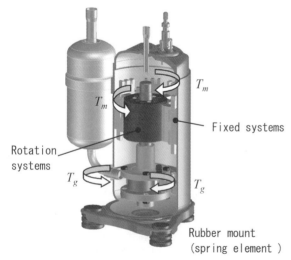

Fig. 4.52 Toruque and vibration reduction system

symmetrical manner, the instantaneous motor output torque T_m and the compression load torque T_g will work on the rotational and the fixed systems in opposing directions around the rotating axis. The governing equations of motion can be derived at as follows:

$$J_R \frac{d\omega_R}{dt} = T_m - T_g = \Delta T_r \tag{4.38}$$

$$J_s \frac{d^2\theta_s}{dt^2} + C_s \frac{d\theta_s}{dt} + K_s \theta_s = -\Delta T_r \tag{4.39}$$

Here, J_R and J_S are inertia moments in the rotational and the fixed systems respectively, K_s is the spring constant of the vibro-isolating system, C_s is the attenuation coefficient, ω_R ($=d\theta_R/dt$) is the angular velocity of the rotational system, θ_R and θ_S are the rotation angles of the rotational and the fixed systems, and ΔT_r is the instantaneous torque difference.

Equation (4.39) shows that the rotational-direction vibration that appears on the rotary compressor periphery occurs as the torque difference $\Delta T_r(\theta_R)$ between the instantaneous motor torque $T_m(\theta_R)$ and the instantaneous compression load torque $T_g(\theta_R)$ excites the supporting system with spring constant K_s.

Variation of the compression torque load is determined by the geometry of the compression elements and their operating pressure conditions. As the motor output torque is normally approximately constant, the instantaneous torque differences determine the magnitude of rotational-direction vibration.

To minimize the degree of these instantaneous torque differences to reduce the compressor vibration, the following two methods are currently available. One is to use a multiple-cylinder configuration to better level out the compression load torque variations. This method is actually employed in two-cylinder rotary compressors.

The other method is to regulate the motor output torque so that it will follow the compression load torque pattern as closely as possible, thus minimizing the instantaneous torque differences to reduce vibration. This is an active vibration control, or torque control technique, where the motor plays a role of a vibration-controlling actuator. Its details are given in the next section.

4.8.2 Vibration reduction through torque control

Figure 4.53 shows the relationship between the load torque, motor torque, shaft rotational speed and supply voltage, in a normal drive system and a torque control-based system. In a normal drive system, the magnitude of compression-induced load torque varies, by a relatively

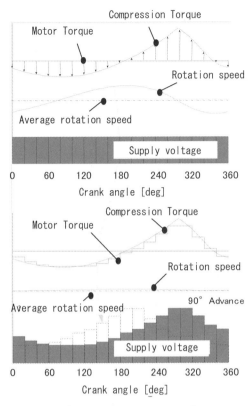

Fig. 4.53 Toruque control

large degree, throughout a crankshaft rotation. The instantaneous differences between the load torque and the motor torque, the latter of which remains constant under the constant supply voltage, generate excitation forces. As a result, the actual rotational speed varies significantly throughout one rotation, deviating from the average rotational speed. In a torque-control-based system, the supply voltage is regulated to vary in such a way that the motor torque will closely follow the load torque pattern, thus reducing instantaneous torque differences. As a result, the excitation forces and the degree of crankshaft rotational speed variation will be smaller.

The followings explain specifically how the motor torque is regulated to follow the load torque pattern. As shown in Fig. 4.53, the effect of instantaneous torque differences between the motor torque and the load torque becomes apparent in the form of rotational speed variation. Detecting this rotational speed variation and adjusting the supply voltage so that the variation becomes the smallest, which is a technique called "feedback control", can provide the desired regulation effect. Such feedback control can be performed by multiple methods. In one method, one rotation of the crankshaft is divided into multiple detection ranges, in each of which angular velocity ω_R is detected, and the motor torque is regulated so that ω_R will remain as constant as possible through all the ranges. In another method, angular rotational acceleration $d\omega_R/dt$ is detected and the motor torque is regulated in a feedback control so that $d\omega_R/dt$ will remain as small as possible[30, 31].

Although the exact variation pattern of the load torque changes according to the geometry of the compression elements and their operating pressure conditions, that degree of change is smaller than the degree of torque variation with shaft rotation. Therefore, almost the same vibration reduction effect as that by the above feedback control can be obtained by providing motor torque regulation based on a pre-assumed load toque variation pattern.

Figure 4.54 shows the vibration amplitude in a single-cylinder rotary compressor in tangential direction, with and without torque control in rotational speed ranges under 50 rps. This indicates that torque control is effective to reduce the compressor vibration, which leads to widening the low speed operation range of inverter-driven compressors.

4.8.3 Noise of rotary compressors

(1) Noise sources by frequency range: Various types of noise are generated from a rotary compressor depending on its capacity or the type of motor used, each in different loudness and coming from a different source in the compressor. The following paragraphs provide classification of rotary compressor noise sources by frequency range.

(i) 50 Hz to 500 Hz

Basic excitation forces, unbalanced inertia forces (occurring to a multiple of the rotational speed) due to torque variation and pressure unbalance resultant of the compression work and electromagnetic motor unbalance (occurring to a multiple of the electromagnetic frequency or the rotational speed).

(ii) 500 Hz to 2 kHz

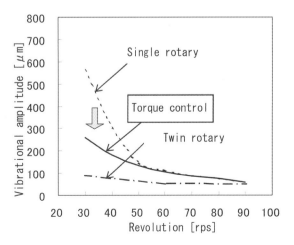

Fig. 4.54 Vibration reduction by toruque control

The dominant source is refrigerant gas pressure pulsation. If peaks in narrowband ranges are observed, fluid resonance in discharge lines or spatial resonance in internal structure of the hermetic casing is suspected. On the other hand, peaks in broadband ranges indicate that structural components are resonance transmission sound due to the excitation forces caused by the above-described fluid or spatial resonance.

(iii) 2 kHz to 4 kHz

The dominant excitation source is the pressure pulsation of high frequency waves that occur in the compression chamber. Noises are amplified by transmission resonance of mechanical components including the cylinder. Electromagnetic noise sources include "slot harmonics" that are determined by the number of slots in the motor stator and rotor and noises that are electromagnetically caused due to the inverter's carrier frequency, the latter type being specific to inverter-driven systems only.

(iv) Over 4 kHz

Moving components such as the discharge valve, vane and rolling piston are potential excitation sources. In higher-speed, higher-compression-ratio operations, noises in this frequency range will have a greater proportion in compressor-generated noise.

(2) Classification of noise sources and respective reduction measures: Reduction of rotary compressor noises should be started by identifying the noise source for each frequency range in question, as described above, and clarifying what factors are contributing to the sound pressure.

(i) Fluid noises

Noises that are amplified by spatial resonance inside the hermetic casing could be reduced by changing the volume and geometry of the expansion-type discharge muffler to improve the acoustic propagation characteristics. Figure 4.55 shows an example of expansion-type discharge muffler.

To reduce the noises caused by pressure waves from the discharge stage, it is known that providing a Helmholtz resonator having a small-volume space and a pressure introduction channel, as shown in Fig. 4.56, attenuates pressure pulsations. This technique allows targeting a specific frequency range to be attenuated by adjusting the volume of the small volume space and changing the pressure introduction channel dimensions[32].

(ii) Electromagnetic noises

Electromagnetic unbalance of the motor causes electromagnetic noises. As a general rule, the basic frequency of such electromagnetic noise will be, from the electromagnetic force frequency determined by fundamental magnetic flux behavior, twice the power supply frequency and will contain the following frequency components:

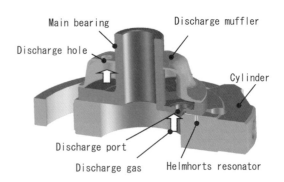

Fig. 4.55 Discharge muffler (expansion-type)

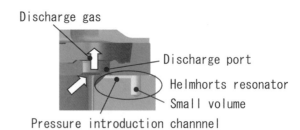

Fig. 4.56 Helmhorts resonator

$$f_e = 2nf \quad (n = 1,2,3,\cdots) \quad (4.40)$$

where f: Power supply frequency

To alleviate electromagnetic unbalance, it is effective to have a uniformly dimensioned air gap between the stator and the rotor. To do that, factors such as stator and rotor roundness and hermetic casing dimensions and geometry must be thoroughly controlled. In a single-phase motor, electromagnetic noise could also be caused by magnetic field unbalance in the motor coil windings. A significant unbalance between the main and auxiliary windings will cause the motor to be electromagnetically out of balance.

(iii) Solid-borne noises

As the compression mechanism of a rotary compressor is directly secured to the hermetic casing, the hermetic casing may, as a solid-borne noise transmission medium, transmit excitation forces from the compression mechanism to the outside or cause resonance in itself or in other components such as the suction muffler. The central tubular section of the hermetic casing, as it needs to be mated with the motor stator by shrink-fitting and therefore is designed to be highly rigid with elevated mechanical impedance, is harder to generate a resonance mode to transmit noise. On the other hand, other parts of the casing such as the top and bottom covers and the suction muffler are more prone to localized resonance. To address this issue, acoustic radiation efficiency must be reduced by any measures, for example by increasing the bending sections in these areas or changing their shapes into spherical form.

To provide a low-noise design of the hermetic casing or the suction muffler, it is critical to fully understand the component's natural frequency mode and mechanical impedance as a solid-borne noise transmission medium. With lower mechanical impedance, the component will tend to vibrate in a specific frequency range.

To improve the mechanical impedance of a tubular hermetic casing and its cover sections, the following measures can be taken:
(a) Increase the wall thickness
(b) Change the geometry
(c) Add vibration-restricting damping material
Of these,
(a) Will help to a certain degree but will have trade-offs such as weight and material cost increase and decrease in machinability.
(b) Highly effective, as shaping the bends and sharp corners as spherical as possible will increase the mechanical rigidity and also lower the acoustic radiation impedance. As completely flat sections are more prone to resonate, it is recommended to add partial convexo-concave on all flat sections. Also avoid, whenever possible, shaping or arranging components in a symmetrical manner.
(c) Practical examples of this technique include gluing a piece of butyl rubber-based damper of a specific mass onto the casing exterior or on piping surface, or fixing the muffler to the hermetic casing with a layer of rubber inserted in between.

4.9 Production Technology

4.9.1 High-precision machining

Components of the rotary compressor's compression mechanism have to be precisely machined on the order of micrometers and then assembled with the clearances in between controlled also to an accuracy of micrometers. It is clear that improvements in machining and assembly precision will directly contribute to the compressor's efficiency and reliability improvement, mainly by reducing leakage and mechanical losses.

The high efficiency and reliability as well as the extended operation range and the diverse applications of today's mass-produced compressors have been made possible not only through advances in design engineering but also through improvements and breakthroughs in production technology. This section explains about high-precision machining, which is an essential part of such advanced production technology.

In a rotary compressor, the vane, reciprocating inside a vane slot in the cylinder wall, divides the space inside the cylinder into compression and suction chambers. Through clearances between the vane and the walls of the slot, refrigerant leaks from the high-pressure side to the low-pressure side. Also, the vane is pressed against the slot by the force of pressure difference while making the sliding action. Therefore, the geometric accuracy of the vane slot as well as the precision of the clearance magnitude between the vane and the slot significantly affects the compressor's efficiency and reliability.

In general, machining of narrow, tight slots like the vane slot used to be done by broaching. By two-step broaching with rough and finishing tools, where the final slot machining accuracy depends on the quality of the cutting tool's ridge line and the straightness of tool motion, the highest level of vane slot flatness achievable in mass production used to be several micrometers with the best possible machining speed.

It was known that grinding process would have a better chance of achieving a superior flatness than broaching. It would require developing the grinding stone drive transmission system that is smaller than the inner diameter of the cylinder, which was practically impossible.

More recently, however, a grinding head system where the grinding stone section and the drive transmission section are separate, as shown in Fig. 4.57, has been newly developed, thus making it possible to apply grinding process to vane slot machining. As shown in the figure, the grinding stone section of this system contains a thin grinding disk that is narrower than the target slot width, while the drive transmission section consists of a drive motor and a pulley.

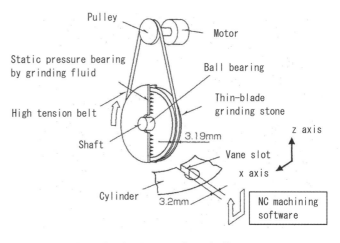

Fig. 4.57 Vane slot grinding

Compressors for Air Conditioning and Refrigeration

The motor power is transmitted to the grinding stone section via a belt. The whole system is controlled by two movable axes, making it possible to accurately grind a narrow-width vane slot of the cylinder. With this advanced grinding system, a flatness accuracy that is three times as good as broaching and a surface roughness that is approximately twice as good has been achieved in vane slot machining on a mass production basis[33].

4.9.2 High-precision assembling

The most important elements of the assembly of rotary compressors are eccentric and concentric assembly steps. "Eccentric assembly" here refers to assembling the crankshaft to its main mating components by having the crankshaft rotation center offset from the cylinder center by a certain amount toward the compression chamber. The reason for doing this is that the compression load generated by the pressure difference between the compression and suction chambers inside the cylinder will work on the rolling piston during compressor operation to cause the assembly gaps to change.

On the other hand, "concentric assembly" refers to assembling the two bearings that support the crankshaft across the cylinder by positioning them fully concentric to each other before fastening them to the cylinder.

Accuracy of these two assembly steps affects the magnitude of gaps between components during compressor operation and greatly influences the degrees of leakage loss and mechanical loss. Accuracy improvement in these assembly steps therefore significantly improves compressor efficiency. Recently, an auto-aligning assembly system using a high-speed measurement unit has been developed, where each workpiece is individually measured before assembly to calculate the best amount of gaps and positioning for the combination of workpieces used. This system was made possible with advances in various supporting technologies including alignment error measurement, counterweight application and high-speed alignment. Figure 4.58 shows such measurement-based auto-aligning assembly system[34].

To secure the compression mechanism to the hermetic casing, providing arc-spot welding on the cylinder used to be the common method. With arc-spot welding, it was unavoidable that welding material that gets between the cylinder and the hermetic casing creates radial reaction force that deforms the vane slot and the inner cylinder circumference, increasing leakage loss and consequently reducing compressor performance as shown in Fig. 4.59.

This issue is now being addressed by various alternative assembly methods, such as providing the arc-spot welding

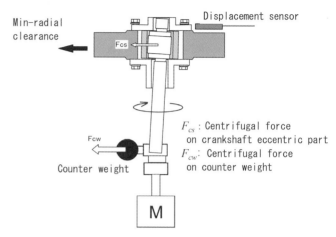

Fig. 4.58 With measureing coaxial assembly method

on non-cylinder components such as the bearings, or use of "thermal caulking" technique, where deformation that has been unavoidable in welding-based assembly could be fundamentally eliminated.

Figure 4.60 shows an example of such thermal caulking-based assembly. Here, the dominant force working on the cylinder is the circumferential binding force created by the thermal shrinkage of thermallly caulked casing parts. With

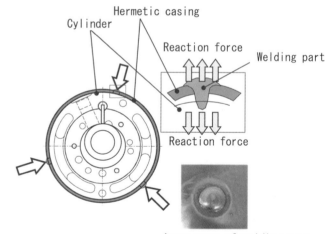

Fig. 4.59 Cylinder fixing by arc-spot welding

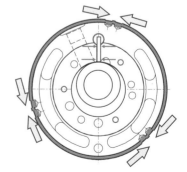

Fig. 4.60 Cylinder thermal caulking

this, radial deformation can be fundamentally eliminated. It is reported that tests using equal-rigidity cylinder specimens have shown that the thermal-caulking technique successfully reduced the degree of vane slot deformation by half [35].

References

1) Japan Society of Refrigerating and Air Conditioning Engineers: "6th EDITION, JSRAE HANDBOOK, II EQUIPMENTS", Japan Society of Refrigerating and Air Conditioning Engineers, Tokyo, p. 10 (2006). (in Japanese)

2) S. Nagatomo: "Recent compressors for refrigeration and air conditioning", Science of Machine, 42 (7), pp. 801-806 (1990). (in Japanese)

3) S. Nagatomo: "Positive Displacement Compressor Technology for Air Congitioners", Transactions of JSRAE, 15 (4), pp. 305-326 (1998). (in Japanese)

4) Y. Sato, Y. Shirafuji: "The Study of Rotary Compressor Driven under Low electric Frequencies", Proc. 1990 Purdue Int. Compressor Tech. Conf., West Lafayette, pp. 548-556 (1990).

5) T. Kato, Y. Shirafuji, S. Kawaguchi: "Comparison of Compressor Efficiency Between Rotary and Scroll Type with Alternative Refrigerants for R22", Proc. 1996 Purdue Int. Compressor Tech. Conf., West Lafayette. pp. 69-75 (1996).

6) Y. Shirafuji, M. Myogahara, T. Saikusa, T. Yamamoto: "Development of Rotary Compressor using Immiscible Oil for R410A Right Commercial Use", Proc. 2002 JSRAE Annual Conference, pp. 229-232 (2002). (in Japanese)

7) K. Tojo: "Development of Compressors for New Refrigerants", Refrigeration, 82 (959), pp. 734-740 (2007). (in Japanese)

8) H. Kobayashi, N. Murata, Y. Ozawa, S. Taniguchi: "Reliability Evaluation of Refrigeration Compressor Journal Bearing", Mitsubishi Heavy Industries Technical Review, 25 (4), pp. 384-389 (1988). (in Japanese)

9) K. Sakaino, S. Muramatsu, S. Shida, O. Ohinata: "Some Approaches Towards a High Efficient Rotary Compressor", Proc. 1984 Purdue Int. Compressor Tech. Conf., West Lafayette, pp. 315-322 (1984).

10) N. Ishii, M. Yamamura, S. Muramatsu, M. Fukushima, K. Sano, M. Sakai: "The High Mechanical Efficiency of Rolling-Piston Rotary Compressors: 1st Report, Experimental Study", Transactions of the Japan Society of Mechanical Engineers (Series C), 54 (507), pp. 2578-2582 (1988). (in Japanese)

11) N. Ishii, M. Fukushima, M. Yamamura, S. Muramatsu, K. Sano, M. Sakai: "The High Mechanical Efficiency of Rolling-Piston Rotary Compressors: 2nd Report, Theoretical Study", Transactions of the Japan Society of Mechanical Engineers (Series C), 55 (511), pp. 687-695 (1989). (in Japanese)

12) K. Imaichi, M. Fukushima, S. Muramatsu, N. Ishii: "Vibration Analysis of Rolling-Piston Rotary Compressor", Transactions of the Japan Society of Mechanical Engineers (Series C), 49 (447), pp. 1959-1970 (1983). (in Japanese)

13) H. Maeyama, N. Hattori, H. Nakao, T. Takayama, E. Sakamoto: "Single-Stage Rotary Compressor for Natural Refrigerant CO_2", Proc. 2008 JSRAE Annual Conference, Osaka, pp.13-16, (2008). (in Japanese)

14) T. Fujiwara: "Technical History of Rotary Compressor", Proc. 2006 JSRAE Annual Conference, Fukuoka, pp.17-20 (2006). (in Japanese)

15) K. Okoma, M. Tahata, H. Tsuchiyama: "Study of Twin Rotary Compressor for Air-Conditioner with Inverter System", Proc. 1990 Purdue Compressor Tech. Conf., West Lafayette, pp. 541-547 (1990).

16) H. Kato, M. Hasegawa, A. Morishima: "Development Of 2-Cylinder Rotary Compressor Series For Light Commercial Use With R410A", Proc. 2002 Purdue Compressor Tech. Conf., West Lafayette, C5-3 (2002). (in Japanese)

17) T. Sato, S. Kitaichi, S. Ide: "DAISEIKAI™ Room Air Conditioner with Advanced Energy-Saving Technologies", Toshiba Review, 64 (11), pp. 23-27 (2009). (in Japanese)

18) S. Ito et al.: "Practical Machine Series, Volume Type Compressor", Sangyo Tosyo, Tokyo, p. 192 (1970). (in Japanese)

19) Japan Society of Refrigerating and Air Conditioning Engineers: "6th EDITION, JSRAE HANDBOOK, II EQUIPMENTS", Japan Society of Refrigerating and Air Conditioning Engineers, Tokyo, p. 12 (2006). (in Japanese)

20) M. Higuchi, H. Taniwa, H. Matsuura, Y. Nabetani, E. Kumakura, H. Gigashi: "Development of Swing Compressor for CO_2 Refrigerants", Proc. International Symposium on HCFC Alternative Refrigerants and Environmental Technology 2002, Kobe, pp. 41-46 (2002). (in Japanese)

21) M. Masuda, K. Sakitani, Y. Yamamoto, T. Uematsu, A. Muto: "Development of Swing Compressor for alternative refrigerant (HFC)", Refrigeration, 71 (821), pp.

230-234 (1996). (in Japanese)

22) F. Wada, K. Asami: "Rotary Compressor for HCFC22", Mitsubishi Denki Giho, 67 (4), pp. 422-426 (1993). (in Japanese)

23) 23) T. Sano: "Capacity Control in Residential Air conditioners", Refrigeration, 74 (863), pp. 777-783 (1999). (in Japanese)

24) I. Onoda, S. Kitaichi, K. Takashima: "Dual-Stage Compressor for Air Conditioner", Toshiba Review, 59 (4), pp. 44-47 (2004). (in Japanese)

25) A. Kubota, M. Nonaka, A. Onuma, Y. Iizuka, Y. Nakada: "Development of 2-stage rotary compressor for gas injection cycle air-conditioner", Proc. 39th Japanese Joint Conference on Air-conditioning and Refrigeration, Tokyo, pp.159-162 (2005). (in Japanese)

26) T. Ito, H. Kobayashi, M. Fujita, N. Murata, T. Shikanai: "Study on the Oil Supply System for Rotary Compressors", Mitsubishi Heavy Industries Technical Review, 29 (5), pp. 458-462 (1992). (in Japanese)

27) H. Kawai: "Horizontal Rotary", Refrigeration, 62 (720), pp. 1090-1099 (1987). (in Japanese)

28) K. Shimamura, Y. Oishi, K. Asami: "Small Size Refrigerator (Rotary)", Refrigeration, 60 (694), pp. 836-842 (1985). (in Japanese)

29) Y. Shirafuji, T. Oikawa, T. Kato, S. Kawaguchi: "Development of High-efficiency Rotary Compressor for Room Air-Conditioner with R410A", Proc. the International Symposium on HCFC Alternative Refrigerants and Environmental Technology 2000, Kobe, pp. 36-41 (2000). (in Japanese)

30) M. Nakamura, H. Hata, Y. Nakamura, T. Endo, K. Iizuka: "Study on Vibration Reduction of a Rolling Piston-Type Compressor by Motor Torque Control: 2nd Report, Experimental Study of Control Effects", Transactions of the Japan Society of Mechanical Engineers (Series C), 56 (527), pp. 1797-1804 (1990). (in Japanese)

31) H. Iwata, M. Nakamura, O. Matsushita, M. Sutou: "Vibration and Noise Reduction of Compressors for Air-Conditioner", Transactions of the Japan Society of Refrigerating and Air Conditioning Engineers, 7 (2), pp. 105-118 (1990). (in Japanese)

32) Y. Yokomizo, Y. Watanabe, M. Hasegawa: "Noise reduction technology of rotary compressor for low temperature service", Toshiba Review, 40 (9), pp. 740-743 (1985). (in Japanese)

33) T. Nakasuji, H. Yokota, Y. Hirai, T. Honoki, T. Hashimoto: "High Efficiency Rotary Compressor by High Accuracy Vane Slot Grinding", Mitsubishi Denki

Ggiho, 75 (10), pp. 663-666 (2001). (in Japanese)

34) S. Hara, T. Iwasaki, T. Mochizuki, M. Ukioka, H. Noda, K. Tominaga: "Coaxial Assembly Method of Rotary Compressor for Refrigerator and Air Conditioner", Mitsubishi Denki Giho, 72 (4), pp. 313-316 (1998). (in Japanese)

35) T. Kato, T. Arai: "Twin Rotary Compressor with Caulking Assembly Innovation", Mitsubishi Denki Giho, 82 (3), pp. 179-182 (2008). (in Japanese)

Chapter 5 Scroll Compressors

5.1 History of Scroll Compressors

The idea of a scroll compressor that transports fluid using an identical pair of spiral-shaped parts, or "scroll wraps", has a relatively long history. From the late 1800s to the early 1900s, several patents related to scroll compressor technology for use in engine or pump mechanisms were applied in Europe and the United States[1-2]. However, none has been put to actual use in those periods, due to the poor production technologies and the lack of practical design concepts which ensure high efficiency and reliability. Full scale research and development oriented toward actual use did not start until 1970s. Since around that period, the structurally simple yet potentially highly-efficient design of the scroll compressor has drawn strong attention worldwide. Active research and development efforts commenced in Japan as well, establishing the fundamental design that still forms the basis of today's scroll compressors. From 1980s onward, significant technological advances were made for commercialization of scroll compressors, where more practical production designs were created, and high precision machining technologies for mass production were developed. Eventually, the world's first scroll compressor for packaged air conditioner and for automotive air conditioner were commercialized both in Japan[3-8].

The scroll compressor has a number of fundamental advantages over other types of compressors, such as that the compression process is performed continuously, that a very smaller amount of leakage occurs between compression chambers, and that better reliability is attainable due to the smaller number of parts and the lower slip speed between sliding components. Scroll compressors are used extensively in the output range of 0.75 to 30.0 kW, mostly in air-conditioners such as residential and commercial air-conditioners and cold storage systems such as freezers and refrigerators. They compete with rotary compressors for small capacity applications and with screw compressors for larger capacity uses. In addition to the above applications including packaged and automotive air conditioners, there are also instances of more industrial and technological uses such as helium compressors in cryogenic systems[9], air compressors[10] and vacuum pumps. More recently, scroll compressors are actively used in CO_2-based heat pump water heater systems[11] as well.

Fig. 5.2 Automotive scroll compressor[4]

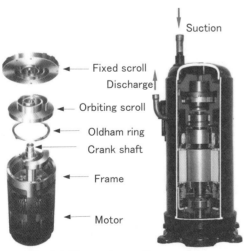

Fig. 5.1 Hermetic scroll compressor

5.2 Operating Principle and Characteristics

5.2.1 Operating principle

Figures 5.1 and 5.2 show examples of hermetic and automotive air-conditioning scroll compressors, respectively. A scroll compressor is a type of rotational compressor that comprises a fixed scroll with a spiral-shaped blade, or "wrap", extending from a flat base plate, and an orbiting scroll which basically is identical with the fixed scroll but is driven by an eccentric crank mechanism. A pair of scrolls is assembled at a relative angle of 180°, so that they touch at several points and form a series of crescent-shaped gas pockets (compression chambers). Relative movement between the two scrolls causes the volume change of these compression gas pockets to compress the fluid inside. The orbiting scroll revolves around the center of the fixed scroll at a certain orbiting radius without rotating (that is, the orbiting scroll maintains the same orientation while orbiting), with the flank of its wrap in contact with that of the fixed scroll.

Figure 5.3 shows the operating principle of a scroll compressor. Between the fixed scroll and the orbiting scroll, several pairs of crescent-shaped gas pockets (compression chambers) are formed. The first figure in the upper left shows the scroll compressor with the suction process just completed. The outer gas pockets are closed, and the refrigerant gas flowed in from the outside of the wraps is trapped in them. The upper right figure shows the scroll compressor 90° ahead from the first one. A new suction process has started in the outer sections while compression and discharge process are taking place in the intermediate and inner sections, respectively. The remaining figures in the lower right and the lower left each show the orbiting wrap position 90° ahead from the previous one, in the order mentioned. With the orbiting motion of the orbiting scroll, each compression gas pocket moves toward the center of the scroll and the pocket volume becomes increasingly smaller with the gas inside more compressed. Finally the gas is discharged through the discharge port at the center of the fixed scroll.

It is clear from this operating principle that the scroll mechanism compresses the working fluid continuously and that it requires neither a suction valve nor a discharge valve. Also, with multiple compression gas pockets working simultaneously from suction to discharge process, the amount of leakage between each compression gas pocket is very small. These characteristics contribute to highly efficient and very quiet operation.

In addition to the above-described scroll operating principle where one of the scrolls is fixed while the other scroll orbits around the center of the fixed one, the two scrolls can also be rotated simultaneously in the same direction (co-rotating) with an offset between the centers of the scrolls by the same amount as the orbiting radius, which provides the same compression effect as the above-described. The co-rotating mechanism has been researched and developed extensively by some companies, resulting in use in vacuum pumps and other applications[12].

5.2.2 Scroll wrap geometry

While the spiral curves that form a scroll wrap can be drawn in several ways, the simplest and therefore most commonly used profile is an involute of a circle. The profile of the fixed scroll wrap can be obtained as an envelope curve of the orbiting locus described by the orbiting scroll. In practice, the fixed scroll is normally shaped as a mirror-image of the orbiting scroll.

The involute curve that represents the outer side of the orbiting scroll wrap can be expressed, on the X_m-Y_m coordinates shown in Fig. 5.4 (with the direction of the starting point of the outer involute curve denoted as X_m), as follows:

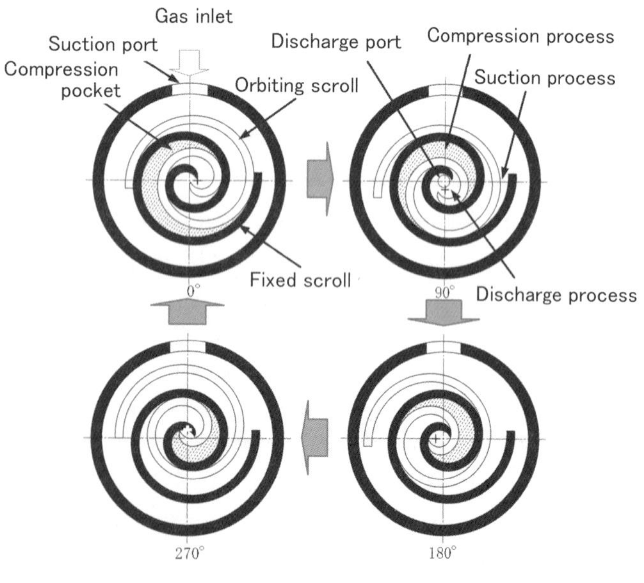

Fig. 5.3 Compression principle

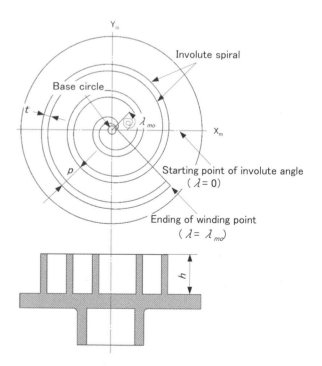

Fig. 5.4 Scroll wrap geometry

$$x_{mo} = a(\cos\lambda + \lambda\sin\lambda)$$
$$y_{mo} = a(\sin\lambda - \lambda\cos\lambda) \quad (5.1)$$

where:

a : Base circle radius of the involute curve
λ : Involute angle of the curve

Similarly, the curve representing the inner side of the orbiting scroll wrap can be obtained as follows:

$$x_{mi} = a\{\cos\lambda + (\lambda-\beta)\sin\lambda\}$$
$$y_{mi} = a\{\sin\lambda - (\lambda-\beta)\cos\lambda\} \quad (5.2)$$

where:

β : Phase difference between the outer side and the inner side of the wrap ($=t/a$)
t : Scroll wrap thickness

Assuming that the thickness of the orbiting and the fixed scroll wraps are identical, the scroll wrap pitch p and the amount of eccentricity (orbiting radius) ε of the crankshaft driving the orbiting scroll have the following relationship:

$$p = 2\pi a = 2(\varepsilon + t) = 2a(\alpha + \beta) \quad (5.3)$$

where:

α : Ratio of orbiting radius to base circle radius ($=\varepsilon/a$)

As shown in Fig. 5.5, each compression gas pocket formed between the inner side of the fixed scroll wrap and the outer side of the orbiting scroll wrap, V_{pi}, is paired with another one formed between the inner side of the orbiting scroll and the outer side of the fixed scroll, V_{qi}. The volume of the i-th compression pocket from the center one ($i = 1$) can be obtained as follows:

$$\begin{aligned}V_i &= V_{pi} + V_{qi} \\ &= 2a^2h\left\{\frac{1}{2}\int_{\lambda_{p(i-1)}}^{\lambda_{pi}}(\lambda_p+\alpha)^2 d\lambda_p - \frac{1}{2}\int_{\lambda_{p(i-1)}}^{\lambda_{pi}}\lambda_p^2 d\lambda_p\right\} \\ &= 2\varepsilon^2 h\left\{2\left(1+\frac{t}{\varepsilon}\right)(\lambda_{pi}-\pi)+\pi\right\}\end{aligned} \quad (5.4)$$

where:

λ_{pi}: Involute angle at the contact point where i-th compression pocket is formed
h : Scroll wrap height

Assuming that the angle of the orbiting scroll center O_m relative to the fixed scroll center O_f is θ ($=$ orbiting angle), θ has the following relationship with the involute angle λ_{pi}:

$$\begin{aligned}\lambda_{pi} &= \left(-\theta+\frac{1}{2}\pi\right)+2(i-1)\pi \\ &\quad \text{if } 0 \le \theta < \frac{1}{2}\pi \\ &= \left(-\theta+\frac{1}{2}\pi\right)+2i\pi \\ &\quad \text{if } \frac{1}{2}\pi \le \theta < 2\pi\end{aligned} \quad (5.5)$$

The displacement volume of a compressor is determined by the volume of space trapped inside it upon suction completion. Assuming that the involute angle at the end point of the orbiting scroll wrap is λ_{mo}, the displacement volume V_{th} is:

$$V_{th} = 2\varepsilon^2 h\left\{2\left(1+\frac{t}{\varepsilon}\right)(\lambda_{mo}-2\pi)+\pi\right\} \quad (5.6)$$

Figure 5.6 shows an example of volume curve, relative to the crankshaft rotational angle, of a compression gas pocket formed between the orbiting and fixed scrolls[7]. One compression cycle from suction to discharge in a given compression pocket is completed after a number of crankshaft rotations. The suction process completes and the compression process starts at point A where the crankshaft completes its first rotation. As the compression process

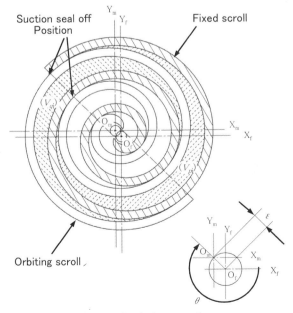

Fig. 5.5 Sealed gas pocket

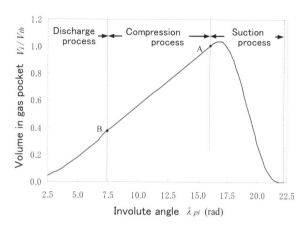

Fig. 5.6 Volume curve

progresses, the volume of the compression pocket decreases at a constant rate as the contact point moves toward the center of the spiral. Gas trapped inside the compression pocket will be discharged when the compression pocket is connected to the discharge port at the center of the fixed scroll.

In scroll compressors, unlike reciprocating and rotary compressors, the compression pocket volume at the start of the compression process and the one at the end of it are determined solely by the scroll wrap geometry. The ratio of the compression pocket volume upon suction process completion (point A) to the one immediately before the start of discharge (point B) is called the compressor's "built-in volume ratio", or V_r, which can be obtained by the following equation:

$$V_r = \frac{2(1+t/\varepsilon)(\lambda_{mo} - 2\pi) + \pi}{2(1+t/\varepsilon)(\lambda_{ms} + \pi) + \pi} \quad (5.7)$$

where:

λ_{ms}: Involute angle at the start point of the orbiting scroll wrap

The scroll wrap geometry cannot be singularly determined by giving displacement volume V_{th} and built-in volume ratio V_r. Other factors, such as the amount of interference between the two scrolls, the outer dimensions of the compressor, the required wrap strength during operation, the manufacturing method and the productivity, must be added to finally determine the scroll wrap profile.

5.2.3 New scroll wrap

The scroll wrap is one of the key elements for improving the performance and reliability of scroll compressors.

An involute curve of a circle is generally adopted in current scroll compressors as the basic volute curve of the wrap, with a uniform wrap thickness all the way from the center to the outer peripheral. New types of scroll wrap profile have been recently developed to cope with dimensional restrictions and operating pressure and compression ratio requirements:

(1) Asymmetric wrap

In order to satisfy the diversified needs, such as higher efficiency, longer life, smaller size and lighter weight, asymmetrical scroll wraps, where the end angle of the fixed scroll wrap is larger than that of the orbiting scroll wrap, was developed[13-15]. In an asymmetrical scroll wrap shown in Fig. 5.7, there is a phase difference between the compression process run by the compression pockets that are formed inside of the orbiting scroll wrap and that run by the compression pockets formed outside of the orbiting scroll wrap. This provides less-pulsating and therefore more smoothly continuous gas flow.

(2) Variable-pitch (variable thickness) wrap

Alternative scroll wraps, such as formed by involute curve based on a circle with a radius varying with involute angle[16], or formed by algebraic-spiral curve[17] enable to gradually vary the wrap thickness or the groove width, as shown in Fig. 5.8. For example, it is possible to gradually increase the wrap thickness toward the center of the scroll where the pressure difference is larger, and decrease the wrap thickness toward the outer peripheral where the pressure difference is smaller. Thus the technology provides dimensional flexibility to realize a scroll design with smaller internal leakage. With the variable thickness or pitch technology, it is also possible to locally thicken the center wrap area where the higher rigidity is required or to increase the curvature radius of the center wrap area, providing a margin in scroll strength in required areas.

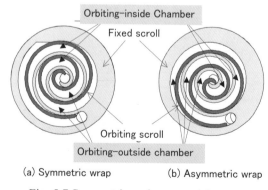

Fig. 5.7 Symmetric and asymmetric wraps

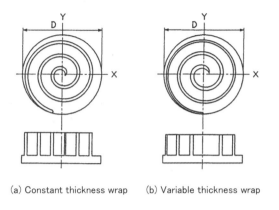

(a) Constant thickness wrap   (b) Variable thickness wrap

Fig. 5.8 Constant and variable thickness wraps

(3) Stepped-height wrap

Another new type of scroll wrap design being used is one where the tip and root of the wrap are stepped (=height changed) so that the outer wrap section is made higher than the inner wrap section, as shown in Fig. 5.9[18]. Making the outer wrap section higher enables to have a larger compressor capacity without changing the outer dimensions of the compressor, thus contributing to size and weight reduction. Also, having volumetric changes not only in the radial direction but also in height direction realizes an even higher compression ratio.

5.3 Structure and Design of Compressor Components

5.3.1 Compression mechanism
(1) Drive mechanism

For a compressor to perform fluid compression efficiently, it is necessary to maintain sealing of the compression gas pockets while driving the orbiting scroll in an orbiting motion. To realize that, such five components as the orbiting scroll, the fixed scroll, the Oldham ring, the crankshaft and the frame are required (see Fig. 5.10). The orbiting and the fixed scrolls are commonly made of cast iron, while aluminum alloys may also be used for weight reduction in some applications such as automotive air conditioners. The orbital motion of the orbiting scroll is achieved by the combination of a crankshaft and an anti-rotation mechanism. While compressing fluid, various forces and moments will work on the orbiting scroll.

In a symmetrical scroll wrap design like the one shown in Fig. 5.11, the tangential component of gas compression force, F_{pt}, works at the midpoint of the line segment con-

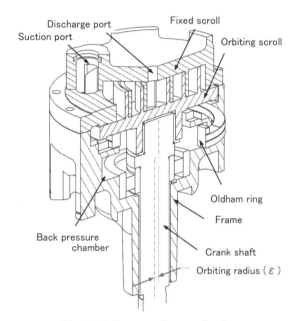

Fig. 5.10 Compression mechanism

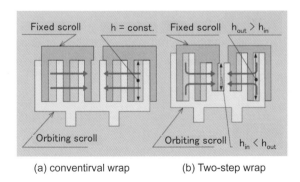

(a) conventirval wrap   (b) Two-step wrap

Fig. 5.9 Two-step wrap

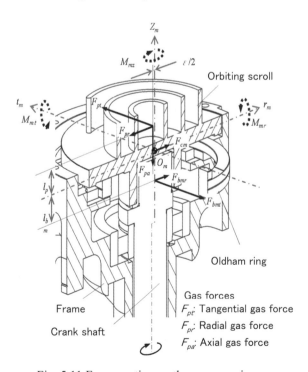

Fig. 5.11 Forces acting on the compression mechanism

necting the centers of the orbiting and the fixed scrolls. Therefore, if the orbiting scroll is driven by the crankshaft at its geometric center, a rotational moment around the geometric center will work in the same direction as the crankshaft rotation, trying to rotate the orbiting scroll about its axis. To prevent such rotation of the orbiting scroll and to convert the crankshaft rotation solely into an orbiting motion, an anti-rotation mechanism like the Oldham ring is required.

(2) Anti-rotation mechanism

An Oldham ring mechanism consists of an Oldham ring that has an orthogonal pair of projecting keys on its top and bottom, and the corresponding key ways provided on the back of the orbiting scroll and on the frame (or, alternatively on the fixed scrolls). As the crankshaft rotates, the Oldham ring reciprocates along the key way of the frame in the same amplitude as the orbiting radius, and the orbiting scroll orbits while reciprocating along the key of the Oldham ring in the orthogonal direction.

Another type of anti-rotation mechanism that is used for air compressors is a set of multiple crankpins with the same amount of eccentricity as that of the crankshaft[10]. For automotive compressors, a ball coupling mechanism, shown in Fig. 5.12, that also works as a thrust bearing is commonly used[4]. A movable ring and a fixed ring are attached to the back of the orbing scroll and to the compressor casing respectively. Steel balls held in ball pockets of each ring roll along the inside of the respective pockets as the orbiting scroll orbits, thus preventing the orbiting scroll from rotating about its own axis. The ball coupling mechanism also serves as a thrust bearing, supporting the axial force working on the orbiting scroll.

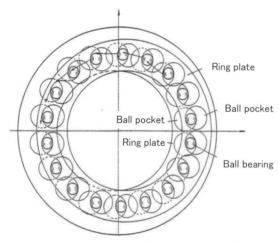

Fig. 5.12 Ball coupling mechanism[19]

5.3.2 Forces and moments working on the drive mechanism

Due to the compression of gas, fluid forces act on the orbiting scroll as shown in Fig. 5.11. These forces are a radial component F_{pr} working in the direction of crankshaft eccentricity, a tangential component F_{pt} orthogonal to F_{pr}, and an axial component F_{pa} orthogonal to the base plate[6]. These fluid forces working on the orbiting scroll can be obtained, based on the suction pressure, as follows:

Assuming that the compression process is adiabatic, pressure P_i in the i-th compression pocket from the center can be expressed as follows using adiabatic index κ:

$$p_i = P_s \left\{ \frac{2(1 + t/\varepsilon)(\lambda_{mo} - \pi) + \pi}{2(1 + t/\varepsilon)(\lambda_{pi} + \pi) + \pi} \right\}^\kappa \quad (5.8)$$

And, the tangential component is:

$$F_{pt} = ah \left\{ (P_d - P_s)(2\lambda_{p1} + \pi + t/a) + 4\pi \sum_{i=2}^{n} (P_i - P_s) \right\} \quad (5.9)$$

The radial component is:

$$F_{pr} = 2ah(P_d - P_s) \quad (5.10)$$

The axial component is:

$$F_{pa} = \sum_{i=1}^{n} (P_i - P_s)(S_{pi} + S_{qi} + S_{tp_i} + S_{tqi}) \quad (5.11)$$

where:

S_{pi}, S_{qi}: The axial projection areas of the i-th compression pockets formed along the respective scroll wraps

S_{tpi}, S_{tqi}: The axial projection areas of the outer wraps surrounding the i-th compression pockets

Also, as the orbiting scroll is driven with an orbiting radius ε, a centrifugal force F_{cm} expressed below works in the orbiting scroll:

$$F_{cm} = G_m \varepsilon \omega^2 \quad (5.12)$$

where:

G_m: Mass of orbiting scroll
ω: Angular velocity of rotation

As the acting point of the fluid forces working on the orbiting scroll differs from that of the driving force of the crankshaft, a moment, shown in Fig. 5.11 and expressed below, works in the orbiting scroll:

$$M_{mr} = F_{pt}l_p + F_{bmt}l_{bm}$$
$$M_{mt} = F_{pr}l_p + F_{bmr}l_{bm} + F_{pa}\frac{\varepsilon}{2} \quad (5.13)$$
$$M_{mz} = F_{pt}\frac{\varepsilon}{2}$$

Sealing mechanism, explained in the next section, that allow closer contact between the orbiting and the fixed scrolls by resisting axial force F_{pa} and moments M_{mr} and M_{mt} have been adopted. Figure 5.13 shows an example of fluid forces working on the orbiting scroll. As indicated in the figure, fluctuation of each fluid force during one rotation is relatively small, due to the multiple gas-pocket compression going on simultaneously.

5.3.3 Sealing mechanism

Compressor performance is directly related to internal leakage and mechanical losses. In a scroll compressor, there are two leakage paths: axial and radial. Radial leakage occurs through the axial gaps between the tip of the wrap of one of the scroll and the flat base plate of the other and circumferential leakage occurs through the gaps between the flanks of the each scroll wrap.

(1) Axial sealing

Axial gaps are closely related to the supporting structure of the orbiting scroll. Pressure of the gas in the compression pockets generates axial forces and moments that try to separate the orbiting scroll away from the fixed scroll (see Fig. 5.11). The axial sealing mechanism, or "axial compliance", that overcomes these forces and moments to provide a closer contact between the orbiting scroll and the fixed scroll can be classified into the following two types, illustrated in Fig. 5.14[20-21].

a) Axial sealing mechanism with a thrust bearing provided behind the orbiting scroll plate and tip seals mounted on the tip of the fixed and orbiting scroll wraps.

In this sealing mechanism, axial force working on the orbiting scroll is received by the thrust bearing behind the orbiting scroll. Therefore, the bearing must be designed to have sufficient yield strength. The geometry and structure of the tip seal also must be carefully designed so that it floats in the tip-seal groove as close contact with the sides of the groove and also with the flat base of the other scroll.

b) Axial sealing mechanism with a back pressure chamber provided behind the orbiting or fixed scroll so that pressure in this chamber push one of the scrolls to shift minutely in the axial direction against the other scroll.

In this sealing mechanism, a small hole ("back pressure port") drilled in the base plate of the scroll provides a connection between the back pressure chamber and the intermediate compression gas pocket. With this, pressure in the back pressure chamber is automatically regulated to an intermediate level between the suction and discharge pressures. If the back pressure port opens to the compression pocket during the orbiting angle from θ_1 to θ_2, pressure in the back pressure chamber can be obtained by the following equation:

$$P_b = \int_{\theta_1}^{\theta_2} P_t d\theta / (\theta_2 - \theta_1) \quad (5.14)$$

Pressure in the back pressure chamber pushes one of the scrolls against the other scroll to provide a closer contact, thus assuring sealing at the tip of the scroll wrap[6]. However, the actual back pressure is also affected by the rotational speed and the amount of lubricating oil fed to the back pressure chamber. To assure both performance and reliability, it is necessary to

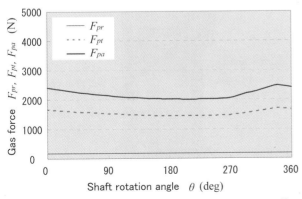

Fig. 5.13 Gas forces acting on the orbiting scroll

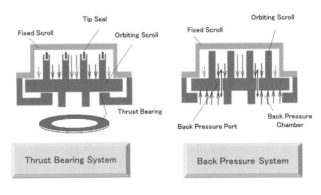

Fig. 5.14 Axial compliance mechanism

minimize leakage through axial gaps and sliding losses induced by thrust loads. Therefore, the size and location of the back pressure port and the gas-oil ratio inside the back pressure chamber must be optimized.

Similar sealing mechanism with a compliant frame that zones the space behind the orbiting scroll into two independent back pressure chambers has been developed[22].

(2) Radial sealing mechanism

Radial gaps along the flanks of the scroll wraps are determined by the scroll wrap geometric accuracy, the amount of crankshaft eccentricity and the relative position between the two scrolls.

In a scroll compressor where the crankshaft eccentricity is fixed, the flank gaps are designed to be sealed by lubricating oil. To make this sealing dependable, the scroll wraps must be machined highly accurately. The simple structure of the mechanism with minute gap between the wraps throughout compressor operation avoid the contact-induced noise and vibration. This makes the compressor highly suitable for inverter-controlled variable speed operation.

Alternatively, a mechanism where the driving crank radius is designed to be minutely variable so that the flank of the orbiting scroll is allowed to move smoothly along the flank of the fixed scroll may be used. This sealing mechanism can work without such a high scroll machining accuracy as described above[23] (see Fig. 5.15). In such features as an eccentric bushing, a linkage and a slider crank, this alternative mechanism allows the orbiting scroll to shift in the direction of crankshaft eccentricity as required (the orbiting radius can minutely change). With this, centrifugal and gas pressure forces radially push the orbiting scroll against the fixed scroll, and reduce the leakage through the gaps between the flanks of the scroll wraps.

5.3.4 Structures of scroll compressor

Typical scroll compressors have the compression mechanism and the motor inside a hermetic casing (shell), with the compression mechanism placed above the motor and with a reservoir of lubricating oil inside the bottom of the casing. The hermetic casing types can be divided into the high-pressure casing type where pressure inside the casing is equal to the discharge pressure and the low-pressure casing type where the pressure is equal to the suction pressure.

(1) High-pressure casing type

An example of high-pressure casing type scroll compressor structure is shown in Fig. 5.16. Inside the hermetic casing, the compression mechanism is located above the motor, with a suction pipe directly connected to the compression mechanism. Refrigerant gas flows into the suction chamber directly through the suction pipe. The gas is compressed in the compression pockets and is discharged through the port provided at the center of the fixed scroll and flows into the upper space inside the hermetic casing. The discharged refrigerant gas is then directed toward the bottom of the casing, cooling the motor along the way and also having oil separated, and finally flows out of the casing through the discharge pipe connected to the casing. This high-pressure casing type structure has the following features.

a) As the gas compressed by the scroll mechanism is ejected into the space inside the hermetic casing before finally discharged out of the casing, pressure pulsation in the discharge pipe will be much smaller than if the

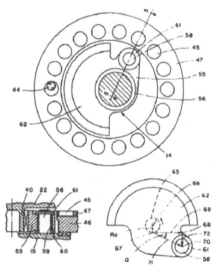

Fig. 5.15 Radial compliance mechanism

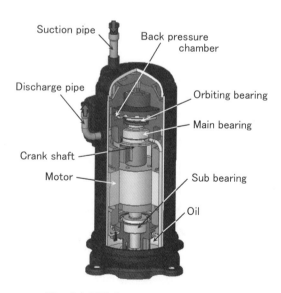

Fig. 5.16 High pressure casing type

gas is directly discharged to the discharge pipe. This contributes to very quiet operation.
b) As the refrigerant gas from the evaporator is directly sucked into the suction chamber, a relatively high volumetric efficiency is attainable.
c) No pressure-drop-induced foaming occurs, which is often experienced upon startup in the low-pressure casing types of compressor structures. This leads to superior lubrication oil retainability.

(2) Low-pressure casing type

An example of low-pressure casing type scroll compressor structure is shown in Fig. 5.17. Inside the hermetic casing, the compression mechanism is located above the motor, with a reservoir of lubricating oil inside the bottom of the casing. The space located above the compression mechanism is a discharge chamber, separated from the motor chamber below. A discharge pipe is connected to the discharge chamber and a suction pipe is connected to the motor chamber.

The refrigerant gas flows into the hermetic casing through the suction pipe, cooling the motor before flowing into the compression mechanism. After being compressed inside the compression pockets, the refrigerant gas is ejected into the discharge chamber above the compression mechanism and then the gas is finally discharged out of the casing through the discharge pipe. This low-pressure-casing type structure has the following features.
a) As the refrigerant gas from the evaporator is sucked into the space inside the hermetic casing before entering the compression mechanism, pressure pulsation in the suction pipe will be much smaller than if the gas is directly sucked into the suction chamber.
b) Cooling of the motor by the suction gas keeps the motor at relatively low temperature.
c) The casing only needs to withstand the suction pressure and thus can be made thinner. The light weight of the casing is favorable for the medium and large capacity compressors.

5.4 Kinematic Analysis

To help the reader better understand the dynamic behavior of the compression mechanism and mechanical efficiency in a quantitative manner, this section provides a kinematic analysis of the scroll compressor[24-25].

5.4.1 Working forces and equations of motion

(1) Orbiting scroll

Figure 5.18 illustrates the forces and moments working on the orbiting scroll. To evaluate the thrust forces, an axial cross section of the scroll mechanism is also provided. It is shown that pressure from the back pressure chamber provided behind the orbiting scroll plate presses the orbiting

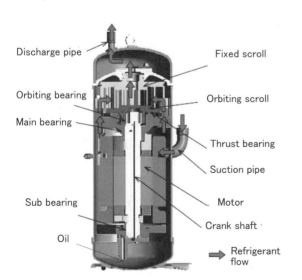

Fig. 5.17 Low pressure casing type

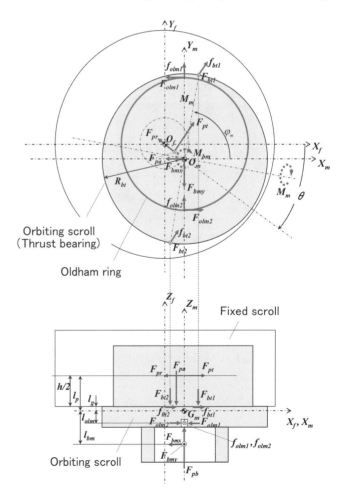

Fig. 5.18 Dynamic model of orbiting scroll

Compressors for Air Conditioning and Refrigeration

scroll against the fixed scroll to assure sealing between the scrolls.

The X_f-Y_f coordinate system shown has its origin at the center of the frame (fixed scroll) while the X_m-Y_m system has the origin at the center of the orbiting scroll. Here, F_{pt}, F_{pr} and F_{pa} are the tangential, radial and axial components of the gas force given by Eqs. (5.9) to (5.11) respectively, working on the midpoint of orbiting radius ε connecting the fixed scroll center O_f and the orbiting scroll center O_m. F_{pb} is the gas force behind the orbiting scroll, working on the orbiting scroll center O_m. Its thrust forces, F_{bt1} and F_{bt2}, work on the thrust face of radius R_{bt} where the orbiting and the fixed scrolls slide against each other, while friction forces f_{bt1} and f_{bt2} work in the direction orthogonal to the orbiting radius and resisting the orbiting motion.

F_{bmx} and F_{bmy} are the forces working on the orbiting bearing to drive the orbiting scroll into motion. M_{bm} is a frictional torque caused by the relative slip at the orbiting bearing. F_{olm1} and F_{olm2} are forces from the Oldham ring with radius R_{old}, while friction forces f_{olm1} and f_{olm2} work on the slide face against the Oldham ring (key way face).

The center of the orbiting scroll, O_m (x_m, y_m), is defined by the following equation:

$$x_m = \varepsilon \cos\theta, \; y_m = -\varepsilon \sin\theta \qquad (5.15)$$

The equations of motion of the orbiting scroll in each coordinate axis direction can be given as the following equations:

$$
\begin{aligned}
(G_m \ddot{x}_m) &= -F_{bmx} + (F_{pt} + f_{bt1} + f_{bt2})\sin\theta \\
&\quad - F_{pr}\cos\theta - F_{olm1} + F_{olm2} \\
(G_m \ddot{y}_m) &= -F_{bmy} + (F_{pt} + f_{bt1} + f_{bt2})\cos\theta \\
&\quad + F_{pr}\sin\theta + (f_{olm1} + f_{olm2}) \\
(G_m \ddot{z}_m) &= -F_{pa} + F_{pb} - G_m g - (F_{bt1} + F_{bt2})
\end{aligned}
\qquad (5.16)
$$

Moments around X_m and Y_m axes working on the orbiting scroll can be obtained by the following equations:

$$
\begin{aligned}
M_{mx} &= -F_{pa}\cdot\frac{y_m}{2} + (F_{pt}\cos\theta + F_{pr}\sin\theta)l_p \\
&\quad + (f_{olm1} + f_{olm2})l_{olm} + F_{bmy}l_{bm} + (f_{bt1} + f_{bt2})l_g\cos\theta
\end{aligned}
$$

$$
\begin{aligned}
M_{my} &= F_{pa}\cdot\frac{x_m}{2} - (F_{pt}\sin\theta - F_{pr}\cos\theta)l_p \\
&\quad - (F_{olm1} - F_{olm2})l_{olm} - F_{bmx}l_{bm} - (f_{bt1} + f_{bt2})l_g\sin\theta
\end{aligned}
$$

$$(5.17)$$

Therefore, moment M_m resultant of moments M_{mx} and M_{my} can be given by the following equation:

$$M_m = \sqrt{M_{mx}^2 + M_{my}^2} \qquad (5.18)$$

The direction of this moment, with its counterclockwise angle from the X_m axis denoted as φ_m, can be given as follows:

$$\varphi_m = \tan^{-1}(M_{my}/M_{mx}) \qquad (5.19)$$

Here, the acting points of the thrust forces are the two points on the line which is equal to the working direction of moment M_m passing through the orbiting scroll center O_m. Assuming that thrust forces F_{bt1} and F_{bt2} are generated at these two points, such thrust forces can be given by the following equations:

$$
\begin{aligned}
F_{bt1} &= \frac{1}{2}\cdot(-M_m/R_{bt} - F_{pa} + F_{pb}) \\
F_{bt2} &= \frac{1}{2}\cdot(M_m/R_{bt} - F_{pa} + F_{pb})
\end{aligned}
\qquad (5.20)
$$

If one of these thrust forces is negative, it indicates that normal contact will not be maintained on the thrust bearing face.

Moment around Z_m axis working on the orbiting scroll will be offset by the Oldham ring. The moment balance equation around Z_m axis can be given as follows:

$$
\begin{aligned}
&F_{pt}\cdot\frac{\varepsilon}{2} + M_{bm} + (f_{bt1} - f_{bt2})R_{bt}\cos(\theta + \varphi_m) \\
&\quad = F_{olm1}(R_{old} - y_m) + F_{olm2}(R_{old} + y_m)
\end{aligned}
\qquad (5.21)
$$

(2) Oldham ring

Figure 5.19 gives a dynamic model of the Oldham ring. The directin of force arrows in th figure indicate the positive direction of each force at the same instance shown in Fig.5.18. The Oldham ring's X-direction key engages with the key way on the frame while its Y-direction key engages with the key way on the back of the orbiting scroll, thus preventing the orbiting scroll from rotating about its own axis while the ring reciprocates in the frame key way. With the Oldham ring's mass denoted as G_{old} and assuming that its center of gravity matches the geometric center of the ring, the coordinates for the ring's center of gravity, O_{old}, can be given as (x_m, 0). Working on the Oldham ring, in the directions respectively shown, are reaction forces F_{olm1} and F_{olm2} and friction forces f_{olm1} and f_{olm2} from the orbiting scroll, and also reaction forces F_{olf1} and F_{olf2} and friction forces f_{olf1} and f_{olf2} from the frame key way along the X-axis.

94

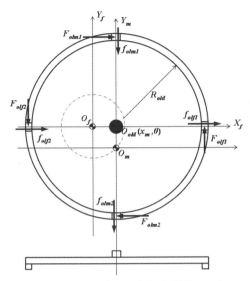

Fig. 5.19 Dynamic model of Oldham ring

With these, the equation of motion of the Oldham ring in the X-axis direction can be given as follows:

$$(G_{old}\ddot{x}_m) = F_{olm1} - F_{olm2} + f_{olf1} + f_{olf2} \qquad (5.22)$$

The force balance equation in the Y-axis direction and the moment balance equation can be given as follows:

$$F_{olf1} - F_{olf2} - f_{olm1} - f_{olm2} = 0$$
$$F_{olm1} + F_{olm2} - (F_{olf1} + F_{olf2}) = 0 \qquad (5.23)$$

By organizing the above equations, the defined loads and moments can each be obtained as a function of the orbiting angle θ (including frictional forces and torques). Of those, loads F_{bmx} and F_{bmy} on the orbiting bearing can be given by the following equations:

$$F_{bmx} = -(G_m + G_{old})\ddot{x}_m + (F_{pt} + f_{bt1} + f_{bt2})\sin\theta$$
$$\quad - F_{pr}\cos\theta + f_{olf1} + f_{olf2} \qquad (5.24)$$
$$F_{bmy} = -G_m\ddot{y}_m + (F_{pt} + f_{bt1} + f_{bt2})\cos\theta$$
$$\quad + F_{pr}\sin\theta + f_{olm1} + f_{olm2}$$

(3) Crankshaft

Figure 5.20 gives a dynamic model of the crankshaft. With the crankshaft inertia moment around Z-axis denoted as I_c, and assuming that motor drive torque $N_{mo}(\dot{\theta})$, the orbiting bearing frictional torque M_{bm} and the crank bearing frictional torque M_{bc} work in the directions respectively shown in Fig. 5.20, the equation of motion of the crankshaft will be:

$$I_c\ddot{\theta} = F_{bmx}\varepsilon\sin\theta + F_{bmy}\varepsilon\cos\theta$$
$$\quad + N_{mo} - M_{bm} - M_{bc} \qquad (5.25)$$

By substituting loads F_{bmx} and F_{bmy} from Eq. (5.24) working on the orbiting bearing into the above equation, the following equation governing the rotational motion of the crankshaft can be obtained:

$$\left(I_c + G_m\varepsilon^2 + G_{old}\varepsilon^2\sin^2\theta\right)\ddot{\theta}$$
$$\quad + G_{old}\varepsilon^2\sin\theta\cdot\cos\theta\cdot\dot{\theta}^2$$
$$= N_{mo} - F_{pt}\varepsilon - (f_{bt1} + f_{bt2})\varepsilon - M_{bm} - M_{bc} \qquad (5.26)$$
$$\quad - \{(f_{olf1} + f_{olf2})\varepsilon\sin\theta + (f_{olm1} + f_{olm2})\varepsilon\cos\theta\}$$

The left side of the equation is inertia terms, while the first term on the right side represents motor torque and the second and subsequent terms represent load torques. Of the load torques, $F_{pt}\varepsilon$ is the torque used for gas compression and all the other torques are those induced by mechanical friction.

5.4.2 Equation of energy and mechanical efficiency

By multiplying Eq. (5.26) of motion by the angular displacement and then integrating the result over one crankshaft rotation, an equation of energy can be obtained. As

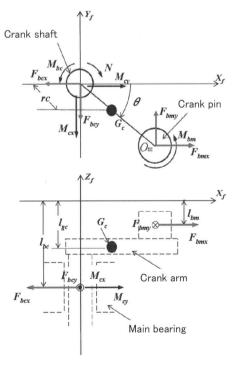

Fig. 5.20 Dynamic model of crank mechanism

the integral for the inertia term will be zero in a steady-state operation, the energy equation can be derived as:

$$E_{motor} = E_{gas} + E_{bt} + E_{bc} + E_{bm} + E_{old} \quad (5.27)$$

where

$$\begin{aligned}
E_{motor} &= \int_{1rev} N_{mo} \dot{\theta} dt \\
E_{gas} &= \int_{1rev} F_{pt} \varepsilon \dot{\theta} dt \\
E_{bt} &= \int_{1rev} (b_{bt1} + f_{bt2}) \varepsilon \dot{\theta} dt \\
E_{bc} &= \int_{1rev} M_{bc} \dot{\theta} dt \\
E_{bm} &= \int_{1rev} M_{bm} \dot{\theta} dt \\
E_{old} &= \int_{1rev} \left\{ \begin{matrix} (f_{olm1} + f_{olm2}) \varepsilon \cos\theta + \\ (f_{olf1} + f_{olf2}) \varepsilon \sin\theta \end{matrix} \right\} \dot{\theta} dt
\end{aligned} \quad (5.28)$$

Mechanical efficiency η_m can be calculated by the following equation:

$$\eta_m = \frac{E_{motor} - (E_{bt} + E_{bc} + E_{bm} + E_{old})}{E_{motor}} \quad (5.29)$$

5.5 Other Scroll Compressor Mechanisms

5.5.1 Capacity control

In pursuit of better air conditioning comfort and further energy saving, there is now an increasing demand for capacity control functionality where the discharge volume of the compressor can be varied and controlled according to the thermal load.

In scroll compressors, such capacity control function can be generally provided by a bypass control method where part of the gas being compressed can be returned to the suction side by a switching valve or a rotational speed control method where the compressor drive frequency can be regulated to vary by inverter control. Besides the above two, newer capacity control methods include one where a back pressure chamber provided behind the fixed scroll can be connected to the suction side or the discharge side through a switching valve so that the fixed scroll shifts slightly up and down in the axial direction to alter the compression/non-compression frequency, resulting in changes in the compressor capacity[26].

Furthermore, the multiple-compressor system is being used for larger-capacity air conditioning or refrigeration applications, where two or more scroll compressors are provided so that the required number of compressors can be activated or deactivated according to the load (Fig. 5.21). In the multiple-compressor system, various engineering efforts are made in the compressor structure and also in the operating method so as to prevent mal-distribution of lubricating oil and to store the oil evenly in the casing of each compressor.

5.5.2 Injection

As the scroll compressor forms in itself a number of compression pockets that will be simultaneously engaged in different stages of compression process from suction to discharge, liquid or gas refrigerant can be easily injected through an injection port into one of these pockets when it is halfway through the compression process. In low-temperature application, liquid injection method is used in order to restrict increase in the discharge gas temperature[27]. As shown in Fig. 5.22, the amount of injected liquid refrigerant can be regulated according to the discharge gas temperature to control temperature rise of the motor wiring. With this, operating with the evaporating temperature as low as -45°C is possible in systems using R404A or other similar refrigerants. Alternatively, gas refrigerant can be injected so as to create an economizer cycle for capacity increase or performance improvement. The latter method is implemented in some residential and commercial air conditioners.

5.6 Uses and Features

As obvious from its operating principle, the scroll compressor has a number of superior characteristics that other types of compressors do not offer. Such characteristics include the followings:

(1) Small torque variation; low vibration and low noise:

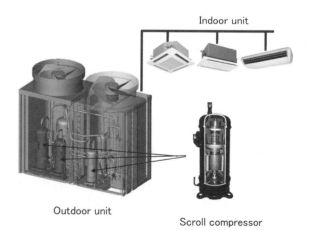

Fig. 5.21 Multiple-compressor system

Chapter 5 Scroll Compressors

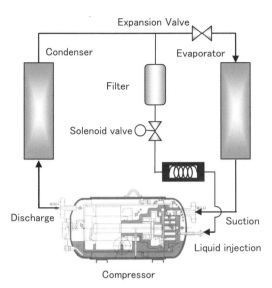

Fig. 5.22 Liquid injection system

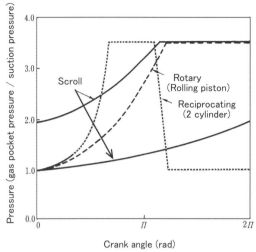

Fig. 5.23 Pressure change in compression process

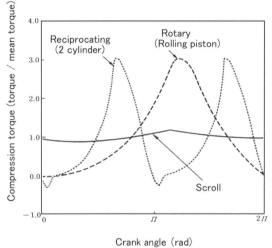

Fig. 5.24 Compression torque

As multiple compression pockets are simultaneously engaged in different stages of compression process from suction to discharge, the torque fluctuation by compressing the refrigerant gas is exceptionally small compared to other types of compressors. Pressure pulsation is also small due to the gas being suctioned and discharged continuously, which makes low-noise, low-vibration operation easily achievable. Figures 5.23 and 5.24 show comparisons of typical variations of compression chamber (or pocket) pressure and gas compression torque between reciprocating, rotary and scroll compressors against the crankshaft rotational angle. The figures reveal the following features concerning the scroll compressor. As one compression cycle from suction to discharge in a given compression chamber will be completed after a number of crankshaft rotations, the rate of pressure variation with shaft rotational angle is low. The degree of torque fluctuation by compressing the refrigerant gas is approximately only one tenth that of conventional reciprocating or rotary compressors. Inertia forces can also be balanced, leading to smaller compressor vibration.

(2) Small amount of leakage throughout the compression process; higher efficiency:

In all positive-displacement type compressors, it is critical to minimize the amount of leakage between the compression side and the suction side in order to achieve high compression efficiency. In a scroll compressor, where multiple compression pockets are formed between the suction pockets and the discharge pockets, the pressure difference between immediately adjacent pockets is relatively small and therefore the amount of leakage-induced efficiency reduction is low. Another structural advantage of a scroll compressor is that no direct connection is made between the discharge pocket and the suction side. These advantages lead to a very high volumetric efficiency. Figures 5.25 and 5.26 show typical efficiency curves of a hermetic scroll compressor[8]. Due to superior compression pocket sealability plus the absence of clearance volume that would lead to reduction in volumetric efficiency by causing re-expansion when the next suction process starts, very good volumetric efficiency is maintained even in high pressure ratio ranges. The degree of efficiency variation from low-speed to high-speed ranges is also small.

(3) Smaller orbiting radius of movable parts and low sliding speed; high reliability and better suitability for high-speed operation:

The orbiting radius of a scroll compressor is typically in the range of several millimeters. As a result, the relative speed between sliding parts is approximately only 2 m/s even in 60 Hz operations, which is exceptionally slow compared to other conventional compressors.

Compressors for Air Conditioning and Refrigeration

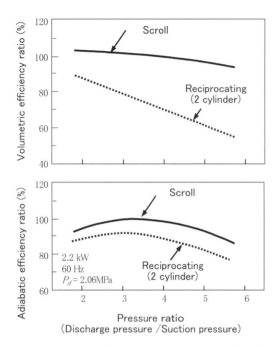

Fig. 5.25 Relation between pressure ratio and efficiency

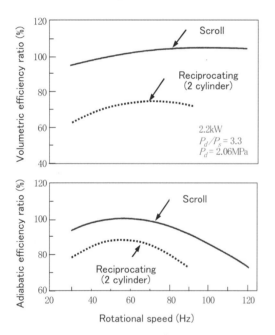

Fig. 5.26 Relation between rotational speed and efficiency

(4) Smaller number of mechanical parts; compact and lightweight

The number of mechanical parts in a scroll compressor mechanism is only one fourth that of a reciprocating compressor.

It has been thirty years since the scroll compressor was first commercialized. Since then, its use has spread widely, mainly as refrigeration and air conditioning compressors. The factors contributing to such high market penetration of scroll compressors include: high-efficiency, low-noise, low-vibration; very small efficiency reduction even in high-speed ranges, making the compressor suitable for the trend of multi-function of inverter-driven heat-pump-based and multi-split air conditioners; less severe lubrication condition in sliding areas, which is advantageous for the use of R410A and other HFC refrigerants. Also unlike other types of compressors, the scroll compressor mechanism is believed to have more potential for future improvements. Due to significant and still-continuing machining accuracy improvement as a result of advances in production technology, operating characteristics that are even better than now are considered achievable. Also, further structural refinements for specific applications are considered possible.

5.7 Production Technology

5.7.1 Scroll machining

Among various methods proposed for scroll wrap machining, the most commonly used one is numerically controlled end-mill cutting (see Fig. 5.27). Machining of the wrap-flank contour forming an involute curve requires very high accuracy as any geometric error in there will directly lead to leakage during gas compression. As an alternative to the general-purpose X-Y controlled milling, R-θ controlled machining that can offer higher speed and better accuracy has been proposed[28]. Machining units that are dedicated to scroll production have also been developed.

Scroll wraps are commonly made of cast iron and various machining-related improvement efforts continue, including those for cutting allowance reduction through higher-precision casting, for faster cutting speed and for longer-lasting end mills. For automotive applications where weight reduction and higher operating speed are of greater importance, aluminum alloys are widely used as scroll compressor material.

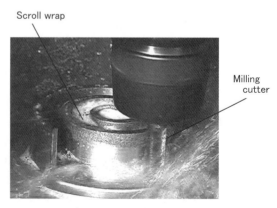

Fig. 5.27 Precision milling

Chapter 5 Scroll Compressors

5.7.2 Machining accuracy control

As scroll wraps need to be produced to an accuracy of micrometers, geometric measurement is one of the essential elements of scroll compressor production technology. While the most commonly used geometric measurement for scroll wraps is a high-precision three dimensional coordinate measuring machine like the one shown in Fig. 5.28, geometric measurement systems dedicated to scroll wrap have been developed to provide faster and simpler high-precision measurement.

Also, efforts are underway to develop a system where geometry measurement results can be immediately fed back to the machining process to improve accuracy.

5.8 Selection Criteria of Scroll Compressor

The followings are some of the most important points to consider when selecting a scroll compressor for a refrigeration or air conditioning system:

1) First, make sure that the system is designed for a refrigerant that best suits its application and purpose. Select a compressor that is appropriate for such refrigerant.

2) The scroll compressor may be horizontal or vertical, may be constant-speed or inverter-regulated, and may be housed in a high-pressure casing or low-pressure casing. Exhaustively investigate what operating characteristics are available with which type of compressor so as to select the optimum one that offers the temperature and other characteristics that are best suited for the intended use.

3) Design the system after carefully checking what operation range is attainable and suitable with the proposed compressor. Do not allow excessively high pressure ratios, and avoid vacuum operations.

Fig. 5.28 Profile measurement by 3-D coordinate measuring machine

4) Make utmost efforts to avoid conditions where too much liquid refrigerant gets sucked into the compressor. If it is anticipated that such liquid refrigerant suction would be unavoidable, consider having an accumulator on the suction side.

5) Reverse rotation will not only result in non-compression but will also lead to excessive noise and component damage due to insufficient lubrication of the sliding parts. Make absolutely sure that the compressor is built into the system with correct rotation direction.

6) As scroll compressors are made of precision-machined components, implement sufficient measures to prevent foreign particles from getting into the compressor. Be very careful when flushing the refrigerant system and its piping or when parts are joined by brazing.

References

1) G. Pelizzola: "Apparecchio Rotatorio a Spirali (Spiral Rotary Machine)", Italian Patent, No. 383 (1886).
2) L. Creux: "Rotary Engine", U.S. Patent, No. 801182 (1905).
3) K. Tojo, M. Ikegawa, M. Shibayashi, N. Arai, A. Arai: "A Scroll Compressor for Air Conditioners", Proceeding of the 7th International Compressor Conference at Purdue, West Lafayette, pp. 496-503 (1984).
4) M. Hiraga: "The Spiral Compressor – An Innovative Air Conditioning Compressor for the New Generation Automobiles", SAE Technical Paper, 830540 (1983).
5) K. Terauchi, Y. Tsukagoshi, M. Hiraga: "The Characteristics of the Spiral Compressor for Automobile Air Conditioning", SAE Technical Paper, 830541 (1983).
6) M. Ikegawa, E. Sato, K. Tojo, A. Arai, N. Arai: "Scroll Compressor with Self-Adjusting Back-Pressure Mechanism", ASHRAE Transactions, 90 (2), 2846 (1984).
7) K. Tojo, M. Ikegawa, N. Maeda, S. Machida, M. Shiibayashi, N. Uchikawa: "Computer Modeling of Scroll Compressor with Self-Adjusting Back Pressure Mechanism", Proceedings of the 8th International Compressor Conference at Purdue, West Lafayette, pp. 872-886 (1986).
8) N. Uchikawa, H. Terada, T. Arata: "Scroll Compressors for Air Conditioners", Hitachi Review, 36 (3), pp. 115-162 (1987).
9) K. Tojo, T. Arata, Y. Tomita, M. Shiibayashi, N. Uchikawa: "Scroll Compressors for Air Conditioning and Refrigerating", Proceedings of the International Congress of Refrigeration, Wien (1987).
10) S. Machida, K. Shinoki: "Large-capacity oil-free type

scroll compressor development technology", Electricity, 613 pp. 10-11 (1999). (in Japanese)

11) H. Kato, H. Kamiya, N. Akiyama, K. Uchida, H. Saikawa, T. Kobayakawa: "Development of CO_2 Compressor for Hot Water Supply", Proceedings of JSRAE Annual Conference, Tokyo, (2001). (in Japanese)

12) E. Morishita, Y. Kitora, T. Suganami, S. Yamamoto, M. Nishida: "Scroll Vacuum Pump", Mitsubishi Electric Technical Paper, 62 (5), pp. 423-426 (1988). (in Japanese)

13) M. Ikegawa, K. Tojo, M. Shibayashi: "Scroll Fluid Apparatus", Japan Patent Application, 54-95032 (1979).

14) S. Hagiwara, S. Ueda, Y. Shibamoto, T. Yoshii, T. Toyama, M. Omodaka, S. Jomura, K. Hori: "Development of Scroll Compressor of Improved High-Pressure Housing", Proceedings of the 14th International Compressor Conference at Purdue, West Lafayette, pp. 495-500 (1998).

15) M. Matsunaga, S. Nakamura, Y. Otawara, T. Tsuchiya, K. Fujimura, Y. Matsunaga: "Development of R410A Scroll Compressor", Proceedings of the JSRAE Annual Conference, Hamamatsu, C307 (2004). (in Japanese)

16) K. Tojo, H. Ueda: "New Wrap Profile for Scroll Type Machine", Proceedings of the International Congress of Refrigeration, (1995).

17) H. Kosokabe, M. Takebayashi, Y. Kunugi, Y. Ohshima, H. Hata: "New Scroll Profiles Based on an Algebraic Spiral and Their Application to Small Capacity Refrigeration Compressors", ASHRAE Trans., 102 (2), 3982 (1996).

18) H. Sato, T. Ito, S. Matsuda, M. Fujitani, H. Kobayashi, H. Mizuno: "High Efficiency 3-D Scroll Compressor", MHI Technical Report, 43 (2), pp. 10-13, (2006). (in Japanese)

19) T. Iimori, K. Terauchi, M. Hiraga: "Scroll Compressor", Japan Patent Application. No. 56-33644 (1981).

20) K. Tojo, I. Hayase: "Improvement of Self-Adjusting Back Pressure Mechanism for Scroll Compressors", Proceedings of the 14th International Compressor Conference at Purdue, West Lafayette, pp. 901-906 (1998).

21) I. Tsubono, M. Takebayashi, I. Hayase, I, Inaba, K. Sekigami, A. Shimada: "New Back-Pressure Control System Improving the Annual Performance of Scroll Compressors", ASHRAE Transactions, 104 (1), (1998).

22) T. Fushiki, F. Sano, K. Ikeda, T. Nishiki, T. Sebata, M. Tani, S. Sekiya: "Development of Scroll Compressor with New Compliant Mechanism", Proceedings of the 16th International Compressor Conference at Purdue, West Lafayette, C18-3 (2002).

23) J. E. McCullough, J. T. Dieckmann, T. P. Hosmer: "Compact Scroll-Type Fluid Compressor with Swing-Link Driving Means", U.S. Patent, No.4892469 (1990).

24) E. Morishita, M. Morishita, T. Nakamura: "Scroll Compressor Dynamics", Transactions of the Japan Society of Mechanical Engineers (Series B), 51 (466), pp. 1981-1987 (1985). (in Japanese)

25) T. Ishi, M. Fukushima, K. Sawai, K. Sano, K. Imaichi: "Dynamic Behavior of A Scroll Compressor", Transactions of the Japan Society of Mechanical Engineers (Series C), 53 (491), pp. 1368-1376 (1987). (in Japanese)

26) S. Wang, A. Majumdar: "Digital Scroll Technology", Proceedings of the International HCFC Alternative Refrigerant Symposium, Kobe, (2002).

27) K. Tojo, S. Sado, N. Hagita, Y. Igarashi, K. Suefuj: "Scroll Compressors for Low Temperature Use", Proceedings of the 14th International Compressor Conference at Purdue, West Lafayette, (1992).

28) M. Masuda: "Processing Machine for Scroll Wrap", Japan Patent Application. No. 60-194798 (1985).

Chapter 6 Twin Screw Compressors

6.1 History of Twin Screw Compressors

A twin screw compressor compresses gas by rotating a pair of screw rotors that have convex- and concave-shaped helical lobes. The history of twin screw compressors began by the discovery of its basic theory by Heinrich Krigar in 1878[1-2]. However, it was practically impossible by the technology of the time to accurately produce the complex-shaped screw rotors. It was many years afterward that the twin screw compressor was actually put to use.

The first-ever operational twin screw compressor was built by Ljungström Steam Turbine Co. (hereinafter "LST") a Swedish company, in 1937. A. J. Lysholm, an engineer at LST who had been working on the development of a gas turbine unit since early 1930s, became interested in the theory of twin screw compressor and realized that it, if realized, would conveniently provide a compact and light-weight compression mechanism that will not surge and can be directly driven by a gas turbine. As twin screw compressors existed only in theory at the time, Lysholm started working on developing them[3]. Progress in the development of twin screw compressor technology and its later implementation owe much to the pioneering work of Lysholm and his employer, LST. Because of this, twin screw compressors are sometimes referred to as "Lysholm compressors". The first operational twin screw compressor was an oil-free type having a four-lobe male rotor and a six-lobe female rotor. These two rotors were synchronized by a timing gear so that they rotated together at high speed without actually touching each other. Subsequent development led to the implementation of an oil-free twin screw compressor as air compressor in the middle of 1940s. After that, manufacturers in various countries started producing twin screw-type air compressors under LST license.

In early 1950s, LST renamed itself as "Svenska Rotor Maskiner AB" (hereinafter "SRM"), and started developing oil-injected type twin screw compressors, which are now dominantly used as the twin screw compression mechanism for refrigeration and air conditioning systems. Oil-injected type twin screw compressors, which are simpler in structure and are superior in many performance and reliability aspects than the oil-free type, were put to use first as air compressors and then as refrigeration system compressors. During this period, a number of refrigeration system manufacturers from various countries entered into a license agreement with SRM to start producing oil-injected twin screw compressors. The first oil-injected twin screw compressors for refrigeration systems were built in 1964.

Before dedicated machining units for screw rotor production were developed in early 1950s, screw rotors used to be fabricated by general-purpose milling machines with subsequent finishing done by manual filing or polishing. Therefore, the manufacturing cost was high and the compressor performance varied significantly with the quality of individual rotors. Then in 1952, British machine tool maker Holroyd successfully developed a dedicated milling machine for screw rotor production, after which manufacturing productivity and accuracy for screw rotors have drastically improved.

Starting from the late 1960s, many engineering efforts were made in the theoretical analysis of the rotor profile, which led to further advances in rotor machining technology. With the development of asymmetric profile in early 1970s which led to significant reduction of internal leakages, twin screw compressor efficiency has improved to be almost the same as that of reciprocating compressors. Also introduced during the same period were the first operational gear hobbing machines, which enabled producing complex-shape rotors more easily and in large quantities.

In Japan, production of oil-free twin screw compressors started in 1950s using SRM technology. Then in 1960s, oil-injected twin screw compressors started to be produced for air compression and refrigeration purposes. In the late 1970s, compressor manufacturers started their own research and development concerning twin screw compressor technology, which led to a number of pioneering achievements including, among others, a semi-hermetic structure where the oil separator is fitted inside the casing together with the motor and the compression mechanism[4], new rotor profiles that provided better efficiency and easiness of machining[5], and performance improvement with the use of inverter drive[6].

Due to their compactness, lightweight and high reliability, twin screw compressors have been applied to various uses such as refrigeration and air conditioning, air compression and process gas compression. A wide variety of twin screw compressor research and development efforts

101

still continue today in companies, universities and research institutions, for alternative rotor profiles and machining methods and for numerical simulation-based analysis techniques. There is a good chance that further improvements in efficiency, reliability and productivity will lead to new innovative applications in the future.

6.2 Working Principle and Basic Structure

6.2.1 Working principle

The basic compression elements in a twin screw compressor comprise a pair of rotors with helical lobes that mesh with each other, a rotor casing enclosing the rotors with a very small clearance around them, and suction and discharge ports. The volume of each space surrounded by the lobes of the two rotors and the casing (hereinafter "compression chamber") increases and decreases as the rotors rotate, compressing the gas inside. With this, screw compressors are classified as a positive displacement-type compressor. As twin screw compressors have a fewer number of parts and are simpler in structure than other types of compressors, a very high reliability is obtainable with an appropriate design. Also, the absence of reciprocating mechanical elements that are necessary for almost all the other types of positive displacement compressors enables high-speed rotation to provide a large compression volume with a compact body.

Figure 6.1 shows the twin screw compressor elements that are directly involved in the gas compression process. (a) is a view from the suction port side (low pressure side) and (b) is a view from the discharge port side (high pressure side). A pair of rotors are fitted inside the rotor casing, with discharge ports and suction ports provided at the ends and the sides of the rotors. Of the two rotors, one with convex-shaped thicker lobes is called the male rotor and the other is called the female rotor. The space between the rotors and the casing form a series of compression chambers along the rotor axis direction.

Figure 6.2 is a schematic presentation of the gas compression mechanism of a twin screw compressor. Figures 6.2 (a) to (d) show progress of the suction process viewed from the suction port side. The darker shaded area represents the gas trapped inside one specific groove (compression chamber). Figure 6.2 (a) is the condition immediately after the suction process started in this groove. As the male and female rotors turn in the respective arrow directions and the compression chamber starts moving from the suction end toward the other end, the compression chamber volume gradually increases and is filled with gas sucked in from the suction port. Around the time that both rotors have completed their first rotation, the compression chamber reaches its maximum volume. Just at the same time, the suction port in the casing is closed off as shown in Fig. 6.2(d), completely trapping the gas inside the chamber.

Figures 6.2 (e) to (h) show the subsequent process viewed from the discharge port side (high pressure side). Figure 6.2 (e) is the condition immediately after gas trapping, the same condition as the one shown in Fig. 6.2 (d). This is where the compression process starts. As the rotors turn, the compression chamber volume gradually decreases as shown in Fig. 6.2 (f), increasing the gas pressure inside. The discharge process starts from where the compression chamber becomes connected to the discharge port as shown in Fig. 6.2 (g). The ratio of compression chamber volume V_s, shown in Fig. 6.2 (d), when suction gas has just been trapped, to the compression chamber volume V_d, shown in Fig. 6.2 (g), when the compression chamber is connected

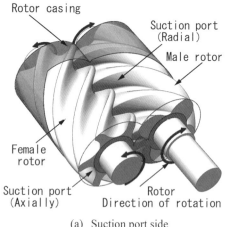

(a) Suction port side

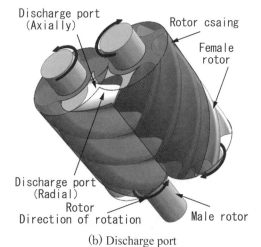

(b) Discharge port

Fig. 6.1 Typical twin screw compressor rotors showing key functions and elements

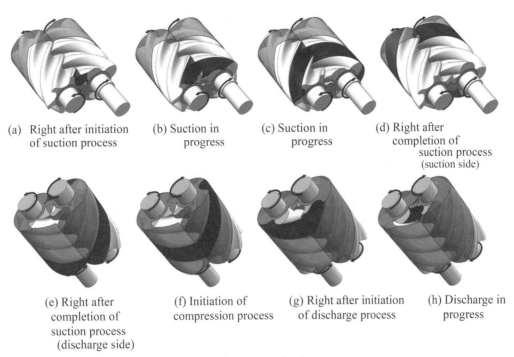

(a) Right after initiation of suction process
(b) Suction in progress
(c) Suction in progress
(d) Right after completion of suction process (suction side)
(e) Right after completion of suction process (discharge side)
(f) Initiation of compression process
(g) Right after initiation of discharge process
(h) Discharge in progress

Fig. 6.2 Fundamental process of twin screw compressor

to the discharge port out of the casing, or V_s/V_d, is called "built-in volume ratio", or V_i. Normally, the built-in volume ratio is designed so that the pressure inside the compression chamber when the discharge process starts will match the discharge pressure required by the system. In other words, the system is designed so that the compression chamber will be connected to the discharge port just when the compression chamber pressure reaches the required level, allowing the compressed gas to be smoothly discharged. Figure 6.2 (h) shows further progress of the discharge process, where the compression chamber volume becomes even smaller as the rotors turn, discharging the gas out of the compression chamber.

This is a complete twin screw compressor cycle from suction to discharge. The whole cycle is driven by rotor rotation and is consecutively repeated in each of the series of compression chambers. Due to such screw compressor working characteristics, gas flow through the compressor is continuous flow even with a certain amount of pulsation.

6.2.2 Basic structure

The previous section 6.2.1 explained about the basic twin screw compressor elements that are directly involved in the gas compression process. In addition to these basic elements, an actual compressor also has bearings and other mechanical components that allow the compression elements to work together. The following paragraphs explain about the typical structure of a twin screw compressor.

Figure 6.3 shows the cross section of an open-type oil-injected twin screw compressor. Each of the male and female rotors has one end rotationally supported by a radial bearing attached to the inlet housing and the other end similarly supported by the combination of a radial and a axial bearings attached to the outlet housing. The two rotors are fitted inside a working space enclosed by the rotor casing and the inlet and the outlet housings, with a very small clearance around the rotors.

Inside this working space, a series of compression chambers is formed by and between the inner wall of the casing and the lobes of the rotors. Each of these chambers repeats a compression cycle as the rotors turn, as explained in Section 6.2.1. The inlet housing has a suction channel and a suction port, through which gas is sucked into the working space. The outlet housing has a discharge port and a discharge channel, through which compressed gas is discharged into the working space. Most oil-injected twin screw compressors have its male rotor driven by a motor and the female rotor driven by the rotating male rotor. With an open-type unit like the one shown in Figure 6.3, one end of the male rotor shaft extends out of the unit, with the penetration area sealed in an airtight manner with a mechanical seal or other shaft-sealing device. The extended end of the rotor shaft is driven by a motor or an internal combustion engine.

In the same compressor shown, the other end of the male rotor shaft is fitted with a balance piston to compensate for the axial load generated with compressor operation. Also fitted in this area is a slide valve for capacity control.

Compressors for Air Conditioning and Refrigeration

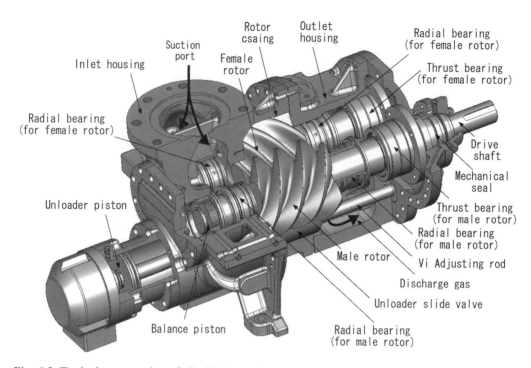

Fig. 6.3 Typical construction of oil oil-injected screw compressor with mechanical seal (open type)

The slide valve works to change the gas-trapping completion timing to adjust the effective displacement volume. A V_i variation mechanism also works together with this slide valve to change the discharge port opening timing in order to control the built-in volume ratio according to the specific operating conditions. However, there are many inverter-driven compact compressors and air compressors that do not have such system of balance piston, capacity control and V_i variation mechanism.

6.2.3 Oil-free type and oil-injected type

Twin screw compressors are mainly classified into the oil-free type and the oil-injected type. The oil-free type works without introducing oil into the compression chambers throughout the whole compression cycle from suction, compression through to discharge, so that no oil gets mixed with the compressed gas (see Figure 6.29). Figure 6.4 shows the cross section of an oil-free twin screw compressor. The oil-free type employs torque transmission from the drive shaft to the driven shaft through a timing gear fitted outside the working space, so that the male and the female rotors turn together without actually contacting each other. To prevent lubrication oil that is fed to the timing gear and the radial and axial bearings from getting into the working space, the male and the female rotor shafts require a total of four shaft seals to separate the working space from other spaces inside the compressor. The shaft seals may be composed of a mechanical seal or carbon packing. In the oil-free type unit, gas temperature tends to rise significantly as the pressure ratio increases. Owing to this tendency, the maximum allowable pressure ratio where a single-stage oil-free twin screw compressor can work efficiently is approximately 3:1. Also, internal leakage has a significantly negative effect on compressor efficiency (see Section 6.3.6). Because of these restricting factors, the oil-free type is not suitable for high differential pressure applications.

An oil-injected type twin screw compressor introduces oil into the compression chambers so as to utilize the lubrication, liquid-sealing and cooling properties. Oil that has lubricated the mechanical components such as the bearings, the balance piston and the unloader cylinder flows toward the compression chambers. In addition to that oil, some amount of oil is injected into the compression chamber at an appropriate timing so that just the required amount of oil is introduced into the compression chambers (see Figs. 6.27 and 6.28).

Figure 6.5 shows the cross section of an oil-injected twin screw compressor. The oil-injected type has a number of advantages compared to the oil-free type. One advantage is that it is permissible for a certain limited amount of oil to get inside the working space from the bearings, allowing the shaft seal to be omitted or be provided by a light-duty sealing feature such as labyrinth-type seals. Another is that no timing gear is required as the oil introduced into the working space lubricates the rotors so that torque can be directly transmitted between the rotors. Owing to these ad-

Chapter 6 Twin Screw Compressors

6.2.4 Semi-hermetic type and open type

The twin screw compressors can be built as a semi-hermetic type or an open type. The semi-hermetic type has the compression mechanism and the motor housed in a common casing, sealed by gaskets and O-rings. Figure 6.6 shows examples of semi-hermetic twin screw compressors; (a) is a single-stage compression unit containing an oil separator and (b) is a two-stage compression unit. On the other hand, an open type has the drive shaft extending out of the compressor casing, which is driven by an external motor. Figure 6.7 shows an example of open-type twin screw compressor unit. The casing section penetrated by the drive shaft is sealed with a mechanical seal or other sealing feature (see Section 6.4.9).

Compared to the open type, the semi-hermetic type is less prone to gas leakage and will not suffer as much mechanical loss because it does not require any mechanical seal or other shaft sealing feature. However, because of the

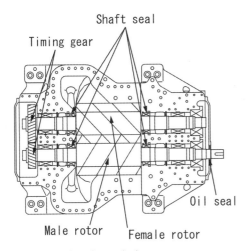

Fig. 6.4 Section through dry screw compressor

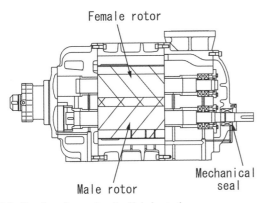

Fig. 6.5 Section through oil oil-injected screw compressor

vantages, the oil injected type can be designed to a simpler and more robust structure. In addition, the lubrication oil's sealing effect contributes to internal leakage reduction. This leads to a higher volumetric efficiency, making the compressor suitable to operate with larger pressure differences. For the same reason, the compressor has good operating efficiency even with low rotational speeds, which makes the oil-injected type superior in terms of bearing reliability and vibration and noise reduction compared to the oil-free type. Furthermore, the oil injection feature itself serves to reduce compression gas temperature, enabling high-efficiency operations with a single stage compression in a pressure range of up to approximately 10:1.

To summarize, the oil-free and the oil-injected twin screw compressors should each be selectively used for their suitable applications, the oil-free type for such applications where avoidance of oil contamination is a high priority and the oil-injected type for almost all the other general applications including refrigeration and air conditioning. For specific application examples, refer to Section 6.6.4.

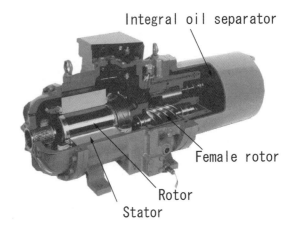

(a) Semi-hermetic twin screw compressor (single-stage)

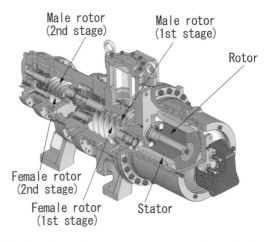

(b) Semi-hermetic twin screw compressor (two-stage)

Fig. 6.6 Semi-hermetic twin screw compressor

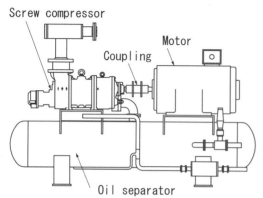

Fig. 6.7 Example of twin screw compressor unit (monocoque type)

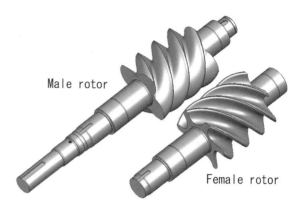

Fig. 6.8 Twin screw rotors

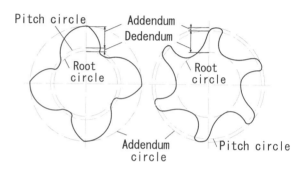

Fig. 6.9 Rotor profile in transverse section

whole motor being exposed to the refrigerant or process gas, types of materials that can be used as motor or coil windings and insulations may be limited depending on the type of gas used.

Advantages of an open-type twin screw compressor include that any drive power source can be selected. For example, a motor or an internal combustion engine can be used as the drive source depending on the amount of shaft drive power required. In addition, the fact that the compressor and the motor are separate from each other can be beneficial from maintenance perspective because each can be individually fixed or replaced if a failure occurs in either of them. In large-scale systems where the rotor outside diameter exceeds 200 mm, the open-type is selected more often because such a large system requires a larger motor as well, where a semi-hermetic structure may not be advantageous from the production engineering and manufacturing cost perspectives.

Due to the above-described characteristics, the semi-hermetic type is dominantly used for small-scale systems in the rated power range of up to 50 kW. Although the open type is still most commonly used for all larger systems, recent increase in social awareness for the need of refrigerant gas leakage prevention due to safety and environmental concerns is leading to more use of semi-hermetic twin screw compressors even for larger systems in the 100 kW class.

6.3 Rotor Profiles

6.3.1 Basics and history of rotor profiles

Figures 6.8 and 6.9 show the male and female rotors of a twin screw compressor and their sections perpendicular to the axis. Of the pair of rotors, the one with a thick, convex shape, with the majority of the profile located outside the pitch circle (addendum) is called the male or main rotor, and the one with a thinner, concave shape, with the majority of the profile located inside the pitch circle (dedendum) is called the female or gate rotor. Both rotors have helical lobes and grooves, which mesh together to form compression chambers.

The rotor profile is one of the most important factors of a twin screw compressor that influences almost all the characteristics of the compressor. The ideal profile is considered to have grooves with a large cross section area, a short meshing seal-line and a small blowhole area (for more about meshing seal-lines and blowholes, refer to Section 6.3.6). These attributes will together provide a larger displacement volume and lesser internal leakage for the same compressor size, which means better compression efficiency. In practice, however, the inherent geometric characteristics of a profile makes it difficult to fully attain all of the above attributes at the same time, and therefore the design engineer's approach should be to aim at the best possible mix. The rotor profile also significantly affects torque transmissibility between rotors, which must be considered in the profile design as well. In addition, the rotor profile for an oil-injected type must also consider the Hertz stress, pressure angle and sliding velocity that occur in the drive band section of the rotor profile (see Section 6.3.2).

As a general rule, the rotor profile is designed by defining a curve on one of the rotors and generating its corresponding curve on the other rotor according to an appropriate criteria. The first curve to be defined can take any form as long as an exactly corresponding curve can be generated. In practice, various geometric curves such as circular arcs, parabolas, ellipses, and hyperbolas have been used. Another profile design method is to define the profile on a coordinate system which is independent from either rotor. This method, generally referred to as "rack generation", has been proposed by Menssen[7], Rinder[8] and Stosic[9].

Figure 6.10 shows the chronological list of twin screw compressor rotor profiles that have been implemented. The rotor profile has evolved from (a) Lysholm profile[10], which was used for the first ever operational twin screw compressor, to (b) symmetrical circular arc profile[11] and then to (c) asymmetric profile[12], finally attaining (d) currently used profile. The leading side of a Lysholm profile was a circular arc while the trailing side was a cycloid generated by the curve of the opposing rotor. However, it was extremely difficult to accurately machine this profile by the technology of the time. As a result, twin screw compressors produced during this period often were not very precise or reliable.

To address this issue, a symmetrical circular arc profile, that is composed of a pair of opposed identical circular arcs on the leading and trailing sides and is easier to machine than the Lysholm profile, was developed to become the most popular rotor profile of the time. For many years until the later-described asymmetric profile was developed, the symmetrical circular arc profile remained the dominantly used rotor profile for twin screw compressors. However, one major disadvantage of the symmetrical circular arc profile was that it has a large blowhole, which made it difficult to attain higher compressor efficiency. Then an asymmetric profile, (c) of Fig. 6.10, was developed in early 1970s. The asymmetric profile has a generated curve on the trailing side, which significantly reduced the blowhole area on the compression side. The introduction of this asymmetric profile improved the efficiency of twin screw compressors to be approximately the same as that of other types of compressors, leading to their increasingly widespread use.

Starting from the later 1970s and continuing into the 1980s, compressor manufacturers actively worked on their respectively unique research and development programs related to twin screw compressors. These efforts collectively led to the creation of the currently most popular profile that provides both superior performance and high productivity, bringing on further efficiency improvement. However, it is generally believed that further improvement of the twin screw compressor rotor profile, both in terms of definition and generation, is still possible through more research and development. If realized, it will certainly lead to a much higher operating efficiency.

6.3.2 Rotor profile modification

A twin screw compressor rotor profile defined and generated by a number of curves is called a "theoretical profile", which, from the purely geometric perspective, allows the male and female rotors to mesh together with no clearance in between. However, considering the anticipated machining errors and thermal expansion that occurs during compressor operation, the actually produced screw rotor profile needs to be designed with some amount of profile modification applied to the theoretical profile so as to provide the realistically required amount of backlash and clearance between the male and female rotors.

Especially with the oil-injected type where drive torque needs to be directly transmitted from the male to the female rotors, the rotor profile must be modified from the theoretical shape so that the male and the female rotors will not contact each other except in specific areas ("drive band") where the contact is required for drive power transmission.

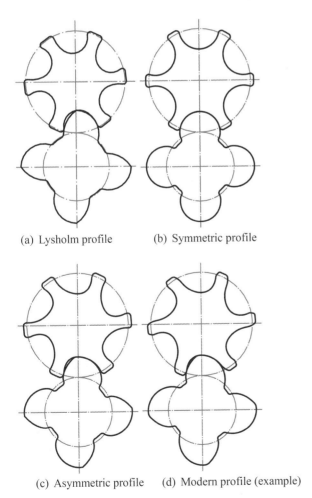

(a) Lysholm profile (b) Symmetric profile

(c) Asymmetric profile (d) Modern profile (example)

Fig. 6.10 Brief transition of rotor profile

However, it must be noted that excessive backlash or clearance will lead to an increase in the internal leakage passage area (see Section 6.3.6), which contributes to compressor efficiency reduction. On the other hand, lack of enough backlash and clearance will prevent successful absorption of machining errors, and as a result contact between male and female rotors in undesirable areas may be caused with thermal expansion during compressor operation, leading to more noise and vibration or even rotor seizure. To assure an optimum efficiency, reliability and quietness in twin screw compressor operation, it is critical, as described above, to provide just the right amount of backlash and clearance in the rotor profile design and profile modification.

6.3.3 Number of lobes

Just as important to the twin screw compressor's operating characteristics as the rotor profile is the number of lobes on the male and the female rotors. The larger the number of lobes is, the smaller the pressure difference between adjacent compression chambers will be, leading to less internal leakage and better compressor efficiency. Also, the larger number of lobes will increase the number of compression chambers formed at a time, which decreases pressure pulsation in the gas being discharged, reducing noise and vibration. On the other hand, the larger number of lobes means decrease in the cross section area of the grooves, reducing the displacement volume per drive shaft rotation for the same rotor length and diameter. These advantages and disadvantages must be carefully considered to determine the best combination of the numbers of male and female rotor lobes.

Since the first ever operational twin screw compressors, a combination of four male rotor lobe and six female rotor lobe, like the one shown in Fig. 6.10 (a) has been most commonly used rotor lobe configuration. However, as more efficiency improvement is demanded as a result of recent increase in social awareness for environmental issues and energy saving, compressor manufacturers have started to develop and introduced compressor design with the number of lobes optimized for individual applications. Figure 6.11 shows examples of such alternative rotor lobe configurations. Compared to the traditional four-male lobe and six-female lobe configuration, the five-male lobe and six-female lobe configuration will, in addition to providing the above-described general advantages for having more lobes, better level out the gas load working on the rotor-supporting radial bearings to lower the bearing load. Therefore, the five-male and six-female configuration is suitable for compressors that are operated under high pressure ratios.

Another advantages of having one more female rotor lobe than the male lobe like the five-male and six-female configuration is that "closure delay", which refers to a condition where the discharge-end groove is not fully closed when the suction port closes and as a result becomes connected to open-to-suction grooves, is eliminated. The four-male and five-female combination, shown in Fig. 6.11 (b), is not only free from the closure delay like the five-male and six-female combination but also provides a larger cross section area of the grooves. Due to these advantages, the four-male and five-female combination is increasingly selected for use in oil-injected type compressors.

For compressors that are used at the discharge pressure of 5 MPa or higher pressure conditions, rotor configurations with larger numbers of lobes such as the six-male and eight-female combination shown in Fig. 6.11 (c) are favored because they provide a larger root diameter, which is advantageous for shaft stiffness, while minimizing the pressure difference between adjacent grooves. On the other hand, such compressors as air compressors for lower pressure-ratio applications may be designed with much fewer number

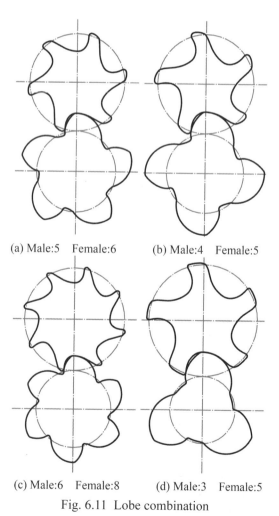

(a) Male:5 Female:6 (b) Male:4 Female:5

(c) Male:6 Female:8 (d) Male:3 Female:5

Fig. 6.11 Lobe combination

of lobes, such as the three-male and five-female configuration shown in Fig. 6.11 (d), to provide a larger displacement volume for the same compressor size to improve efficiency.

6.3.4 Wrap angle

"Wrap angle" in a twin screw compressor refers to the amount of angular travel made by a groove helix about the rotor shaft while it axially travels over the length of the rotor. Figure 6.12 shows a comparison of two male rotors with different wrap angles. The larger the warp angle is, the thinner the lobe thickness in the section perpendicular to the helix will be, causing mechanical strength reduction in the lobe. Compressors that are operated under high pressure conditions are typically designed with a small wrap angle so as to provide a sufficiently thick lobe profile. Considering this factor and also that excessively large wrap angles make the rotor machining process difficult, the appropriate range of wrap angle is relatively limited.

The amount of wrap angle also affects the volume change curve profile and other operating characteristics. If a smaller wrap angle is used, a compression process from compression start to discharge will be completed in a smaller rotation angle. This will reduce the time required to complete a compression process for the same rotation speed. Therefore, reducing the wrap angle by an appropriate amount can contribute to internal leakage reduction. On the other hand, a smaller wrap angle will reduce the number of compression chambers formed, which results in a larger pressure difference between adjacent compression chambers and may possibly reduce compressor efficiency instead of improving it depending on the compressor design and application. A smaller wrap angle will also make the pressure rise in the compression chamber steeper, which increases excitation forces, possibly causing more noise and vibration. However, an excessively large wrap angle will lead to the closure delay, reducing the effective compression chamber volume and thereby decreasing the displacement volume. The wrap angle should be designed with all the above-described factors taken into consideration. The commonly used wrap angle range for the male rotor is around 300°.

6.3.5 L/D_m and center distance

Screw rotor dimensions are generally determined by the ratio of rotor length L to the male rotor outside diameter D_m, namely L/D_m. Figure 6.13 shows a comparison of two male rotors with the same wrap angle and different L/D_m.

The lobe profile, the number of lobes, the wrap angle, L/D_m and the center distance are the basic rotor parameters that determine the displacement volume, which is the most important specification of a compressor. Of these rotor parameters, L/D_m and center distance are particularly important as they affect the internal leakage passage area, lobe thickness and strength, applicable bearing dimensions and bearing load, and therefore must be selected based on exhaustive design study.

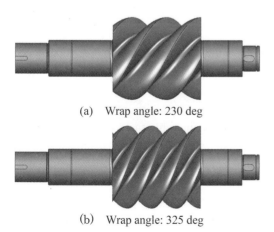

(a) Wrap angle: 230 deg

(b) Wrap angle: 325 deg

Fig. 6.12 Difference in male rotor form due to change in wrap angle

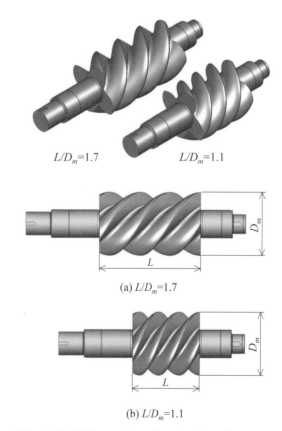

(a) L/D_m=1.7

(b) L/D_m=1.1

Fig. 6.13 Difference in male rotor form due to change in L/D_m (length-diameter ratio)

6.3.6 Sealing line and internal leakage passage

With the basic rotor parameters determined, the shape and length of the sealing line can be obtained. Figure 6.14 shows an example of sealing line geometry for four-male and six-female lobe configuration. In the figure, (a) represents the sealing line on the compression side and (b) those on the suction side, with the diagrams on each side showing, from top to bottom, projections onto x-z, y-z and x-y planes respectively (see Fig. 6.16 for the definitions of the coordinate axes). The sealing line of a compression chamber is formed along the areas of engagement between the male and female rotors, between the rotor tip and the rotor casing, at the blowhole and at the end plane clearances between the rotor suction end and the housing and between the rotor discharge end and the housing. As the rotors rotate, the sealing line travels from the suction end toward the discharge end while compression progresses. In the diagrams,

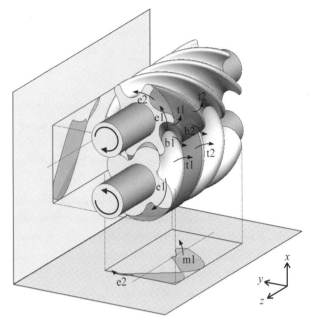

Fig. 6.16 Internal leakage path

C3 - C4 - C5 - C6 - C7 and S3 - S4 - S5 - S6 - S7 represent the rotor engagement areas, C2m - C3 - C2f and C8m - C7 - C8f the blowhole rim on the compression side, S2f - S3 - S2m and S8f - S7 - S8m the blowhole rim on the suction side, and C1m - C9m and C1f - C9f the clearance between the rotor discharge-side end plane and the outlet housing. Here, a "blowhole" refers to the opening formed between the cylindrical intersection of the rotor casing, or "cusp", and the male and female rotor tips. While the nominal sealing line of a single compression chamber passes along the rim of the blowhole, there is no actual "sealing line" there. In reality, adjacent compression chambers are connected with each other through a blowhole.

The boundary framed by the sealing line is the internal leakage passage, which is illustrated in Fig. 6.16. The plus direction of z axis in the figure is the direction toward the discharge side. The figure also shows the male rotor sealing line as projected on the x-z and the y-z planes. In the figure, one specific chamber where compression is underway (shown shaded) receives leakage t1, across the sealing line along its leading rotor tip, from the chamber immediately in front of it where compression process has progressed further. At the same time, the same chamber lets out leakage t2, across the sealing line along its trailing rotor tip, into the chamber immediately behind it where pressure is still lower. These are the leakages that occur between compression chambers across rotor tip. Similarly, leakages e1 and e2 occur between compression chambers through the gap between the rotor discharge end plane and the outlet housing. The above-described blowholes also provide channels

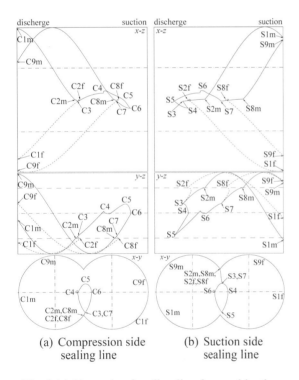

Fig. 6.14 Example of sealing line for combination of 4 male and 6 female lobes

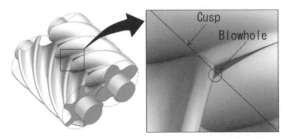

Fig. 6.15 Blowhole (compression side)

for inter-chamber leakages such as b1 and b2. In addition to these inter-chamber leakages, internal leakage also include leakages across areas of engagement between male and female rotors, such as m1 shown projected on the *y-z* plane, where gas flows from the groove where compression is ongoing into another groove into which gas is still being sucked. Similarly, leakages that occur in the vicinity of the discharge port through the gap between the rotor discharge end plane and the outlet housing may flow back to the suction side.

Internal leakage from the compression side to the suction side brings about reduction in volumetric efficiency, while the one between compression chambers leads to increase in the gas compression power requirement and reduction in adiabatic efficiency. Internal leakage is one of the most influential factors on the twin screw compressor efficiency. For efficiency improvement, it is imperative to design a rotor profile that will reduce these internal leakages to an absolute minimum. It is also important to thoroughly study and optimally control other factors that influence leakage passages, such as rotor profile modification (see Section 6.4.2), machining and assembly accuracy and parts deformation due to heat and gas force during operation.

6.3.7 Displacement volume

As the compression chambers in a twin screw compressor are in very complex three-dimensional forms, analytical techniques such as theoretical volume determination from the sealing line geometry projected on a plane based on the principle of virtual work[13] have been proposed. Figure 6.17 shows volume and pressure changes in a compression chamber, or groove, of a male-rotor-driven compressor with a four-male and six-female lobe configuration. The horizontal axis represents the male rotor rotational angle α_M. The vertical axis represents the ratio of groove volume V to maximum groove volume V_{max} and also the ratio of in-groove pressure P to discharge pressure P_d. As the rotors rotate, the groove volume gradually increases and becomes the largest at rotational angle α_{Vmax}, upon which the suction stage is completed. As soon as the groove volume reaches the maximum value, the suction port closes to start the compression stage. As the compression stage progresses, the groove volume decreases to increase the pressure inside. When the rotational angle reaches α_D, the groove becomes connected with the discharge port to start the discharge stage.

With the maximum groove volume V_{max} obtained from the sealing line geometry, the displacement volume per rotor rotation, V_{th}, can be expressed by the following equation:

$$V_{th} = V_{max} \cdot Z_l = C \cdot D^2 \cdot L \qquad (6.1)$$

where:
  C: Profile coefficient
  D: Rotor outside diameter
  L: Rotor length
  Z_l: Number of lobes on the driving rotor

C is a constant determined by the rotor profile and is equivalent to the ratio of the displacement volume to the cylindrical volume expressed by the rotor length and diameter. As long as the rotor profile is geometrically similar, the profile coefficient remains constant even when the rotor diameter changes.

6.4 Structure and Design of Compressor Components

6.4.1 Screw rotors

To design a rotor for a twin screw compressor, it is necessary to first determine the anticipated load and torque working on the rotor shaft under the expected operating conditions based on the gas force, and then to conduct evaluations related to lateral and torsional critical speeds (see Section 6.5.4) before determining the required shaft diameter. The amount of gas force that works on the rotor shaft varies depending on the rotor profile and the number of lobes. Also, the rotor shaft diameter is restricted by the root diameter of the rotor profile. Therefore, the rotor profile and the number of lobes must be investigated together in designing the rotor shaft. As a screw rotor is a moving component that rotates at high speed, sufficient stiffness and superior dynamic balance is required to assure reliability and low noise.

Rotors can be built with various cast irons, forged steels,

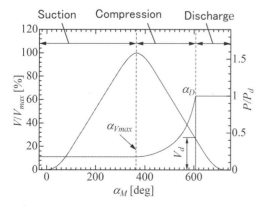

Fig. 6.17 Volume curve and indicator diagram

Compressors for Air Conditioning and Refrigeration

high-strength special steel or free-cutting steel. Induction hardening, plating or other surface treatment may be provided on the journals as required. As the accuracy of male and female rotor groove machining directly affects compressor performance, it is important to finish the rotors by precision milling machine or grinding machine and to ensure dependable accuracy control. More information about rotor machining is given in Section 6.8.

6.4.2 Rotor casing

The rotor casing is a structural component that accommodates the screw rotors, constitutes the outer wall section of the compression chambers and also part of the suction and discharge ports. The rotor casing also joins other structural components together, such as the inlet and outlet housings. The rotor casing has an opening that is shaped like two intersecting cylinders, where a pair of intermeshing rotors is housed to form the compression chambers. The intersection line of partially overlapping two cylinders is called the "cusp". The rotor casing must be designed as a pressure vessel. Its wall thickness and reinforcing rib locations must be determined based on the amount of stress in various locations of the casing under the design pressure. FEM analysis techniques are now being used in rotor casing design. As any gap between the rotor casing's inner wall and the male and female rotor tips will constitute an internal leakage passage, it is necessary to make all such gaps as small as possible. For that purpose, high precision machining techniques such as boring are being used. The rotor casing also joins inlet and outlet housings that support the rotor shafts, and therefore the casing's finish accuracy affects bearing alignment. Parallelism between the casing end planes and their perpendicularity of bore to the end planes are particularly important. The rotor casing is typically built with cast iron or cast steel. Similarly to the previously described rotors, high rigidity is required to assure reliability and low noise.

6.4.3 Inlet housing

The inlet housing closes off the suction end of the working space, and provides a gas suction channel and a suction port in such a way that gas will be sucked in and trapped inside in the best timing. The inlet housing is also a structural component that supports the radial bearings for the male and female rotors. The gas suction channel must have a sufficient cross section area to accommodate the incoming gas flow and also must be shaped and designed to reduce fluid loss that occurs when the gas is sucked in. The end surface mating with the rotor casing and the portion to house the bearing must be very precise, as accuracy in these areas affects rotor bearing alignment and the magnitude of gaps between the casing and the rotors. The inlet housing is typically built with cast iron or cast steel. High rigidity is required to assure reliability and low noise. In some product models, the inlet housing is integrated with the rotor casing.

6.4.4 Outlet housing

The outlet housing closes off the discharge end of the working space and provides a discharge channel and a discharge port. The discharge port serves to connect the compression chamber to the discharge channel so as to release the compressed gas in the best timing. The outlet housing is also a structural component that supports the radial and thrust bearings for the male and female rotors. Due to the slower gas flow speed, fluid loss through the discharge channel does not have as significant effect as that through the suction channel. However, the discharge channel still needs to have a sufficient cross section area to accommodate the outgoing gas flow. Like the rotor casing and other structural components, the outlet housing is typically built with cast iron or cast steel, and must be machined and produced with high precision and rigidity. In smaller compressors, the outlet housing may be integrated with the rotor casing.

6.4.5 Suction port

A twin screw compressor has either an axial suction port in the inlet housing or a radial suction port in the rotor casing. Sometimes the suction port is a combination of axial and radial ones (see Fig. 6.1 (a)). In a commonly used oil-injected twin screw compressor, the suction port is normally designed to a geometric shape so that gas will be trapped inside the compression chamber exactly at such rotational angle where the compression chamber volume is the largest. However, an angular closure delay may be provided on the male rotor side in oil-free high rotation models so as to allow a sufficient time to have the groove filled with gas, or the closure angle may be advanced in slow-rotation models to reduce the effect of internal leakage.

6.4.6 Discharge port

A twin screw compressor typically has an axial discharge port provided in the outlet housing and also a radial discharge port provided in the rotor casing (see Fig. 6.1 (b)). The discharge port opening timing has a significant effect on the compressor's discharge loss. Therefore, the basic discharge port design is normally carried out by obtaining the built-in volume ratio (V_i) from the following equation so that the operating pressure ratio and V_i will have the best possible relationship under the most frequently employed operating conditions.

$$\frac{P_d}{P_s} = \left(\frac{V_s}{V_d}\right)^\kappa = V_i^{\ \kappa} \qquad (6.2)$$

P_d: Discharge pressure
P_s: Suction pressure
κ: Specific heat ratio of the gas

The relationship between the operating pressure ratio and V_i and its potential effects on the compressor performance are explained further in Section 6.6.1. In compressor models with a capacity control or variable-V_i mechanism, the radial discharge port is often provided in a slide valve or similar feature instead of in the rotor casing so that the discharge port will move along with the operation of the above-mentioned mechanisms. But the basic design remains the same.

6.4.7 Other ports

In addition to the above-described suction ports, the working space of a twin screw compressor may also have other ports to take in oil or gas or liquid refrigerant from the outside. For example, oil-injected models may have an oil injection port that supplies oil into the compression chamber and also an oil return port to return the lubrication oil that has lubricated the bearings to the compression chamber. These ports will be located in the respectively suitable positions considering the lubrication oil flow path and the refrigerant gas characteristics. Oil-free models may also have ports to introduce liquid into the compression chambers to seal or cool the gas inside.

Also, refrigerant compressors with economizer or liquid injection features may have an economizer port that introduces intermediate pressure gas into rotor grooves or a liquid injection port to introduce liquid refrigerant into the grooves. See Section 6.6.1 for more information about economizer and liquid injection features.

6.4.8 Bearings

One of the dominant loads that work on the bearings of a twin screw compressor is the gas load on the rotors. Gas pressure inside a rotor groove increases as the groove moves toward the discharge port, where the pressure difference causes axial and radial loads on the rotor shaft. In oil-injected models, torque transmission between the two rotors causes both radial and axial loads. In timing gear-fitted oil-free models or speed increasing gear-fitted models, forces generated by their torque transmission act on the bearings.

In start-up and other transient condition where the ro-tational speed is not stable, moment of inertia of the rotor causes the transmission torque to increase, and its reaction force acts on the bearings in both axial and radial directions. In most semi-hermetic models, the extended portion of the drive shaft is coupled with the motor rotor, where all of the rotor mass as well as the reaction forces to all the forces generated in the motor work on the drive shaft bearings. In addition, any slight unbalance in the shaft or in the entire rotational system including the subsidiary components will cause a corresponding load.

To be able to support all these radial and axial loads, the male and the female rotors of a twin screw compressor are each supported by both radial and thrust bearings. The functionality of the bearings of a twin screw compressor includes not only supporting the above-described various loads that work on the bearings but also retaining the rotors in the correct position both radially and axially. The bearings are required to provide all these functions with high reliability and minimum loss, while generating smallest possible noise.

As long as both male and the female rotors can be maintained in their correct positions during compressor operation, only the smallest amount of design clearance needs to be secured between components, enabling higher compressor efficiency. To realize that, it is critically important to precisely calculate all the anticipated loads and select bearings that are sufficiently strong and rigid to support them. The radial positioning accuracy is affected by the bearing's radial clearance and rotational accuracy, while the axial positioning accuracy is affected by the bearing's axial clearance and the bearing preload. Other factors, such as the bearing's fitting accuracy with the shaft or the housing and the degree of assembly deformation, also significantly affect positioning accuracy. Sufficient care must be taken to select, design or fabricate the components and features related to these factors.

Bearings can be mainly grouped into sliding and rolling bearings, either of which may be used for twin screw compressors depending on the case. While technical details of each group of bearings should be obtained from academic literature, the slide and the rolling-element bearings have the following characteristics from the perspective of twin screw compressor design.

Rolling bearings, when used as a twin screw compressor component, enables more accurate positioning and thereby facilitates the control of clearances and gaps during the assembly process. Running clearance can also be reduced to an absolute minimum. Also, rolling bearings are generally less prone to friction losses regardless of the operating conditions such as rotational speed or lubrication oil temperature and can operate with only a small amount of oil to lubricate and cool the bearings. This characteristic contributes to reduction in mechanical and pumping losses. The rolling bearings can withstand poor lubricating conditions such as dissolution of refrigerant into the lubrication oil or momentary lack of lubrication and also do not need to be primed with oil for start-up. Therefore, compressors with rolling bearings can be lubricated by the force of pressure difference only, which enables the system to be built without an oil pump.

In spite of offering so many advantages, rolling bearings are disadvantageous in that they can only last for a specific lifetime and therefore compressor maintenance must be scheduled in time for bearing service, and also that a shaft sealing feature is required between the bearing and the compression chambers. These factors complicate the compressor structure and also possibly increase cost. In addition, inherently large outside diameter of the bearing may restrict the choice of products that can be used, especially if the distance between the rotor centers is fixed based on the rotor profile and the displacement volume. Figures 6.18 and 6.19 show examples of rolling bearings that can be used in a twin screw compressor. Cylindrical roller bearings and duplex angular-contact ball bearings are most commonly selected as radial and thrust bearings, respectively. Needle roller bearings, tapered roller bearings and four-point-contact ball bearings may also be used depending on the application.

Sliding bearings as a twin screw compressor component last for a long time as they are supported by oil film pressure and are operated under fluid lubrication condition. It is generally believed that sliding bearings, if used under ideal condition, can have an almost indefinite service life. Another advantage of sliding bearings is that their load capacity increases with speed, which makes their use quite beneficial in applications where the amount of load increases with speed. The absence of rolling elements between races also contributes to quiet operation. An important characteristic of slide bearings is that the oil film formed inside it has an impact absorption effect, which makes this kind of bearings especially suitable for twin screw compressors where the amount of bearing load changes in a cyclic manner. From the design and manufacturing perspective, sliding bearings are advantageous in that they do not require a large installation space and therefore less subject to layout restriction by the rotor center distance. Under such operating conditions that a sufficiently larger oil film pressure is obtained, the amount of bearing loss will be either equivalent to or less than that of rolling bearings. On the other hand, disadvantages of sliding bearings include potential deterioration of lubricating condition when the oil film pressure decreases, for example during startup or low speed operation, and susceptibility to damages caused by changes in the lubrication oil temperature, refrigerant resolution and mixing of liquid refrigeration. The sliding bearings must always be primed with oil for start-up, and therefore an oil pump is required in the system. Figure 6.20 shows an example of sliding bearing that is used as a radial bearing (see Fig. 6.3 as well). For special applications where an exceptionally large amount of static or variable load in axial direction is anticipated, a tilting pad bearing that is often used in gas turbine systems, like the one shown in Fig. 6.21, may be used as an axial bearing.

As described above, a wide variety of bearings may be employed in twin screw compressors. It is necessary to fully understand the advantages and disadvantages of each type of bearing and to select an optimum one by considering the characteristics of the entire system in which the relevant compressor are used.

When selecting a rolling bearing, select one that will satisfy the service life requirement under the bearing load, rotational speed and other operating conditions anticipated, while taking into consideration the dimensional limitations such as the rotor profile and center distance.

In the case of sliding bearings, dimensional restrictions like the rotor profile and center distance are not likely to be

Fig. 6.18 Roller bearing

Fig. 6.19 Angular contact ball bearing (double-acting type)

Fig. 6.20 Sliding bearing

Fig. 6.21 Tilting pad bearing

a major problem. In the design process, the shaft diameter, the bearing width and other factors should be determined based on the anticipated bearing load and rotational speed as well as the allowable surface pressure, sliding velocity and other characteristics of the bearing material. At the same time, evaluate such factors as the Sommerfeld number, the minimum oil film thickness and oil film temperature to make sure that no conflicting issues occur. For further details of sliding beating design, numerous academic publications are available.

While selection and design of rolling and sliding bearings can as a general rule be made in the same way as for general machinery, an important point specific to compressor bearings is the bearing material's chemical resistance to the process gas or refrigerant. Special consideration is required for applications that employ a gas or refrigerant that is corrosive.

6.4.9 Shaft seals

Open-type compressors, where the drive shaft must penetrate the compressor casing to be extended outside, requires a shaft sealing feature in the area of penetration. Functional requirements for a shaft seal include dependable separation of spaces of different pressures, sealing of the fluid inside the casing and enabling the shaft rotation with a minimum loss. While many types of shaft sealing features

Chapter 6 Twin Screw Compressors

provide the above-described functions, the most commonly used type of shaft sealing device for open-type compressors is mechanical seals. Mechanical seals allow an extremely small amount of leakage compared to other sealing devices such as gland packings. They also do not incur as much power loss or heating. Other advantages of a mechanical seal include the smaller degree of abrasion incurred in the sliding components and superior adaptability to demanding operating conditions such as high pressure and high speed.

Figure 6.22 shows an example of mechanical seal structures that are used in open-type twin screw compressors. As shown, a mechanical seal comprises a rotary ring that turns together with the shaft and a fixed ring that is attached to the casing. The sealing faces on the two rings are mated with each other by the internal fluid pressure plus the force of a coil spring or bellows as they rotate and slide against each other, thus restricting fluid leakages. A mechanical seal requires its own sealing elements between the fixed ring and the casing and also between the rotary ring and the shaft, which are provided by an O-ring or V-ring. These sealing elements are commonly referred to as "secondary seal". The sliding components of a mechanical seal are typically built with a combination of synthetic carbon, graphite, special cast iron, silicon carbide, tungsten carbide and Hastelloy.

The mechanical seal is not intended for complete blocking of fluid leakage. A very small amount of fluid enters between the sealing faces of the rotary and fixed rings sliding against each other, and part of it leaks outside. This minute amount of leakage is indispensable for lubricating and cooling the sliding faces. Either the seal manufacturer or the compressor manufacturer normally specifies a range for appropriate amount of leakage. As the mechanical seal of

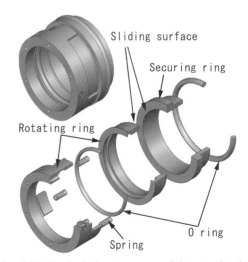

Fig. 6.22 Typical elements comprising mechanical seal assembly

a twin screw compressor needs to be lubricated and cooled by the lubrication oil, areas on the compression mechanism side of its sliding plane must always be filled with oil while the compressor is operating. To achieve this, the areas immediately around the mechanical seal and other areas inside the compressor must be partitioned by an oil seal or other sealing feature so that only the areas surrounding the mechanical seal will be filled with oil. This oil-filled space is often referred to as the "seal box".

Mechanical seals are divided into several types according to their structure. For technical details of each type of seals, refer to academic publications or professional information available from the manufacturers. The following paragraphs explain about a number of mechanical seal types and their structures that are commonly used in twin screw compressors.

One of the criteria that are used for mechanical seal classification is the balance ratio. The sliding faces of the fixed and the rotary rings of a mechanical seal is pressed together, mainly by the internal fluid pressure. The higher this fluid pressure gets, the larger the surface pressure on the sliding faces will be. If this surface pressure rises above a certain limit, the lubrication film formed by fluid between the sliding faces will break, possibly leading to damages such as abrasion, seizure or gas leakage. To prevent this, the mechanical seal may be designed so that the ratio of the internal fluid pressure application area A_0 to the sliding face area A will be smaller, thus reducing the surface pressure on the sliding faces if the application entails high fluid pressure. This A_0/A ratio is called the "balance ratio" of a mechanical seal. As a general rule, mechanical seals with a balance ratio of 0.7 to 1.0 are called "balance type mechanical seal" and those with a higher balance ratio is called "unbalance type mechanical seal". Figure 6.23 shows an unbalance type mechanical seal and Figs. 6.24 and 6.25 show balance type mechanical seals, each in a cross section view.

Mechanical seals can also be grouped into rotary and stationary types. Mechanical seals that have a coil spring or bellows attached to their rotary ring are called "rotary type". The rotary type can be built with a simpler sealing structure and therefore can be designed smaller. For this reason, rotary-type mechanical seals are most commonly used for twin screw compressors. However, mechanical seals may have a centrifugal force generated in its coil spring or bellows when it is rotating at high circumferential speed, which may deteriorate the rotary ring's mechanical compliance to the mating ring, causing larger amounts of leakage. Stationary type mechanical seals are the ones with a spring or bellows attached to the fixed ring. The stationary

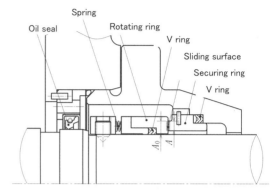

Fig. 6.23 Unbalance type mechanical seal with rotating ring

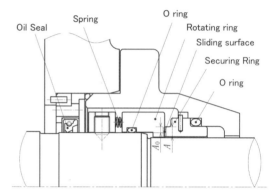

Fig. 6.24 Balance type mechanical seal with rotating ring

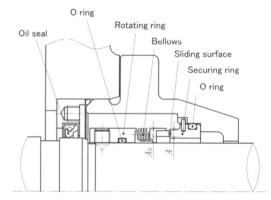

Fig. 6.25 Unbalance type mechanical seal with bellows rotating ring

type is less susceptible to the effect of centrifugal force and is therefore suited for compressors used under high speed rotation condition.

Another method of classifying mechanical seals is by the configuration of sliding faces, into single, double and tandem types. The single type, shown in Figs. 6.23 to 6.25, provides sealing function with a single pair of sealing faces and is the most commonly used one. For applications with tighter leakage restriction criterion, the double type or the tandem type, both of which use two pairs of sliding faces, may be selected. The double type, shown in Fig. 6.26, have two pairs of sliding faces one behind the other and in opposing orientations (fixed face – rotating face – rotating face – fixed face). To enhance gas leakage control reliability, fluid is normally injected between the two seals for sealing and cooling. The double type may also be employed for better compatibility with high circumferential speed. While the single type uses an oil seal to partition the seal box inside the compressor, lip abrasion may occur where the circumferential lip speed exceeds the allowable value. Therefore, compressors that are operated at high circumferential speed may use a double seal configuration. In that case, the above-mentioned oil seal inside the compressor is replaced with a mechanical seal for better leakage control reliability. A tandem type mechanical seal has two pairs of sliding faces one behind the other and in the same orientation (rotating face – fixed face – rotating face – fixed face), to increase reliability by providing two layers of sliding seals. For further details about mechanical seals, including their structure, performance and how they keep internal fluid from leaking into the atmosphere, refer to API (American Petroleum Institute) Standard API 682 or JIS (Japanese Industrial Standards) B 2405 "Mechanical seals - General requirements".

Very few compressor manufacturers produce mechanical seals in-house. Mechanical seals for compressors are selected based on the seal manufacturers' information and according to compressor operating conditions including the gas type to be compressed, the lubrication oil type, pressure and rotation speed. Shaft sealing features other than the mechanical seals include carbon packings and non-contact dry gas seals, some of which may be employed in oil-free compressors and chemical plant gas compressors. However, use of those seals are not common for twin screw compressors and therefore they will not be explained in any further details here.

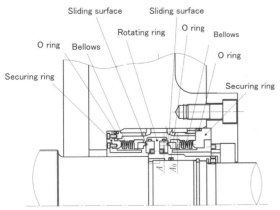

Fig. 6.26 Stationary type mechanical seal – double arrangement

6.4.10 Oil supply system

Figures 6.27 and 6.28 each show the oil flow of a standard oil supply system used in an oil-injected twin screw compressor. An oil-injected twin screw compressor may use either a pressurized (Fig. 6.27) or differential-pressure-based (Fig. 6.28) oil supply system. Selection between the two types should be made based on the operating conditions and system requirements such as the compressor application, pressure difference between high pressure and low pressure sides and the gas temperature.

Pressure at the oil feed header is normally maintained at approximately the discharge pressure plus 0.3 MPa in a pressurized oil supply system and the discharge pressure minus 0 to 0.1 MPa in a differential-pressure-based system. Pressure difference between the oil feed header pressure and the groove pressure is utilized to feed oil to various compressor components. Oil is supplied from the oil feed header to the radial and thrust bearings of the male and the female rotors, the balance piston and the mechanical seal to lubricate and also take heat away from these components, before being returned to a groove in which suction gas has just been trapped. In addition to lubricating these components, the oil feed header injects oil into an appropriate area of the rotor groove in which compression is underway. The injected oil in the groove lubricates the rotor surface, seals the leakage passage and cools the compressed gas, while getting pressurized and discharged through the discharge port together with the compressed gas. The stirred mixture of the compressed gas and oil flows into the oil separator, where gas and oil contents are separated from each other. The separated oil has heat taken away at the oil cooler and then gets filtered through the oil filter, before returning to the oil feed header and repeating the oil circulation cycle. In a pressurized-type system, oil is pressurized to the required level by an oil pump before being returned to the oil feed header.

In compressor models with a slide-valve-based capacity control system (see Section 6.4.11), oil needs to be fed to the unloader cylinder as well. In a pressurized oil supply system, oil to the unloader cylinder is supplied by the oil pump and then through a control-purpose solenoid valve. Even in models with a differential-pressure-based oil supply system, oil for the capacity control system may need to be separately pressurized by an oil pump. Also, compressors using sliding bearings need to have the slide bearings primed with oil for start-up. For this purpose, even differential-pressure-based oil supply systems may have an oil pump and an oil feed line connected to it, as shown in Fig. 6.28. In the example shown, a solenoid valve closes

Compressors for Air Conditioning and Refrigeration

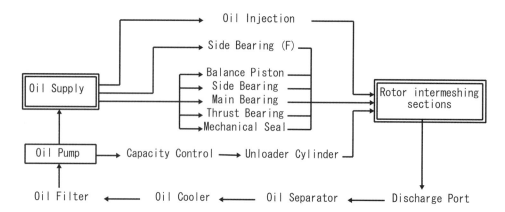

Fig. 6.27 Typical pressurized oil supply system

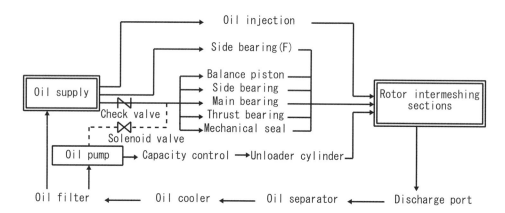

Fig. 6.28 Typical oil supply system using differential pressure

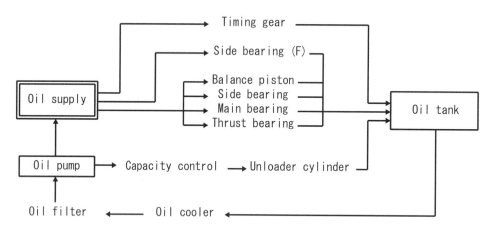

Fig. 6.29 Typical oil supply system for oil free compressor

off all the oil feed lines except for the one to the capacity control system in steady-state operation so that all the other components and areas will be lubricated by the force of differential pressure only.

On the other hand, an oil-free type compressor model, which does not have a differential-pressure oil supply system, have lubrication oil pressurized through an oil pump and fed to the compressor unit. Figure 6.29 shows an example of oil supply system for an oil-free compressor model. Oil is supplied from the oil feed header to the male and the female rotor bearings, the timing gear and other components for cooling and lubrication before flowing into an oil tank. After exiting from the oil tank, oil flows through an oil cooler and then through an oil filter before being returned to the oil feed header by the oil pump.

Figure 6.30 illustrates the oil flow through an oil-injected type twin screw compressor. Oil is supplied from the oil feed header to the mechanical seal, the thrust and radial bearings of the male and female rotors and the balance piston before being directed into the groove in which compression is taking place.

Various types of oil may be used as lubrication oil for twin screw compressors, including mineral oils, alkyl benzene (AB) oils, polyalkylene glycol (PAG) oils, polyol ester (POE) oils and polyvinyl ether (PVE) oils. Selection of the lubrication oil must be based on careful study as the oil's physical and chemical properties may affect the compressor performance, system efficiency and reliability. Oil viscosity varies significantly with temperature. Too much viscosity leads to increased mechanical loss while insufficient viscosity reduces bearing reliability and, in oil-injected type compressors, lowers compressor efficiency. Considering these, oil selection should be done by checking the information available from the oil manufacturer and selecting one that will dependably provide an appropriate level of viscosity in the anticipated in-operation oil temperature range. Also, oil for an oil-injected compressor must be chemically stable in the presence of the process or refrigerant gas as the oil and the gas will be mixed and compressed together. Stability under temperature variation must also be considered if the gas in the discharge stage is expected to become very hot. In cases where an oil-injected compressor is used in a closed cycle system like refrigeration systems, oil solubility with the gas can be critically important. Whether to select a soluble oil or insoluble oil influences the compressor efficiency (see Section 6.6.1 for details). When selecting a soluble oil, possible changes in the oil viscosity and other physical properties when the gas dissolves into the oil should also be considered. Note also that O-rings and other elastomer parts used in the lubrication system may detrimentally expand or harden depending on its compatibility with the lubrication oil. Furthermore, electrical insulation characteristics of the oil is an important factor in a semi-hermetic compressor where lubrication oil gets inside the motor casing. All of the above-described factors must be comprehensively studied to select the best possible oil for the system.

6.4.11 Capacity control mechanism

In a refrigerant compressor used for refrigeration and air conditioning purposes, capacity control function is an absolute necessity to accommodate thermal load variations. Also in compressors for other applications, capacity control features may be provided to alleviate start-up load torque or for other purposes. Requirements for a capacity control system to be used in a twin screw compressor include that power demand will be reduced during part load operation and that the effect of capacity control operation on the built-in volume ratio V_i will be as small as possible. It is also important that the capacity control system will not have detrimental effects on the noise and vibration characteristics of the machinery.

Various types of mechanisms have been developed for capacity control of a twin screw compressor. Each type has a specific set of characteristics and should be selected depending on the intended use. One of the simplest capacity control methods is to restrict the suction channel by a throttle valve. While the throttle valve method is not used very commonly as it entails more loss than other methods, it may be employed in some applications with a large V_i where a relatively large amount of loss is acceptable. Another capacity control method is a bypass method where the discharge gas is returned to the suction side, but it is very rarely employed on its own as this method does not provide any power requirement reduction effect. However, the bypass method may sometimes be used in conjunction with other capacity control methods, for special applications where very precise capacity control is required.

The most commonly used capacity control method in a twin screw compressor is the slide valve method. In this method, a slide valve that can move along the rotor axis is provided on the high pressure side of the rotor casing so that part of the rotor length which will be effective for compression is adjusted by valve movement to control the compressor capacity. Figure 6.31 schematically shows how this capacity control mechanism works. Under a full load operating condition, gas will be trapped in the suction port area which is positioned against the suction end of the rotor casing so that the groove volume gradually decreases to compress the gas as the groove moves over the full ro-

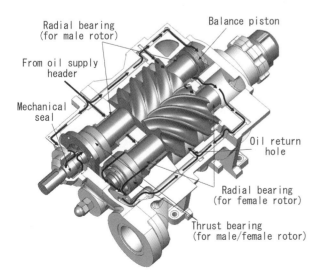

Fig. 6.30 Example of lubricant flow within compressor

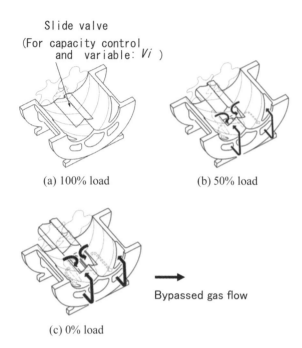

Fig. 6.31 Capacity control using slide valve

(a) 100% load
(b) 50% load
(c) 0% load

Bypassed gas flow

tor length. Under a part load operating condition, the slide valve shifts toward the discharge side to open the bypass channel that connects to the suction side. This causes a delay in gas trapping timing so that the effective displacement volume is decreased. The more the slide valve's suction end moves toward the discharge side, the more delay will be caused in gas trapping timing, further reducing the displacement volume. With this, the displacement volume is continuously adjustable from the full load to the minimum-load conditions.

In a commonly used structure, a radial discharge port is provided at the cusp section on the high pressure side of the slide valve so that compressor performance under the unload operating condition will be optimized while restricting the effect of capacity control on the built-in volume ratio V_i to a minimum. The opening of the axial discharge port under full load condition is designed to occur later than that of the radial discharge port, thus reducing the effect of V_i inconsistency during low-load operation. However, as the relative position between the suction-side end of the slide valve that regulates the displacement volume and the radial discharge port on the discharge-side end that regulates V_i is fixed by the slide valve length and cannot be individually changed, it is unavoidable that the capacity control action causes a certain amount of changes in V_i. Figure 6.32 shows the relationship between the amount of load and power requirement while slide-valve-based capacity control is operating. The vertical axis represents the ratio of part load shaft power requirement L to the full load shaft power requirement L_{max}, and the horizontal axis represents the ratio of the part load displacement volume Q to the full load displacement volume Q_{max}. The shaft power requirement for gas compression decreases as the displacement volume decreases. However, as the amount of mechanical loss does not decrease as much, the ratio of mechanical loss to the shaft power requirement will become larger. Also, the larger the degree of V_i inconsistency is, the greater the discharge loss will be. To summarize, the amount of shaft power requirement does not decrease as much as compressor load does when slide-valve-based capacity control is active, and as a result compressor efficiency becomes somewhat lower.

An advanced version of the slide valve method runs compressor capacity control and the adjustment of built-in volume ratio V_i separately. In this method, the capacity control slide valve is positioned, as shown in Fig. 6.33, away from the high-pressure side cusp section where the radial discharge port is located. Changes in V_i that occur as a result of the movement of the capacity control slide valve are adjusted by a separate slide valve dedicated to V_i adjustment, thus always maintaining an appropriate level of V_i even under part load operating conditions. However, this method is not actually employed very often due to the complex structure, manufacturing cost increase and potential reduction in reliability.

Other types of capacity control mechanisms include a slot valve method, shown in Fig. 6.34, and a lift valve method, shown in Fig. 6.35. The slot valve method uses multiple slots arranged in a row along the inner wall of the rotor casing and a turn valve provided also in the casing with a spiraled opening. As the turn valve is rotated, the slots will become open one by one to return the intermediate pressure gas to the suction side. Capacity control provided by this

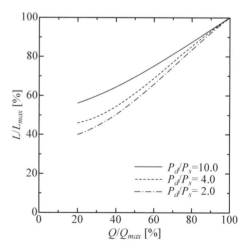

Fig. 6.32 Characteristics of capacity control with slide valve

Chapter 6 Twin Screw Compressors

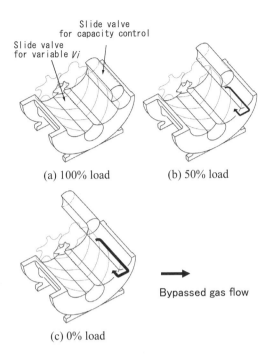

Fig. 6.33 Simultaneous control of capacity and built-in volume ratio

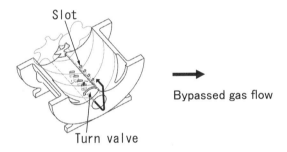

Fig. 6.34 Capacity control with slot gate valve

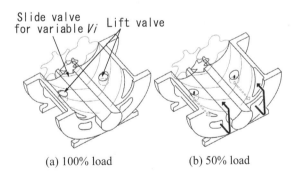

Fig. 6.35 Capacity control with lift valve

method, which depends on the phased opening of multiple slots, will be discontinuous and stepped in contrast to the continuous control provided by the slide valve method. The number of steps will depend on the number of slots provided in the casing. Due to structural restrictions, this mechanism cannot be easily accommodated with the built-in volume ratio V_i adjustment mechanism, and as a result performance under the part load operating condition will be lower than the slide valve method.

The lift valve method uses a plug-shaped lift valve provided on both the male and the female sides of the casing. These valves open and close to return the gas in the groove to the suction side for capacity control. The lift valve method provides a discontinuous stepped control similar to that by the slot valve method. As the lift valve method is simpler in structure, under certain conditions a combination of the lift valve method and a V_i adjustment mechanism may be able to offer part load performance similar to that by the slide valve method. There have been instances where such combination is actually employed in a compressor product.

Most of the capacity control methods that are commonly employed in actual products entail increase in built-in volume ratio (V_i) inconsistency under low-load operating conditions. This may lead to compressor efficiency reduction as well as increase in noise and vibration as a result of greater degrees of discharge pressure pulsation and shaft torque variation. Therefore, capacity control mechanism for a twin screw compressor must be one that provides an appropriate displacement volume for the intended application and is designed so that the system will run at full load under normal operating conditions. Make sure to avoid usages where the compressor may be operated for long period of time under an extremely low-load condition.

An inverter-based rotational speed control can also be used for twin screw compressor capacity control purpose. For further details about inverters, refer to Chapter 10, "Motors and Inverters". As capacity control by this method is provided solely by electrically controlling the rotational speed of the compressor drive motor by an inverter, no special mechanism is required on the compressor side. Also, fully continuous capacity control can be obtained by continuously varying the rotational speed. Another advantage of this rotational speed control method is that it does not cause built-in volume ratio (V_i) inconsistency like the slide valve and other mechanical capacity control methods do. As long as the rotational speed is controlled within an appropriate range, compressor efficiency during part load operation can be maintained at as good a level as that of full load operation.

However, there are a number of cautions concerning the use of rotational speed control for the capacity control of a twin screw compressor. Due to the existence of internal leakage passages, performance of a twin screw compressor can be significantly affected by the rotational speed of its drive shaft (for further details concerning this point, refer

Compressors for Air Conditioning and Refrigeration

to Section 6.6.1). Therefore, it is necessary to check before implementing a rotational speed control if the required efficiency can be obtained throughout the intended rotational speed range. In addition, considerations related to noise and vibration is important. A compressor and its system are composed of various structural elements, each of which has its own natural vibration frequency. When rotational speed control is active, the vibration frequency of the compressor unit, which is a main excitation source, may vary in a wide range, and therefore significant resonance may be caused depending on the relationship between the vibration frequency of the compressor and the natural vibration frequencies of other structural elements. Before implementing a rotational speed control on a twin screw compressor, exhaustively study the mass and stiffness of each structural element to make sure that no detrimental resonance may be caused in any part of the anticipated rotational speed range.

6.4.12 Balance piston

As already explained, when a twin screw compressor is operating, axial load caused by gas pressure will work on the male and female rotors in the direction from the discharge end toward the suction end. Due to the geometric properties of the rotor lobes, load on the male rotor will be greater than that on the female rotor. As the male rotor has less lobes than the female rotor, the male rotor's rotational speed will also be greater than that of the female rotor in proportion to the ratio of the numbers of their lobes. Due to these factors, load on the male rotor thrust bearing will be significantly larger than that on the female rotor thrust bearing. One method that is sometimes used to alleviate this thrust bearing load is to fit a balance piston at the end of the main rotor shaft so as to offset, by the force of oil pressure, part of the axial load working toward the suction side. The compressor shown in Fig. 6.3 has a balance piston fitted to the suction-side end of the male rotor shaft. With oil pressure applied to the space enclosed by a balance piston sleeve and a cover, the male rotor is pushed toward the discharge side so as to reduce axial load on the bearing.

When designing a balance piston, select a piston diameter that is sufficient to offset the amount of axial load that will work on the male rotor under the anticipated operating conditions based on the oil pressure on the piston. When selecting the piston diameter, note that the axial load on the male rotor is not constant and will vary according to the operating conditions. Under certain operating conditions such as part load operation, the balance piston force could become excessively high and may potentially generate a reverse-direction axial load on the bearing. Due to these factors, balance piston design must be done after fully examining the anticipated operating conditions.

6.5 Dynamic-mechanical Analysis

6.5.1 Motion of the mechanism, constraint forces and motion equations

When designing a twin screw compressor, it is critically important to estimate the anticipated load torque, bearing load and other operating quantities through a dynamic-mechanical analysis of design parameters including the ones explained in Section 6.3. This section takes an example of an oil-injected twin screw compressor, where torque is directly transmitted between the male and the female rotors, to explain the basic knowledge required for conducting such dynamic-mechanical analysis.

Figure 6.36 shows a typical example of perpendicular-to-axis cross section of a twin screw compressor mechanism. The male rotor shown on the left rotates clockwise by being driven by motor torque M_{mo}. The drive torque is transmitted to the female rotor to rotate it counterclockwise. As shown here, a twin screw compressor is a purely rotational mechanism with no reciprocating motion elements.

To facilitate the dynamic-mechanical analysis, a pair of Cartesian coordinate systems, (x_1, y_1, z_1) and (x_2, y_2, z_2), with their respective origins located at the two rotors' rotational axes, O_1 and O_2, is defined here. All coordinate values with a subscript 1 are associated with the male rotor and those with a subscript 2 are associated with the female rotor. Axes z_1 and z_2 each match the rotational axis of the respective rotor shaft, as shown. Components of the gas force-generated radial load in the (x_1, y_1) and (x_2, y_2) directions are presented as (F_{1x}, F_{1y}) and (F_{2x}, F_{2y}) while the load torques are presented as M_{g1} and M_{g2}.

The amount of bearing load caused by gas pressure P in the compression chamber shown shaded can be obtained relatively easily, as long as the coordinates for point A on the seal line, (x_{1A}, y_{1A}) and (x_{2A}, y_{2A}), and also for intersection B with the rotor casing, (x_{1B}, y_{1B}) and (x_{2B}, y_{2B}), are obtained, by integrating the forces working on the projected plane. Analysis concerning the micro-elements in the depth direction, dz_1 and dz_2, can be done as follows: The radial load components in the (x_1, y_1) and (x_2, y_2) directions, namely (dF_{1x}, dF_{1y}) and (dF_{2x}, dF_{2y}), can be obtained by the following equations:

$$dF_{1x} = \int_B^A P dy_1 dz_1 = P(y_{1A} - y_{1B})dz_1 \\ dF_{1y} = \int_B^A P dx_1 dz_1 = P(x_{1A} - x_{1B})dz_1 \Bigg\} \quad (6.3)$$

$$dF_{2x} = \int_B^A P dy_2 dz_2 = P(y_{2B} - y_{2A})dz_2 \\ dF_{2y} = \int_B^A P dx_2 dz_2 = P(x_{2B} - x_{2A})dz_2 \Bigg\} \quad (6.4)$$

Load torques, dM_{g1} and dM_{g2}, can be obtained as follows[14]:

$$dM_{g1} = -\int_B^A x_1 \cdot P dx_1 dz_1 + \int_B^A (-y_1) \cdot P dy_1 dz_1 \\ = \frac{P}{2}\{(-x_{1A}^2 + x_{1B}^2) + (-y_{1A}^2 + y_{1B}^2)\}dz_1 \quad (6.5)$$

$$dM_{g2} = -\int_A^B (-x_2) \cdot P dx_2 dz_2 + \int_A^B y_2 \cdot P dy_2 dz_2 \\ = \frac{P}{2}\{(x_{2B}^2 - x_{2A}^2) + (y_{2B}^2 - y_{2A}^2)\}dz_2 \quad (6.6)$$

By integrating the above equations in the z_1 and z_2 axis directions and summing up the results for all the chambers, the final radial load components in the (x_1, y_1) and (x_2, y_2) directions generated by gas pressure, that are (F_{1x}, F_{1y}) and (F_{2x}, F_{2y}), and also gas-compression load torques, M_{g1} and M_{g2}, can be obtained.

To facilitate power transmission from male rotor rotation θ_1 to female rotor rotation θ_2, a commonly employed technique is to modify the rotor profile so that transmission will take place in the vicinity of the intersection between the seal line and the pitch circle (see Section 6.3.2). Figure 6.37 (a) shows a cross section of the power transmission mechanism between the two rotors. Figure 6.37 (b) shows the same mechanism, with the two rotors shown separately to clarify the motion equation. Constraint forces that appear on the male rotor shaft are represented as R_{1x} and R_{1y} and those that appear on the female rotor shaft are represented as R_{2x} and R_{2y}. Also, constraint forces that appear on power

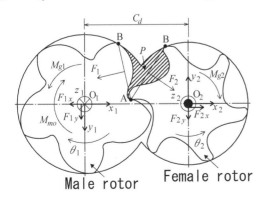

Fig. 6.36 Rotor profile in transverse section

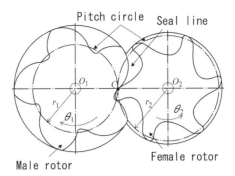

(a) Work transmission mechanism

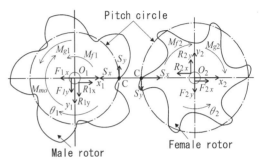

(b) Forces and moments on each rotor

Fig. 6.37 Forces and moments acted on male and female rotor

transmission point C are represented as S_x and S_y. From the force balance equations related to the (x_1, y_1) and (x_2, y_2) axis directions, the following can be obtained:

$$R_{1x} - F_{1x} - S_x = 0 \\ -R_{1y} + F_{1y} - S_y = 0 \Bigg\} \quad (6.7)$$

$$-R_{2x} + F_{2x} + S_x = 0 \\ R_{2y} - F_{2y} - S_y = 0 \Bigg\} \quad (6.8)$$

And, from the moment balance equations about the rotational axes, the following can be obtained:

$$-I_1 \ddot{\theta}_1 + M_{mo} - M_{g1} - M_{f1} - S_y \cdot r_1 = 0 \quad (6.9)$$

$$-I_2 \ddot{\theta}_2 + S_y \cdot r_2 - M_{g2} - M_{f2} = 0 \quad (6.10)$$

Here, I_1 and I_2 are moments of inertia about the rotational axes of the male and the female rotors respectively, and r_1 and r_2 are the male and female rotors' pitch circle radii. M_{f1} and M_{f2} are frictional torques working on the rotational axes, including those occurring at the thrust bearings.

A condition concerning S_x and S_y is that the resultant force of the two will as a general rule work in the direction normal to the lobe surface, as expressed in the following

Compressors for Air Conditioning and Refrigeration

equation:

$$S_x = S_y \tan \alpha \qquad (6.11)$$

where α is the inclination angle of the lobe surface. Furthermore, the relationship between θ_1 and θ_2 is determined by the ratio of pitch circle radii, r_1 / r_2, as follows:

$$\theta_2 = \frac{r_1}{r_2} \theta_1 \qquad (6.12)$$

S_y can also be obtained by using Eqs. (6.10) and (6.12), as follows:

$$S_y = \frac{I_2 \dfrac{r_1}{r_2} \ddot{\theta}_1 + M_{g2} + M_{f2}}{r_2} \qquad (6.13)$$

By substituting the above into Eq. (6.9), the following equation of motion about male rotor can be obtained:

$$\left\{ I_1 + I_2 \left(\frac{r_1}{r_2} \right)^2 \right\} \ddot{\theta}_1 =$$
$$M_{mo} - \left\{ M_{g1} + M_{f1} + \frac{r_1}{r_2} (M_{g2} + M_{f2}) \right\} \qquad (6.14)$$

The left side of the equation is inertia terms, while the first term on the right side represents the motor torque and the second and subsequent terms represent load torques. Of the load torques, M_{g1} and M_{g2} are the torques for gas compression and all the other torques are those induced by mechanical friction.

The male rotor's rotational behavior θ_1 can be obtained from the above equation of motion. Constraint forces (S_x, S_y) at the power transmission point C can be obtained from Eqs. (6.13) and (6.11). The male and female rotor bearing loads, (R_{1x}, R_{1y}) and (R_{2x}, R_{2y}), can be obtained from Eqs. (6.7) and (6.8) respectively.

$$\left. \begin{array}{l} R_{1x} = F_{1x} + S_x \\ R_{1y} = F_{1y} - S_y \end{array} \right\} \qquad (6.15)$$

$$\left. \begin{array}{l} R_{2x} = F_{2x} + S_x \\ R_{2y} = F_{2y} + S_y \end{array} \right\} \qquad (6.16)$$

6.5.2 Equation of energy and mechanical efficiency

By multiplying Eq. (6.14) of motion by the angular velocity and then integrating the result over one rotor shaft rotation, an equation of energy can be obtained. As the integral for the inertia term will be zero in a steady-state operation, the energy equation can be derived as:

$$E_{motor} = E_{gas} + E_{shaft} \qquad (6.17)$$

where

$$E_{motor} = \int_{1rev} M_{mo} \dot{\theta} dt$$
$$E_{gas} = \int_{1rev} \left(M_{g1} + \frac{r_1}{r_2} M_{g2} \right) \dot{\theta} dt \qquad (6.18)$$
$$E_{shaft} = \int_{1rev} \left(M_{f1} + \frac{r_1}{r_2} M_{f2} \right) \dot{\theta} dt$$

Equation (6.17) indicates that motor supply energy E_{motor} will be consumed as gas compression energy E_{gas} and as frictional energy E_{shaft} at the male and female rotor and thrust bearings. Mechanical efficiency η_m can be calculated by the following equation:

$$\eta_m = \frac{E_{motor} - E_{shaft}}{E_{motor}} \qquad (6.19)$$

6.5.3 Mechanical excitation force and vibration of compressor body

With all the forces and moments working on the compressor unit expressed as x- and y-axis direction forces F_x and F_y and torque M_z about the z axis, the following equations can be obtained:

$$\begin{array}{l} F_x = 0 \\ F_y = 0 \\ M_z = -\left\{ I_1 + I_2 \left(\dfrac{r_1}{r_2} \right)^2 \right\} \ddot{\theta}_1 \end{array} \qquad (6.20)$$

This is the result of taking the best static and dynamic balances from the principle of partial balancing. Here, all the internal force terms such as the gas force, frictional forces and torques have been eliminated, leaving only inertia forces and torques. Vibration of the compressor unit is caused as a result of excitation induced by these unbalanced inertia forces and torques.

Vibrations of the compressor unit can be determined in exactly the same way as for reciprocating, rolling piston and scroll compressors. The matrix [F] of excitation forces working on the compressor unit about the center of gravity can be given by Eq. (3.37), and the equation of motion that describes the vibration can be given by Eq. (3.33). The numerical solution for the motion equation can also be obtained in exactly the same way.

6.5.4 Lateral and torsional critical speeds

As a twin screw compressor entails a relatively high shaft rotational speed for a positive displacement compressor, rotor rotation behavior must be analyzed in the design stage to fully evaluate lateral and torsional critical speeds. In all machines with a rotating shaft system like a screw rotor, a noticeably greater level of vibration sometimes occurs in a specific rotational speed range when rotational speed is increased or decreased. This is the effect of resonance that occurs when the natural vibration frequency of the rotating shaft coincides with the rotational speed, and such speed is called "critical speed". However, if other excitation sources than the rotational speed exist in the operating mechanism, such as rotor-meshing collision of a twin screw compressor, the vibration frequencies of such excitation sources can also be a resonance factor.

A rotating shaft system has two types of natural vibration frequencies, that are lateral and torsional vibration frequencies. The lateral vibration frequency relates to the compressor unit only. Take measures in the compressor design stage to avoid such lateral frequency-induced resonance. On the other hand, torsional vibration frequency is a natural frequency of the entire rotational system including motor output shafts and couplings, combustion engines and other drive sources. In the case of open-type compressors, torsional vibration frequency of the compressor unit alone normally will not be evaluated. A standard practice is to identify the moments of inertia and spring constants of the compressor components in the compressor design stage and then select appropriate types of motors and shaft couplings in the system configuration stage so as to avoid or minimize resonance. However, for semi-hermetic and fully hermetic compressor models that have a built-in drive motor, torsional vibration frequency must also be evaluated in the compressor design stage.

(1) Evaluation of lateral critical speed

Analysis of lateral natural vibration frequency is commonly done based on the transfer matrix method or the finite element method. Technical details of these analysis methods can be obtained from rotation dynamics literature. When analysis is done with the bearing support stiffness as a variable, the result will be the natural vibration frequency for each of the anticipated vibration modes. Here, "vibration mode" refers to in what lateral deformation pattern the rotor shaft will turn. While theoretical analysis gives higher vibration modes and their natural frequencies, what really matters in actual resonance experience is the primary to tertiary vibration modes. The "critical speed diagram" of Fig. 6.38 shows changes in the critical speed and vibration modes relative to the level of bearing support stiffness[15]. The horizontal axis represents the rotor support stiffness and the vertical axis represents the critical speed, on which the primary, secondary and tertiary critical speeds are plotted. Also included in the diagram are schematics of vibration modes predicted at various points. Types of shafts where shaft deformation predicted in the ranges up to the compressor's rated rotational speed can be ignored are classified as "rigid shaft". On the other hand, the types of shafts where such deformation cannot be ignored are classified as "flexible shaft". Whether a machine shaft should be considered a rigid or flexible shaft depends on its relative relationship with the critical speed. In the diagram shown, areas below the broken line are the rigid shaft region and areas above the same are the flexible shaft region.

As the rotors of a twin screw compressor need to be designed so that they will operate in the rigid shaft region, it is necessary to plot results of the natural vibration frequency analysis into the critical speed diagram to check if the rotors will always operate in the rigid shaft region under the bearing support stiffness conditions proposed for the project. Determination of whether a shaft is in the rigid or flexible shaft region at the primary and secondary critical speeds should be done based on what vibration mode (deformation pattern) is obtained from analysis. If the shaft remains straight, then it is in the rigid shaft region. If the shaft deflects, it is in the flexible shaft region. If the intended operating speed is close to a critical speed, vibration response analysis or other study should be implemented to

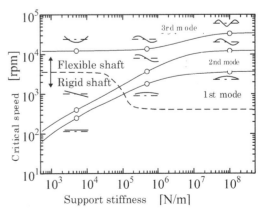

Fig. 6.38 Critical speed diagram

determine how much vibration increase is expected in the vicinity of the critical speed before selecting the allowable rotational speed range. A common practice is to set the operating speed not more than 80 % of the critical speed.

(2) Evaluation of torsional critical speed

As already explained, the torsional natural frequency is the coupled natural frequency of the entire system including motor output shafts and couplings and is determined based on all the moments of inertia and spring constants involved. Figure 6.39 shows an example of torsional vibration analysis model for a motor-driven compressor. Torsional vibration analysis comprises modeling the entire rotational shaft system based on its moments of inertia and spring constants and then carrying out calculation by applying the transfer matrix method or the finite element method in the same way for lateral vibration analysis. The calculation result will be the torsional natural frequency and the vibration mode at that frequency. Here, the "vibration mode" refers to in what torsional deformation pattern the shaft will turn at that vibration frequency.

Torsional resonance occurs when the shaft's natural frequency coincides with the variable torque component. In a male-rotor-driven twin screw compressor with a four-male-lobes and six-female-lobes configuration, four cycles of torque variation occur per one drive shaft rotation and the torsional critical speed will be a rotational speed corresponding to one-fourth of the torsional natural frequency (for further details about torque variation, refer to Section 6.7.1). Also concerning the torsional critical speed, it is necessary to identify the frequency of the variable torque component of the motor and make sure that it does not match the natural frequency of the rotational shaft system.

Except in cases where the vibration level is exceptionally high, torsional vibration is not very apparent to human observation or measurements in any form other than increased torque variation, and therefore it can be easily missed or overlooked by ordinary vibration measurement techniques. This in some cases results in delay in detecting a torsional vibration until it causes noise or damage to the shaft coupling. Therefore it is very important, in the shaft coupling and motor selection stage for an open type compressor and in the design stage for semi-hermetic and fully hermetic compressors, to check whether or not any point of potentially detrimental resonance is present in the system.

6.6 Operating Characteristics and Applications

6.6.1 Performance characteristics

Performance of a twin screw compressor varies significantly with the operating conditions such as the operating pressure ratio, rotational speed and oil supply temperature. Like most other fluid machinery, a twin screw compressor can provide its best performance only when its design specifications are appropriate for the operating conditions. The following paragraphs explain about the generally anticipated performance characteristics of a twin screw compressor commonly used for refrigeration or air conditioning purposes, on the condition that its design specifications are appropriate for the refrigerant gas and the operating conditions.

The built-in volume ratio V_i of a twin screw compressor will as a general rule be fixed by the position and shape of its discharge ports and is not changeable except in models equipped with a variable V_i mechanism. As explained in Section 6.4.6, it is necessary, at the beginning of designing or selecting the compressor, to identify what would be the best V_i for the system requirements and to reflect such V_i in the compressor specifications. Although models with a variable V_i allow some degree of subsequent changes, it is still important to check, when selecting the compressor, that its basic V_i is consistent with the system's operating conditions. Figure 6.40 shows an example of performance characteristics data of twin screw compressors used for refrigeration and air conditioning applications. The horizontal axis represents the operating pressure ratio. The vertical axis represents adiabatic efficiency η_{ad}, volumetric efficiency η_v, and shaft power requirement L. Each of the vertical lines in the graph represents a point where the $(P_d/P_s)^{1/\kappa}$ calculated from the operating pressure ratio coincides with the compressor V_i.

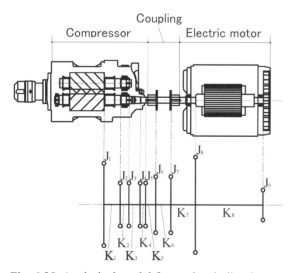

Fig. 6.39 Analytical model for torsional vibration – example

Chapter 6 Twin Screw Compressors

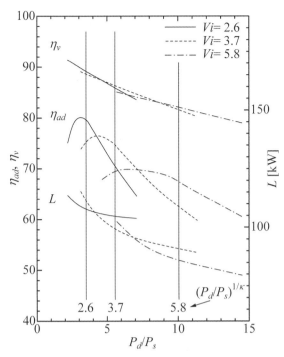

Fig. 6.40 Typical characteristics of twin screw compressor performance

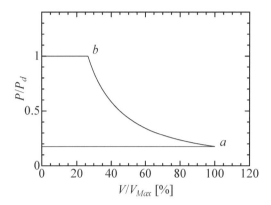

Fig. 6.41 *P-V* diagram under optimum operating condition

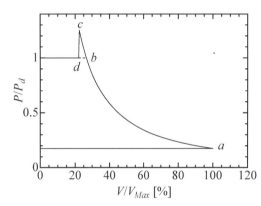

Fig. 6.42 *P-V* diagram for excessive compression (over-compression)

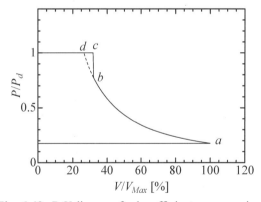

Fig. 6.43 *P-V* diagram for insufficient compression (under-compression)

The adiabatic efficiencies of twin screw compressors of which the built-in volume ratio (V_i) is 2.6, 3.7 and 5.8 respectively are described, in the operating pressure ratio range of up to 15.0, by three overlapping curves shown. Each of these V_i-specific curves indicates high levels of adiabatic efficiency in the ranges where the system-required V_i and the compressor-based V_i are close to each other. The greater the V_i inconsistency (gap between the system requirement and the compressor specification) gets, the less the adiabatic efficiency will be. This is because V_i inconsistency causes over-compression or under-compression of gas in the groove at the time of discharge, leading to greater loss. Figures 6.41 to 6.43 show conceptual *P-V* diagrams showing different states of V_i consistency/inconsistency. Figure 6.41 shows a state of V_i consistency. The discharge port opens just when the system-required discharge pressure is reached, followed by a constant-pressure discharge. On the other hand, Figs. 6.42 and 6.43 show over-compression and under-compression states respectively, both occurring as a result of V_i inconsistency. Figure 6.42 shows a situation where an overshoot in groove pressure occurs as a result that the discharge port does not immediately open when the groove pressure has reached the discharge pressure level. Area "bcd" represents the resulting loss. On the other hand, Fig. 6.43 shows a situation where gas from the high pressure side flows back into the groove as soon as the discharge port opens. The reverse flow increases the groove pressure to the discharge level. As the flowed-back gas needs to be discharged again, area "bcd" represents the resulting loss. As shown above, it is critically important for maintaining good efficiency to make sure that the compressor's internal volume ratio V_i is appropriate for the operating pressure ratio of the system.

When the effect of operating pressure ratios on all the V_i-specific curves shown in Fig. 6.40 is closely examined, it is apparent that, even where an appropriate V_i has been selected for the operating conditions, the compressor's adiabatic efficiency tends to decrease as the pressure ratio increases. A major cause for this is that internal leakage increases as the pressure ratio increases.

Figure 6.44 shows the relationship between the rotational speed and performance. For each of the V_i = 2.6, 3.7 and 5.8 patterns, and given that the operating pressure ratio is appropriate and that all the other conditions remain identical, increasing the rotational speed apparently improves adiabatic and volumetric efficiencies. This is mainly because that increase in the rotational speed causes the amount of internal leakage relative to the displacement volume to decrease. As excessive rotational speed increase will result in efficiency reduction due to increase in discharge and mechanical losses and also will detrimentally affect bearing reliability, there is a limit to how much the rotational speed can be increased in an appropriate manner.

Performance of a twin screw compressor will also be affected by the amount and temperature of oil supplied. This is especially significant in oil-injected type compressors where oil is directly introduced into compression chambers. In an oil-injected type compressor, oil injected into the compression chambers helps moderate gas temperature increase and also reduce internal leakages by its sealing effect. However, increase in the oil temperature will cause the gas temperature to rise as well, leading to more compression loss. The higher temperature will also cause decrease in oil viscosity, which lowers the oil's sealing effect and thereby increases internal leakage loss, resulting in lower efficiency. But a higher oil temperature may also contribute to less mechanical loss as it alleviates the stirring loss that occurs when the injected oil is stirred by rotor rotation and ejected through the discharge port and also the oil friction loss that occurs at the bearings. The amount of oil supply also has similarly conflicting effects: feeding more oil may contribute to efficiency improvement by reducing gas temperature increase and enhancing the oil's sealing effect, while having more oil may also increase mechanical and stirring losses. Therefore, it is critically important that lubrication oil is supplied in the right amount and at the right temperature. Even in the case of an oil-free compressor, it also needs oil to be supplied to its timing gear and bearings, where variation in the degree of stirring loss caused by changes in the oil amount and temperature can also affect compressor performance.

Another critical factor that affects the performance characteristics of a twin screw compressor is the configuration of the system it is installed. Although the following paragraphs may not specifically relate to the compressor unit, they include important information when considering the overall efficiency of a refrigeration system where a twin screw compressor is installed. Figure 6.45 shows the circuit diagram of a refrigeration system with an economizer. An economizer is a feature that enhances refrigeration capacity by directing the condensed high-pressure liquid refrigerant into an intermediate cooler and then routing part of that refrigerant into an expansion valve for depressurization and gasification, so that the evaporator receives refrigerant with lower temperature than otherwise. The part of refrigerant gas that has evaporated in the intermediate cooler will be directed to the economizer port located in the middle of the compression chamber in the compressor, where it is fed into the intermediate pressure region of the system to minimize volumetric efficiency reduction and shaft power requirement increase. Figure 6.46 shows the characteristics of a twin screw compressor that utilizes an economizer feature. The vertical axis represents the rate of change referenced against a single-stage compressor without an economizer system. As shown in the figure, the lower the evaporation temperature gets, the larger the refrigeration capacity will be due to the effect of the economizer system, which more than compensates for the increase in shaft power requirement. As a result, 5 to 18% COP improvement is obtained in the evaporation temperature range of -25 to -50°C.

Figure 6.47 shows the circuit diagram of a refrigeration system with a liquid refrigerant injection feature. In a normal refrigeration system, lubrication oil that has become hot

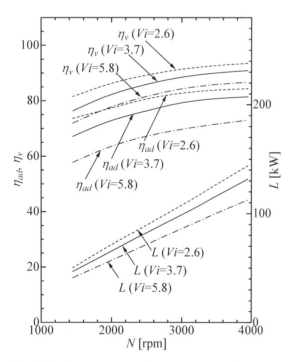

Fig. 6.44 Compressor performance vs running speed

Chapter 6 Twin Screw Compressors

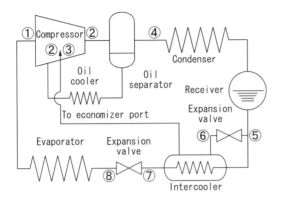

(a) Schematic diagram

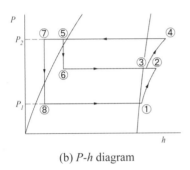

(b) *P-h* diagram

Fig. 6.45 Schematic diagram and *P-h* diagram for economizer system

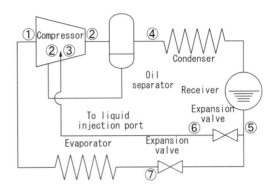

(a) Schematic diagram

(b) *P-h* diagram

Fig. 6.47 Schematic diagram and *P-h* diagram for liquid injection system

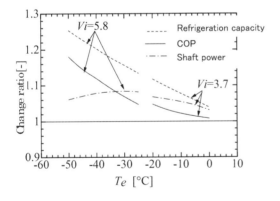

Fig. 6.46 Performance characteristics for economizer system

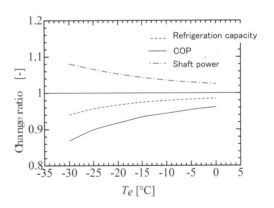

Fig. 6.48 Performance characteristics for liquid injection system

by absorbing heat from the refrigerant gas in the compressor will be separated from the refrigerant gas through the oil separator and will then have its heat taken away in the oil cooler before entering the compressor again (see Fig. 6.27). On the other hand, a refrigeration system fitted with a liquid refrigerant injection feature directs part of the condensed liquid refrigerant to the compressor and have it directly injected into the compression chamber. The injected refrigerant evaporates in the chamber and thereby lowers the gas temperature there. With this, lubrication oil is also lowered

in temperature and does not need to be cooled by an oil cooler. A major advantage of this liquid refrigerant injection feature is that it eliminates the need for an oil cooler and its associated cooling circuit in the refrigeration system. Figure 6.48 shows the characteristics of a twin screw compressor using a liquid refrigerant injection feature. The figure indicates a significant reduction in refrigeration capacity as well as increase in the shaft power requirement as the evaporation temperature gets lower. Under the condition where the evaporation temperature is -25°C, COP drops 10%. Due to

129

this tendency, the liquid refrigerant injection feature is more often employed in air conditioning systems that operate in high evaporating temperature ranges or systems that are used in locations where the supply of cooling water is difficult. Also, users who prioritize equipment introduction cost reduction over running cost reduction may prefer to have a liquid refrigerant injection feature in the system.

Finally, oil type selection could also significantly affect the performance of a twin screw compressor. An important decision concerning an oil-injected twin screw compressor is whether to select an oil that is soluble with the gas to be compressed or to select one that is not. This selection between soluble and insoluble oils results in a significant difference in efficiency, especially in closed-cycle systems like a refrigeration or air conditioning system. An advantage of using a soluble oil in the system is that the oil circulation path can be designed simpler as the oil and the gas circulate in refrigeration cycle together. On the other hand, disadvantages of using a soluble oil include that when oil is injected into or returned to a groove where compression is underway the gas content that has dissolved in the oil will vaporize in the groove due to depressurization. The resulting groove pressure increase will add to compression power requirement and thus may detrimentally affect compressor performance. Therefore, the relationship between the gas and the lubrication oil is a critical factor in oil-injected compressor models.

The above are the major factors, both compressor-specific and system-wise, that influence the performance characteristics of a twin screw compressor for refrigeration and air conditioning purposes. Except for highly advanced models such as those for CO_2 supercritical systems, the above explanations cover most of the important qualitative performance tendencies generally experienced in commercially available twin screw compressors. As there are other product-specific parameters including such as the number of lobes, rotor profile, L/D_m and port shapes and positions, compressor selection must be made by carefully referring to the manufacturer information so as to choose the best product available for the intended application.

6.6.2 Capacity and pressure range

Figure 6.49 shows the general relationship between the capacity and the pressure range of a currently available oil-injected single-stage twin screw compressor. The horizontal axis represents the per-hour displacement volume when the compressor is run at standard rotational speed, and the vertical axis represents the operating pressure range.

A twin screw compressor operates with a relatively high

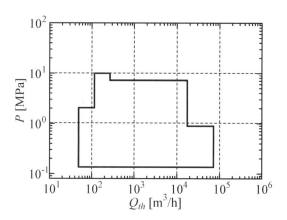

Fig. 6.49 Typical operating range of twin screw compressor

rotational speed of drive shaft for a positive displacement compressor and therefore can provide the same displacement volume with smaller outer dimensions compared to other types of compressor mechanisms such as reciprocating compressors. Due to this characteristic, twin screw compressors are suitable for applications that require a larger displacement volume. Of all the positive displacement compressors, twin screw compressors are the ones that are being employed for the highest capacity ranges. On the other hand, the efficiency of a twin screw compressor is significantly affected by its internal leakage passage area and therefore is not suitable for small capacity applications. In the capacity ranges below what is shown by a solid line in this figure, other positive displacement compressors such as reciprocating, rolling piston and scroll compressors are used more often as they provide higher efficiency in these low capacity regions. On the other hand, in the capacity ranges above those shown in the figure, turbo compressors, such as centrifugal compressors, that are better suited for larger capacities are used more often than positive displacement compressors. However, the capacity range that twin screw compressors are now being applied to is a result of the currently available efficiency, manufacturing cost, performance characteristics and other factors that are chosen over other types of compressors. With new market needs and advances in technology, the capacity range is likely to change further in the future.

While twin screw compressor models with a wide variety of operating pressure ranges are available, the most commonly used pressure ranges are those around 3 MPa for refrigeration and air-conditioning compressors and around 5 MPa for process gas compressors. But models with pressure ranges higher than 10 MPa have been developed in recent years for use in new applications such as CO_2 heat pump systems.

6.6.3 Single-stage compression and multiple-stage compression

As explained in Section 6.6.1, the adiabatic efficiency of a twin screw compressor generally decreases as the operating pressure ratio increases. This is due to a number of factors, including compression power requirement increase caused by more inter-groove leakages inside the compressor and volumetric efficiency reduction caused by the re-expansion of gas that has leaked from the compression side to the suction side through rotor meshing clearances. In addition, as gas temperature gets higher due to increase in the pressure ratio, transfer of heat from the compression gas to suction gas, by thermal conduction through the casing or through internal leakage passages, causes increase in the suction gas temperature as well, which is another potential factor that contributes to volumetric and adiabatic efficiency deterioration. Compression gas heating also accelerates lubrication oil deterioration, which potentially affects component reliability such as that of bearings and motor.

As explained above, operating a twin screw compressor under an excessively large pressure ratio is clearly detrimental to compressor efficiency and reliability. To accommodate very large pressure ratios, compressor designs with two or more compression stages are often employed. Achieving compression in multiple stages helps reduce the pressure ratio per stage, thereby avoiding above-described detrimental effects on compressor efficiency and reliability. Also, additional cooling of the discharge gas in the lower stage helps keep the higher stage discharge temperature from rising too much.

To select between single-stage and multiple-stage compression in the system design phase, start by determining, based on the relationship between pressure ratio and efficiency like the one shown in Fig. 6.40, whether or not the required level of efficiency can be achieved by single-stage compression. If the efficiency achievable by single-stage compression does not seem to be sufficient, first consider adding efficiency enhancement features such as an economizer to a single-stage system. If it is determined that the required efficiency still cannot be achieved by a single-stage system even with additional enhancement features, then opt for a multiple-stage compression system.

As the efficiency requirement varies depending on for what purpose the compressor is used and what are the user requirements, it is not easy to determine the threshold pressure ratio between where a single-stage system can suffice and where a multiple-stage system is more desirable. The commonly applied rule for oil-free twin screw compressors is to select single-stage compression for pressure ratios of up to approximately 3:1, and to select multiple-stage compression for pressure ratios higher than that. In the case of oil-injected models, higher pressure ratios are achievable even with single-stage compression as the lubrication oil has compression gas cooling effect. With oil-injected models, single-stage compression is typically selected for pressure ratios of up to approximately 10:1. For any higher pressure ratios, use of two-stage compression is generally considered desirable even for oil-injected models, but in some specific cases single-stage compression may still be used for pressure ratios of up to approximately 13:1 depending on the system requirements. As an actual example, consider a refrigeration system that uses R717 refrigerant (NH_3) and operates at a condensing temperature of 35°C. In this case, the key to selecting between a single-stage and a multiple-stage systems would be whether the evaporating temperature is higher or lower than approximately -30°C. If the evaporating temperature is expected to be lower, normally a two-stage compression should be selected.

6.6.4 Main applications

As explained in Section 6.2.3, twin screw compressors can be generally classified into oil-free and oil-injected types, each of which is applied to their respectably suitable applications based on their advantages and disadvantages. Oil-free types are mostly employed for applications where oil contamination avoidance is a high priority. Common applications for oil-free twin screw compressors include air compressors in electric and electronic industries like semiconductor manufacturing, pharmaceutical industries and food manufacturing. They are also favored as process gas compressors in chemical plants where oil contamination cannot be tolerated. In addition, with recently increasing focus on energy saving, more oil-free twin screw compressors are now being used for MVR system compressors. MVR, or "mechanical vapor recompression", refers to a technique of compressing vapor generated from industrial processes such as evaporation, drying, rectification and brewing, so as to use the latent heat of vapor more efficiently. The MVR technique is being applied to many manufacturing process including salt and sugar production, beer brewing and whiskey distillation. Figure 6.50 shows a system diagram where the MVR technique is applied to the wort kettle in a beer brewing process. Conventionally, steam from wort kettles used to be either directly emitted to atmosphere or condensed and drained off. Now with an MVR system, steam from the wort kettle is compressed by a twin screw compressor and supplied to the heat exchanger for the kettle in a high saturation temperature state so that the kettle will be

Compressors for Air Conditioning and Refrigeration

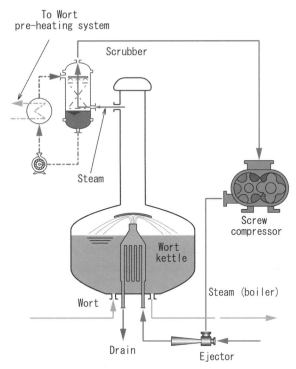

Fig. 6.50 MVR in brewery process

heated with the latent heat of the steam. This system significantly reduces steam consumption in beer brewing process. Figure 6.51 shows a steam-compressing twin screw compressor for an MVR system. As an MVR compressor processes water steam, it needs to be designed corrosion- and erosion-resistant. To satisfy these requirements, the rotors and the rotor casing are typically built with stainless cast steel.

On the other hand, oil-injected twin screw compressors are widely used for applications where a minute oil contamination in the discharge gas is tolerable. Oil-injected twin screw compressors are employed most commonly in general industrial compressed air supply systems as they are more efficient and cheaper to introduce than oil-free models. Twin screw compressors that are employed for refrigeration and air conditioning systems are also dominantly oil-injected. In the fields of industrial refrigeration and cold storage and commercial air conditioning, use of oil-injected twin screw compressors with capacity control function has started to spread gradually almost as soon as they were developed. In North America, Europe and other industrial refrigeration markets in the world, oil-injected twin screw compressors are now fast replacing reciprocating compressors. A great majority of industrial refrigeration compressors that are currently manufactured and employed for 50 kW or higher-capacity applications are oil-injected twin screw compressors. Oil-injected twin screw compressors are also being favored for various gas compression operations where minute oil contamination is tolerable, including gas turbine boosters that pressurize fuel for power generation gas turbines and crude gas compression and recovery systems at oil wells.

Also, even where oil contamination cannot be tolerated, applications that use gases that have a small molecular weight and thus are very leak-prone or those that have a high specific heat ratio and entails significant temperature increase when compressed may use oil-injected models. Use of an oil-injected compressor in non oil-tolerant processes can be achieved by using an advanced multiple-stage oil separation system to reduce oil contamination of the compressed gas to an absolute minimum. Oil-injected twin screw compressors are also employed in the field of aerospace engineering as hydrogen compressors for liquid hydrogen fuel production. Another newer application is in the elementary particle physics, as helium liquefying com-

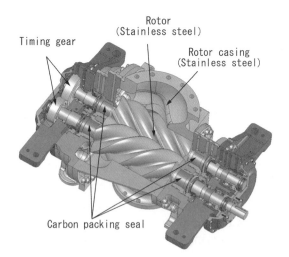

Fig. 6.51 Typical MVR compressor showing key elements

Fig. 6.52 Twin screw compressor unit for helium gas compression

pressors for large-scale particle accelerators. Such compressors are in use at the Fermi National Accelerator Laboratory ("Fermi Lab") in US, High Energy Accelerator Research Organization ("KEK") in Japan, the European Organization for Nuclear Research ("CERN") and other various research institutions throughout the world. Figure 6.52 shows the oil-injected two-stage-compression twin screw compressor system being used for a large scale hadron-collider-type accelerator at CERN.

6.7 Noise and Vibration

6.7.1 Vibration characteristics

Compared to other types of compressors, twin screw compressors are generally regarded as a low vibration machinery as their operating mechanism solely comprises rotational motion elements and do not contain any reciprocating elements. Vibrations from a twin screw compressor can be generally divided into those that originate in the compression mechanism, those that are related to the accuracy of rotary components and those that are related to the operating conditions. In addition, some vibrations are considered to be related to installation and maintenance conditions. When vibration from an actually operating compressor is analyzed, combinations of two or more of these factors are often observed. The following paragraphs explain about the vibration characteristics of a twin screw compressor.

Figure 6.53 shows an example of vibration measurements from a male-rotor driven, four-male-lobe and six-female-lobe twin screw compressor operating at 2,985 rpm. The vertical axis represents vibration acceleration and the horizontal axis represents time. Typical vibrations from a twin screw compressor operating under normal conditions describe a cyclic time-domain waveform like the one shown in the figure.

A twin screw compressor will, due to its structural characteristics, repeat a cycle of suction, compression and discharge the same number of times as the number of its lobes for every drive shaft rotation. The internal casing pressure, the drive shaft torque and the discharge gas pressure also undergo the same cyclic changes. Figures 6.54 and 6.55 show pressure pulsations of a compressor, measured in the casing and in the discharge channel respectively, with the same specifications as those of the one in Fig. 6.53 and operating at 2,985 rpm. In an oil-injected twin screw compressor, which has torque directly transmitted between male and female rotors, rotor meshing occurs the same number of times as the number of its lobes for every drive shaft rotation. The pressure and torque pulsations and the

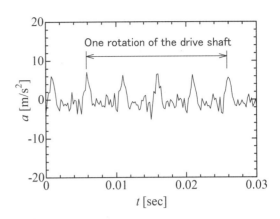

Fig. 6.53 Example of time-domain vibration wave form in axial direction (acceleration)

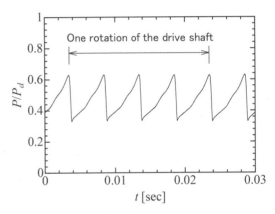

Fig. 6.54 Example of pressure fluctuation within compressor

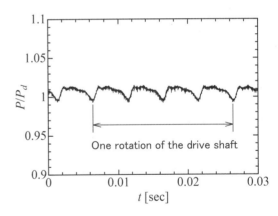

Fig. 6.55 Example of pressure fluctuation in discharge chamber

rotor meshing impacts described above constitute an excitation source, causing vibrations in the casing and the rotors. These are the types of vibrations originating in the compression mechanism and are a major factor determining the vibration characteristics of a twin screw compressor. Based on their generation mechanism, frequency f_n of the basic and harmonic components of vibration can be expressed by the following equation:

$$f_n = n \cdot Z_1 \cdot f_0 \quad (n=1,2,3,\cdots) \quad (6.21)$$

n: A positive integer
Z_1: Number of lobes on the driving shaft
f_0: Rotational frequency of the driving shaft

Figure 6.56 shows the spectrum analysis of the time-domain waveform shown in Fig. 6.53. A typical vibration acceleration spectrum of a twin screw compressor contains the basic component f_1 and the harmonic components f_n ($n \geq 2$) of the frequency expressed by Eq. (6.21). In the horizontal axis range shown, components f_1 to f_9 are observed. Higher harmonic components represent the degree of impact that the time waveform has, which is determined by the impacts caused by pressure variation and rotor engagements that occur in the cycle of suction, compression and discharge. Such harmonic components are clearly observed in normal operations in a frequency region of up to approximately 2 kHz, while harmonic components that are even higher may be observed under inappropriate operating conditions where the degree of pressure variation and the impact of rotor meshing contact are greater. Other excitation sources related to the compression mechanism include gear engagement impacts that are observed in speed increasing- or timing- gear-fitted oil free models. Vibrations caused by these excitations will appear in the vibration spectrum as the basic and higher harmonic components of the frequency obtained as the product of the drive shaft rotational speed and the number of gear teeth on the same shaft.

In addition to the above-described vibrations originating in the compression mechanism, vibrations related to the machining and assembly accuracy of rotating elements may also be observed. As the pair of rotors rotate at high speed,

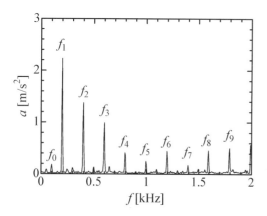

Fig. 6.56 Example of vibration spectrum (acceleration)

any unbalance in either of them will cause an excitation force to occur in time with its rotation cycle, resulting in vibrations. Any minute eccentricity in the rotor, where its rotor profile center does not match its rotational axis, will also be a vibration factor. The frequency of vibrations that occur as a result of defect in these rotational elements can be calculated from the rotation speeds of the driving and the driven shafts, as follows. By identifying the frequency that appears in the vibration spectrum, it is possible to determine whether the vibration originates in the driving shaft or in the driven shaft.

$$f_{drive} = \begin{cases} f_0 & (6.22) \\ n \cdot Z_1 \cdot f_0 \pm f_0 & (n=1,2,3,\cdots) \quad (6.23) \end{cases}$$

$$f_{driven} = \begin{cases} f_0 \cdot (Z_1/Z_2) & (6.24) \\ n \cdot Z_1 \cdot f_0 \pm f_0 \cdot (Z_1/Z_2) & (n=1,2,3,\cdots)(6.25) \end{cases}$$

f_{drive}: Frequency of a vibration originating from a defect in the driving shaft
f_{driven}: Frequency of a vibration originating from a defect in the driven shaft
Z_2: Number of lobes on the driven shaft

Frequencies expressed by Eqs. (6.22) and (6.24) will respectively be the same as the rotational frequencies of the driving and the driven shafts, and Eqs. (6.23) and (6.25) express the sidebands that occur in their basic (f_1) and harmonic (f_n; $n \geq 2$) components as a result of defects in the driving and/or driven shafts like the ones described above[16,17]. The vibration acceleration spectrum shown in Fig. 6.56 reveals a minute peak around 50 Hz. This is the primary rotational component of the male rotor and corresponds to frequency f_0 of Eq. (6.22). As compressors are produced by industrial process under a certain limited level of accuracy control, it is practically impossible to completely eliminate all the vibrations originating from such accuracy errors. However, it is critical, in the machining and assembly of rotors and other rotational elements of a compressor, to maintain a level of accuracy so that primary rotational vibrations like the ones described above are minimized so that they will not be detrimental to the overall compressor vibration quality. In addition to these primary rotational vibrations, vibrations originating from rotational element accuracy errors include those where excitation occurs as a result of rotational transmission errors caused by inaccurate rotor pitch or lead. Each of these exhibits a characteristic vibration spectrum and can be easily identified by spectrum analysis[17].

Operating conditions such as the pressure ratio and the

Chapter 6 Twin Screw Compressors

oil feed amount will also significantly affect the vibration of a twin screw compressor. If a compressor is operated under inappropriate conditions such as significantly mismatched built-in volume ratio V_i or excessive amount of oil injection, the resulting over- or under-compression or oil compression may give rise to excitation factors such as torque or load variations, causing abnormal vibrations. Also, if a compressor fitted with a capacity control mechanism like the slide valve explained in Section 6.4.11 is operated under part load condition, V_i inconsistency is more likely to occur due to its mechanical characteristics. This may result in greater vibration and/or increase in the harmonic components of the vibration spectrum compared to full load operation. Another factor concerning part load operation is that the absolute drive torque of the female rotor will be significantly smaller under part load but the degree of its variation does not decrease as much. This may lead to greater instability of the female rotor, causing rotor bouncing or severer collision between the male and the female rotors and resulting in abnormal vibration. In this type of situation, vibration waveforms with a high degree of impact will be observed, and the vibration spectrum will reveal higher harmonic components.

In addition to the factors described above, misalignment between the compressor drive shaft and the motor output shaft in an open-type compressor may also contribute to vibration. As shown in Fig. 6.7, an open type twin screw compressor has its drive shaft coupled to the motor output shaft by a shaft coupling feature. Coupling of the shafts is achieved by an alignment operation which aims at positioning the axes of the two shafts exactly on a single line. However, a truly perfect alignment is practically impossible and therefore a certain amount of misalignment will remain. Such misalignment between rotational shafts may be in various forms such as parallel (eccentric) or angular (declination) misalignment. Although up to a certain amount of misalignment can be absorbed by the shaft coupling feature, severer misalignment that is caused by alignment operation error or by other post-installation factors and is too great to be successfully absorbed by the shaft coupling may increase in-operation vibration. Misalignment-induced vibrations are often more clearly apparent in the axial direction. The vibration components will mainly be those of primary rotation, but greater misalignment may accompany higher harmonic components as well. Also, different types of shaft couplings may accompany their respectively characteristic spectra. Primary rotational vibration caused by misalignment may sometimes be confused with those caused by unbalance in rotating elements, but they are clearly distin-

guishable; misalignment-induced vibration will not increase or decrease with rotational speed but unbalance-induced vibration will be in proportion to the square of the rotation speed.

Twin screw compressors are basically a rigid and low-vibration machinery, but their use or installation under seriously inappropriate conditions or misalignment due to inadequate maintenance can cause detrimental vibration that affects component reliability. Therefore, compressors must be selected by fully evaluating the required operating conditions and choosing the best available product, and must also be used under appropriate operation and maintenance control.

6.7.2 Noise characteristics

"Noise" refers to vibrations from an acoustic source that are transmitted through air to reach a human ear, causing discomfort. Therefore, almost all the causes of twin screw compressor noise are associated with the types of vibrations explained in Section 6.7.1. Vibrations due to cyclic pressure variation in the grooves or due to rotor meshing impact will be transmitted to the rotor casing and other components exposed to the outside, to radiate from there into the open air as noise. These types of radiation noises will exhibit a noise spectrum including the basic and the harmonic components of the frequency obtained as the product of the drive shaft rotational speed and the number of rotor lobes.

In addition to these, pulsation of the discharge gas is another major noise source. The volumetric flow rate of gas passing through the discharge port can, if simplified with all the subsidiary factors such as discharge pressure overshoots or discharge port pressure loss ignored, be obtained from changing rate of compression chamber volume in the discharge stage. Figure 6.57 shows the volumetric flow rate of discharge gas, calculated by the above-described method, of a male-rotor-driven, four-male-lobe and six-female-lobe compressor operating at 2,985 rpm. Volumetric flow pulsation like the one apparent in this figure converts into a pressure pulsation at a specific impedance and spreads outward, where it is perceived as noise. In the figure, the chronological changes in the volumetric flow of discharge gas describe a saw-tooth waveform. Similarly to the above-described noises originating in the compression mechanism, the cycle frequency f_1 is the product of the drive shaft rotational frequency and the number of lobes. A spectrum analysis of this waveform would reveal the basic component f_1 and the higher harmonic components f_n ($n \geq 2$).

As an example, Fig. 6.58 shows noise measurements on a male-rotor-driven oil-injected twin screw compressor with

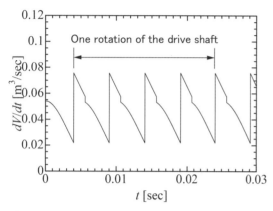

Fig. 6.57 Example of fluctuation in volume flow due to fluctuation in discharge pressure

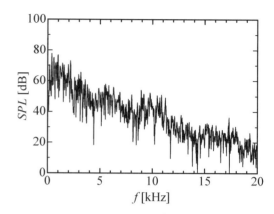

Fig. 6.59 Example of noise spectrum (Range: 20 kHz)

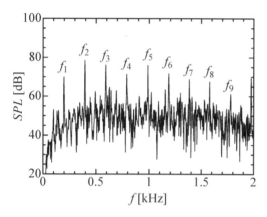

Fig. 6.58 Example of noise spectrum (Range: 2 kHz)

four-male-lobes and six-female lobes and operating at 2,985 rpm. A line spectrum of 199 Hz frequency corresponding to a 199 Hz component and its harmonic components is clearly apparent. Similar to vibration harmonics, these noise harmonic components do not attenuate much through to high frequency levels of over 1 kHz and cause high-frequency noises. As human ears are relatively sensitive to these frequency ranges, the resulting noises may be recognized as highly unpleasant.

In many cases, the actual operating noise will also contain fluid noises that the compressed fluid makes when it passes through suction and discharge channels. These fluid noises typically have characteristics similar to those of pink noise. As shown in Fig. 6.59, fluid noises become apparent as a continuous spectrum component that attenuates in inverse proportion to the frequency. In addition, vibrations that originate from rotor machining and assembly accuracy errors, inappropriate operating conditions or shaft misalignment, will have their corresponding components appear in the noise spectrum.

Twin screw compressors are a relatively low-noise machinery as they do not have reciprocating mechanical elements or repetitively impacting plate valves like those of a reciprocating compressor. However, as the use of twin screw compressors becomes widespread, more units are being installed to residential and commercial buildings in cities where noise and vibration are particularly undesirable and quiet operation is becoming an increasingly higher priority. To fulfill such needs, compressor manufacturers are actively working on better noise reduction, for example by changing the rigidity or weight of structural components and by conducting radiation noise analysis. The design of compressor unit itself, which is the main excitation source, is critically important for noise reduction, while peripheral components such as compressor base, piping and accessories may be equally important. Rigidity and natural vibration frequencies of these peripheral components should be fully examined to make sure that they will not resonate with the compressor unit. Separation of vibration sources by an appropriate use of vibro-isolation rubber or flexible joints can also be effective. With all possible noise reduction features in place, the compressor must be installed, operated and maintained in an appropriate manner so as to minimize noise from the compressor and its associated systems.

6.8 Production Technology

6.8.1 Screw rotor machining

Screw rotors can be classified as a helical gear with special tooth profile and therefore can be machined in generally the same way as gear wheels. Commonly used rotor groove machining techniques include hobbing, milling and grinding, which are conventional and have been widely used for decades. In addition to these, mill-turn, which are a relatively new technology, are being increasingly used in recent years. Each of these techniques are selectively used depending on the case. The following paragraphs explain about these screw rotor machining techniques.

(1) Hobbing machine

A hob, shown in Fig. 6.60, is a machine tool having a shape of a worm with cutting edges formed by several longitudinal grooves and also a relief along its screw thread. These hobs are mounted on a hobbing machine to cut gear teeth on mechanical components.

Figure 6.61 shows a rotor being cut by a hobbing machine, illustrating the machining motion and the position relationship. Rotor machining by hobbing is a type of gear generating process where the hob and the rotor workpiece rotate about their respective axes while the hob also linearly travels along the rotor axis. The combination of these motions creates the specified gear shape.

Hobbing cuts multiple rotor grooves at the same time by a worm-shaped hob. It is a highly productive process and also is less prone to pitch errors compared to other machining methods. A hobbed rotor has a characteristic, scale-like trace on its surface called "cutter marks". Examination of these cutter marks allows the operator to determine whether or not hobbing was done correctly and, in case of a defect, gives a clue to what has been wrong in the process. Hobbing is generally employed to produce rotors for smaller compressors.

(2) Screw rotor milling machine

Screw rotor milling machines are dedicated machines for screw compressor rotor production. While the machines are similar to thread milling machines in appearance, their characteristic feature is an indexer for rotor pitch control. Figure 6.62 shows one of the earliest CNC-controlled screw rotor milling machine developed in early 1980s[18].

A screw rotor milling machine sets the cutting tool at an angle against the rotor workpiece axis. The workpiece is driven to rotate and also to linearly travel along its axis in a synchronized manner to have each helical groove cut. While the workpiece is making this composite movement, the rotor cutter rotates about its axis to cut the required rotor profile. This screw rotor milling technique is classified as a kind of gear forming (as opposed to gear generation) process. Figure 6.63 shows the relative movement of and the position relationship between a rotor workpiece and the

Fig. 6.60 Hobbing tool for screw rotor

Fig. 6.62 Screw rotor milling machine with CNC (by the courtesy of Holroyd Precision Ltd.)[18]

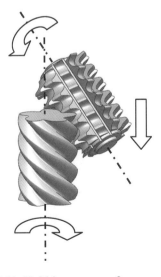

Fig. 6.61 Hobbing process for screw rotor

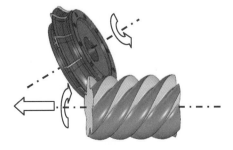

Fig. 6.63 Arrangement and relative motion between rotor blank and cutting tool

cutting tool. Figure 6.64 shows an actual workpiece being cut on a rotor milling machine.

Cutting tools that are used on a screw rotor milling machine can be classified into two types, the formed blade type like the one shown in Fig. 6.65 and the type composed of tungsten-carbide chips fitted to a cutter body like the one shown in Fig. 6.66. The formed blade types are used mostly for finish machining while the tungsten-carbide chips types are used for rough and intermediate machining. The tungsten-carbide chips type has chips fitted to a cutter body that has an axial cross section similar to that of a formed blade, so that the chips on a rotating cutter body will describe an envelope approximately the same as that described by a formed blade. A rotor surface machined by a tungsten-carbide cutter will have specific traces like the ones shown in Fig. 6.67, which are lines between adjacent chip paths. Positioning the chips on the cutter body with a circumferential offset helps mitigating the cutting resistance for efficient machining.

Compared to hobbing where multiple grooves are cut at the same time, a screw rotor milling machine cuts only one groove at a time and therefore requires angular indexing upon each groove cut. As a result, screw rotor cutters are inherently more prone to pitch errors than hobbing machines. However, recent improvements in technology have enabled highly accurate rotor machining by these screw rotor milling machines.

(3) Screw rotor grinding machine

As increasing demands for screw compressor performance calls for more precise rotors, use of screw rotor grinding machine is becoming more widespread. A screw rotor grinding machine uses a formed grinding wheel like the one shown in Fig. 6.68 to cut grooves on a rotor workpiece. The cutting edge of a grinding wheel is composed of an agglomeration of numerous abrasive grains. The amount of cut produced by a single grain is quite small, resulting in finer machined surface and making this technique suitable for finish machining.

The relative movement and the position relationship be-

Fig. 6.64 Actual machining process on milling machine

Fig. 6.65 Formed cutter for finishing of male rotor

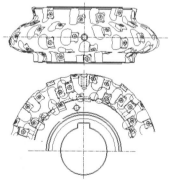

Fig. 6.66 Carbide cutter for rough milling of female rotor

Fig. 6.67 Tooth surface after rough milling

Fig. 6.68 Grinding wheel for male rotor

tween the rotor workpiece and the grinding wheel, shown in Fig. 6.69, are similar to those with the screw rotor milling machine. The screw rotor grinding machine can be understood as a screw rotor milling machine with a grinding wheel in the place of the cutting tool.

There are mainly two types of grinding wheels that can be used on a screw rotor grinding machine, that are standard-material grinding wheels and CBN (cubic boron nitride) wheels. A standard-material grinding wheel for screw rotor production is built by truing and dressing a general-purpose wheel into a formed profile, with abrasive grains exposed on the surface. In the screw rotor grinding process using a standard material wheel, the rotor workpiece can be measured after grinding to feed back the measurement data to the grinding system, to correct the grinding wheel profile to offset any errors and deviations found to provide more accurate grinding quality. On the other hand, CBN wheels are made by setting CBN grains on a formed tool body, either by gluing or by electrodeposition, and have more cutting power than standard-material wheels, which contributes to higher grinding efficiency and provides a longer tool life. However, one drawback of the grinding process with a CBN wheel, where the tool body shape is fixed and is not changeable, is that it is not compatible with freely reshaping the wheel or measurement data feedback for wheel profile correction as is done in the standard-material wheel process.

(4) Mill-turn

Due to advances in machine tools and CAD/CAM technologies, general-purpose mill-turns are now capable of machining highly complex shapes relatively easily and are increasingly used for screw rotor machining.

Mill-turns use multiple tools that are automatically changed over by ATC (automatic tool changer). As a result, rotor shafts, key slots and profile can all be machined by a single mill-turn. For example, it is technically feasible to complete all the profile machining steps from rough cutting through to finish cutting, each of which can be performed by a special tool on the same mill-turn. In consideration of productivity, however, it is currently more common to rough-machine the rotor groove on a mill-turn and then finish-machine using a screw rotor milling machine or screw rotor grinding machine. Figure 6.70 shows how a screw rotor can be machined by a general-purpose ball end mill.

6.8.2 Machining accuracy control
(1) 3-D coordinate measurement and pairing inspection

On a twin screw compressor, where the minute gaps between rotor lobes must be dependably sealed to create compression chambers, enlargement of these inter-rotor gaps due to rotor machining inaccuracy would directly lead to compressor performance deterioration. As rotors are a kind of gear element, their geometrical error will inevitably cause greater noise and vibration. Therefore, it is absolutely necessary and critically important in twin screw compressor manufacturing to appropriately control rotor machining accuracy. Techniques for rotor machining accuracy control include rotor geometry measurement (single piece inspection) using a coordinate measuring machine like the one shown in Fig. 6.71 and pairing inspection, shown in Fig. 6.72, that checks the meshing condition of pairs of male and female rotors.

3-D coordinate measuring machines are used to determine the degree of errors in rotor profile, lead and pitch. Profile errors may occur as a result of inaccurate machining tool geometry, incorrect machining conditions or deterioration of machining tools. Evaluation of rotor profile error distribution will provide a clue to determine what has caused the errors. On the other hand, lead and pitch errors are solely due to inaccuracy and other conditions of the machining equipment and not related to the machining tools.

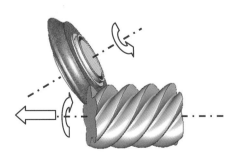

Fig. 6.69 Arrangement and relative motion between rotor blank and grinding wheel

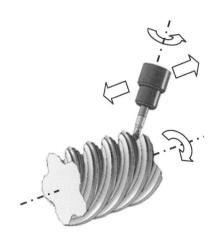

Fig. 6.70 Rough milling on mill-turn

Fig. 6.71 Measurement on coordinate measuring machine

Fig. 6.72 Pairing inspection

Examine the machining equipment based on pitch measurements obtained, and carry out overhaul or retrofits as necessary.

Pairing inspection involves actually meshing each pair of male and female rotors to be installed to a compressor unit on the pairing inspection stand, thereby checking the amount of backlash and clearances. Pairing inspection is also useful to check for abnormal rotor contact or screw lead inconsistency between the male and female rotors, which, if any, will cause noise and vibration in operation.

(2) Dynamic balancing (balance test)

In a twin screw compressor where compression is achieved by a pair of rotors rotating at high speed, unbalance of these rotors, each of which is a heavy object, can lead to not only noise and vibration increase but also mechanical failure. Except for key slots, each rotor is shaped to balance itself about the axis. However, minute geometric deviations within machining tolerance or uneven material density will cause a certain degree of unbalance. To eliminate such unbalance, a balance test followed by balance correction is required as the final step of rotor manufacturing process. In the balancing process, the rotor will be rotated on the balancing machine for measurement to determine how much unbalance is present in what part of it. The measured unbalance will be corrected by adjusting the rotor mass distribution, for example by drilling a hole in the end face.

References

1) H. Krigar: "Schraubengebläse", Kaiserliches Patentamt. Patentchrift No. 4121, (1878). (in German)
2) H. Krigar: "Verwendung eines Schraubengebläse als Gebläse, Pumpe, Motor und Messapparat", Kaiserliches Patentamt. Patentchrift No. 7116, (1878). (in German)
3) A. J. R. Lysholm: "A New Rotary Compressor", Jour. and Proc. I. Mech. E, 150, 11 (1942).
4) S. Nozawa, M. Izushi: Hitachi Review, 36 (3), (1987).
5) N. Arai, K. Matsubara: Hitachi Review, 34 (3), (1985).
6) N. Tsuboi, K. Tanaka, E. Kanki: "Energy Efficient High Speed 2-stage Screw Refrigerators", Kobe Steel Engineering Reports, 56 (2), (2006). (in Japanese)
7) E. Menssen: "Screw Compressor with Involute Profiled Teeth", US Patent 4,028,026, (1977).
8) L. Rinder: "Screw Rotor Profile and Method for Generating", US Patent 4,643,654, (1987).
9) N. Stosic: "Plural Screw Positive Displacement Machines", Patent Application GB9610289.2, (1996).
10) A. Lysholm: "Screw Rotor Machine", US Patent 3,314,598, (1967).
11) H. R. Nilson: "Helical Rotary Engine", US Patent 2,622,787, (1952).
12) C. B. Schibbye: "Screw-Rotor Machine with Straight Flank Sections", UP Patent 4,140,445, (1979).
13) M. Fuziwara, K. Kasuya, T. Matsunaga, M. Watanabe: "Performance Analysis of Screw Compressor", Trans. Jpn. Soc. Mech. Eng., 50 (452), pp. 1027-1033 (1984). (in Japanese)
14) N. Stosic, I. Smith, A. Kovacevic: "Screw Compressors Mathematical Modelling and Performance Calculation", Springer-Verlag, Berlin, pp. 61-64 (2005).
15) K. Shiraki, H. Kanki: "A Vibration Problem in the Machine Industry", Science of Machine, Yokendo, 29 (9), pp. 107-108 (1977). (in Japanese)
16) J. Derek Smith: "Gear Noise and Vibration Second Edition Revised and Expanded", Marcel Dekker, Inc., USA, pp. 161-162 (2003).
17) A. Fujiwara, K. Matsuo, and H. Yamashita: "Vibration analysis of oil-injected twin-screw compressors

using simple simulated waveforms", Proc. Inst. Mech. Eng. Part E: J. Process Mechanical Engineering, 225, pp.105-106 (2011).

18) Holroyd Precision Ltd.: "HOLROYD 2E C.N.C Rotor Milling Machine", (1989).

Chapter 7 Single Screw Compressors

7.1 History of Single Screw Compressors

Single screw compressors belongs to rotational positive-displacement machinery like twin screw compressors, but their structures are significantly different. While a twin screw compressor operates with a pair of parallel screw rotors with spiraled convex- and concave-shaped lobes, a single screw compressor comprises one grooved screw rotor and two gate rotors that mesh with the screw rotor in a perpendicular manner.

The single screw compressor mechanism was first invented by Zimmern of France. A French patent was granted in 1960. This was more than twenty years after twin screw compressors were first put to use in the late 1930s. Zimmern's ambition was to establish the single screw compressor as a technology that can validly compete with twin screw compressors. To that end, Zimmern founded OMPHALE Inc., a French company engaged in the research and development of single screw compressors, and dedicatedly worked on the development and implementation of the technology. OMPHALE later renamed itself as SSCI (Single Screw Compressor Inc.) and relocated to the United States, where it continued its business. SSCI have possessed a large number of patents as well as know-how implementation rights. In many countries of the world, there were licensee companies that operated utilizing SSCI technology.

Similar to twin screw compressors, single screw compressors were first put to use as air compression machinery. But, due to the difficulty to provide accurately machined screw rotors, market penetration did not much progress until high-precision machining systems were developed for screw rotor and gate rotor production. In 1970, Mitsui Seiki Kogyo Co., Ltd. of Japan started developing the single screw compressor technology with an aim at implementing it for air compression. Mitsui Seiki, which originally was a machine tool maker, simultaneously worked on developing dedicated machining systems to produce more accurate screw rotors and gate rotors. The company eventually succeeded in producing air compressors that were superior in quality than the existing products of the time. Their products became very popular and spread widely in the market. Following Mitsui Seiki's success, many refrigerator manufacturers started to adopt single screw compressors as the refrigerant compressor for their products. Production in Europe started during 1970s. Production of single screw compressors as open-type refrigerant compressor was started in Japan in 1982. Production of semi-hermetic refrigerant compressor models followed shortly. The single screw compressor technology still continues to develop today, diversifying into newer models such as refrigerant-injected, refrigerant-cooled models as well as larger-scale and two-stage-compression models.

7.2 Basic Mechanism and Operating Principle

7.2.1 Compression mechanism

Single screw compressors can be theoretically classified into four types according to the combination of the shapes of the screw rotor and the gate rotor, which are the compressor's fundamental compression elements and can be cylindrical or planar. The four types, each shown in Fig. 7.1, are the cylindrical screw and cylindrical gate rotor (CC) type, the plane screw and cylindrical gate rotor (PC) type, the cylindrical screw and plane gate rotor (CP) type and the plane screw and plane gate rotor (PP) type[1]. All of the four types operate with a single screw rotor and two gate rotors that mesh with the screw rotor. Due to screw rotor machinability, however, only two of the four types, that are CP and CC, have been actually implemented to date.

In the fields of refrigeration and air-conditioning, the CP type, which uses the combination of a cylindrical screw rotor and planar gate rotors, is the only type of single screw

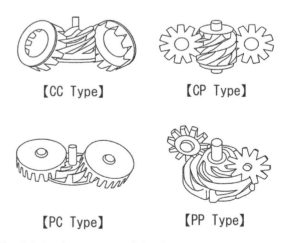

Fig. 7.1 Basic structure of the single screw compressor

compressor implemented.

7.2.2 Operating principle

Figure 7.2 illustrates the operating principle of a single screw compressor. The compressor uses a screw rotor having six grooves and two gate rotors with eleven teeth that mesh with the screw rotor. Each compression chamber is enclosed by the screw rotor, the gate rotors and the casing.

(1) Suction process: As the screw rotor turns the volume of the groove, that is enclosed by the screw rotor and the casing and is connected to the suction side, increases from that of suction chamber "a" to the one of suction chamber "b" shown. Refrigerant flows into the groove from the suction side.

(2) Compression process: As the screw rotor turns further, a gate rotor tooth meshes with the screw rotor groove at the entrance of the groove, partitioning the space from the suction side. At this point, the suction process is completed and the compression process starts. As the screw rotor continues to turn, the volume of the space enclosed by the screw rotor, the gate rotor and the casing ("compression chamber") decreases to compress the refrigerant gas inside.

(3) Discharge process: When the volume of the compression chamber has decreased to a certain level, the compression chamber will open to the discharge port. With this, the compression chamber becomes connected to the discharge side, and its volume further decreases as the screw rotor continues to rotate, discharging the compressed gas to the outside.

Figure 7.2 presents processes of the compression cycle that proceeds on one side of the screw rotor. On the other side of the screw rotor behind the gate rotors, an identical compression cycle takes place.

Similar to twin screw compressors, each single screw compressor has a built-in volume ratio. The ratio of compression chamber volume V_s, when suction gas has just been trapped and separated from the suction side to start the compression process, to the compression chamber volume V_d, when the compression chamber is connected to the discharge port, or V_s/V_d, is called built-in volume ratio. Normally, the built-in volume ratio is designed so that the compression chamber pressure at the starting of the discharge phase will match the discharge pressure required by the system.

7.2.3 Basic structure of single screw compressors

Figure 7.3 shows the general structure of a typical single screw compressor. It has a motor on the suction side and an

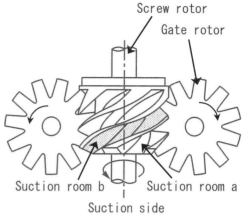

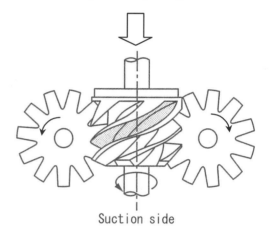

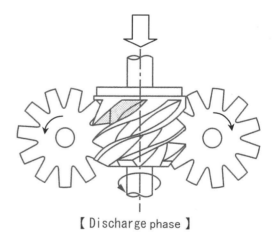

Fig. 7.2 Compression process

oil separator on the discharge side, integrated with the compression mechanism. The refrigerant gas sucked into the compressor first cools the motor and then flows into a space (compression chamber) that is enclosed by the cylindrical screw rotor, the planar gate rotor and the casing. The screw rotor and the gate rotors are each supported by rolling bearings. The compressor also has a slide valve that acts for ca-

Compressors for Air Conditioning and Refrigeration

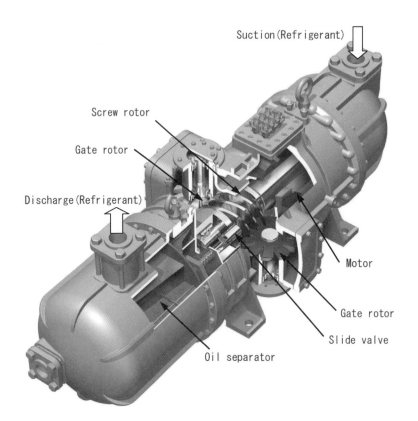

Fig. 7.3 Single screw compressor

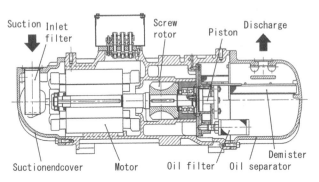

Fig. 7.4 Semi-hermetic single screw compressor

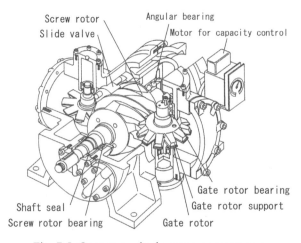

Fig. 7.5 Open type single screw compressor

144

Chapter 7 Single Screw Compressors

pacity control. More explanation about the slide valve will be given in a later section. After compressed in the compression chamber and discharged from it, the refrigerant gas will have its oil content separated by the oil separator and then delivered out of the compressor unit.

While Fig. 7.3 shows a semi-hermetic type, an open type has one end of the screw rotor shaft extended out of the compressor unit with a mechanical seal or other shaft sealing feature fitted to the shaft penetration area, so that the shaft can be driven by an external drive source while keeping the space inside airtight.

7.2.4 Semi-hermetic type and open type

Similar to twin screw compressors, single screw compressors can be classified into semi-hermetic and open types[2]. Figure 7.4 shows a semi-hermetic type and Fig. 7.5 shows an open type.

7.3 Structure and Design of Compressor Components

7.3.1 Shaft

An important characteristic of a single screw compressor is that an identical pair of compression cycles take place on both the top and the bottom halves of the screw rotor, above and below the two gate rotors, and therefore radial gas loads on the screw rotor are balanced and canceled, as shown in Fig. 7.6. As a result, it is generally not necessary to evaluate the lateral deflection strength of the screw shaft and only the torsional strength needs to be evaluated. However, in some semi-hermetic models the motor may be supported on one side only depending on the bearing configuration. In this case, it is necessary to estimate the amount of in-operation shaft deflection by calculating the centrifugal force of the motor rotor and also the electromagnetic force generated by the motor's air gap imbalance.

On the other hand, gate rotor shafts are subject to in-operation deflection which contributes to compression chamber leakage and therefore their strength must be carefully evaluated. Normally, gate rotor shafts should be designed to reduce the amount of in-operation rotor tip displacement to a minimum so as to allow only the smallest possible amount of leakage.

7.3.2 Screw rotor

Screw rotors are typically built with cast iron for the ease of machining. Screw rotor groove machining on a dedicated machining unit is done by using a tool that has an identical shape with the gate rotor. With this, only a minimum amount of gate rotor mesh clearance is secured, allowing high compression efficiency. As shown in Fig. 7.7, a pressure-equalization channel that connects the suction-side end and the discharge-side end of the screw rotor is provided through the core of the rotor.

7.3.3 Gate rotor

Figure 7.8 shows the configuration of the gate rotor section. Each gate rotor system comprises a gate rotor and a gate rotor shaft.

To enhance sealability with and abrasion resistance against the screw rotor, the gate rotor disk is typically built with an engineering plastic material composed of a blend of heat-resistant base plastic, glass fiber and lubricant. As the gate rotors are subject to gas pressure load during compression operation, a cast iron gate rotor support, integrated with the gate rotor shaft, is fitted to the back of each gate rotor.

To be able to withstand impact loads generated by liq-

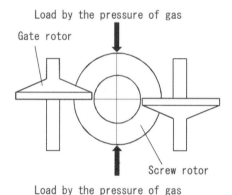

Fig. 7.6 Balanced loading of the screw rotor (radial loads)

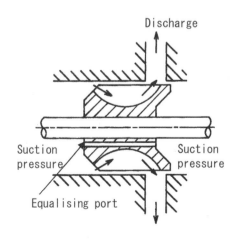

Fig. 7.7 Balanced loading of the screw rotor (axial loads)

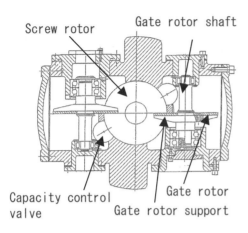

Fig. 7.8 Section of gate rotor

uid compression or other conditions that may occur during compression operation, the gate rotor shaft is typically built with ductile cast iron.

7.3.4 Discharge port

Discharge ports in the casing allow the refrigerant gas compressed in the compression chamber to be discharged in the radial direction. As in the case of twin screw compressors, the discharge port opening timing of a single screw compressor has a significant effect on the compressor efficiency. The discharge port opening timing must be designed so that the operating pressure ratio and the built-in volume ratio of the compressor will be compatible under its most frequently encountered operating conditions. For further information, refer to Sections 6.4.6 and 6.6.1.

7.3.5 Bearings

A single screw compressor is a mechanism that requires three-dimensional accuracy, where the amount of bearing play should be as small as possible. Therefore, all of its six bearings, that are two bearings for screw rotor and four bearings for gate rotor, are commonly composed of rolling bearings. Types of the rolling bearing typically used are cylindrical roller bearings to support radial load and angular contact ball bearings to support thrust load.

As shown in Figs. 7.6 and 7.7, the amounts of both the radial and the axial loads that act on the bearings are small due to the screw rotor's pressure canceling effect. This contributes to a long bearing life.

7.3.6 Shaft sealing feature

Similar to twin screw compressors, open-type single screw compressors commonly use a mechanical seal for shaft sealing. For further information about mechanical seals, refer to Section 6.4.9.

7.3.7 Casing

The casing has a double-layer structure to minimize potential deformation by heat and pressure. In addition, ribs are provided in optimized positions for rigidity enhancement. The shaft cores for screw rotor and gate rotors are machined and assembled with high accuracy using special fixtures to assure precise inter-shaft distance and perpendicularity.

7.3.8 Capacity control mechanism

Similar to that of a twin screw compressor, the capacity control mechanism of a single screw compressor uses a slide valve to bypass part of the gas once trapped in a compression chamber back to the suction side. Figure 7.9 shows how the capacity control mechanism operates.

Under full-load condition, the slide valve is pressed against the stationary portion so that all the suctioned gas is trapped in volume "a" and compressed. Under partial-load condition, the slide valve shifts to create a flow channel between the stationary portion and itself so that part of the gas in the compression chamber flows back to the suction side. Compression of the gas does not begin until position "b", which results in a smaller suction volume.

The slide valve can be driven by various drive sources, including hydraulic, gas pressure and motor drives. Figure 7.10 shows an example where the slide valve is driven by gas pressure[3]. The slide valve is coupled to a piston, and a solenoid valve operates to control pressures on both sides of the piston so as to move the piston and the coupled slide valve back and force. This mechanism relies on pressure difference between the high pressure and the low pressure sides and therefore may not work dependably in the start-up phase or in the defrost mode. To address this issue, a forced "load up" feature that operates by using the system pressure from halfway through the compression cycle will be provided to assure capacity control performance in all conditions.

The mechanism described above provides a discontinuous and stepped capacity control. On the other hand, there are mechanisms, like the one illustrated in Fig. 7.11, which provide non-stepped continuous capacity control. The chamber to the right of the piston is connected to and equalized with the high pressure side, which is the chamber to the left of the piston, through a very small equalization hole. The right chamber has another hole, opened and closed by operation of a pilot valve, which connects it to the low-pressure space on the suction side.

Figure 7.11 also shows that the slide valve constantly receives a leftward force generated by pressure difference

Chapter 7 Single Screw Compressors

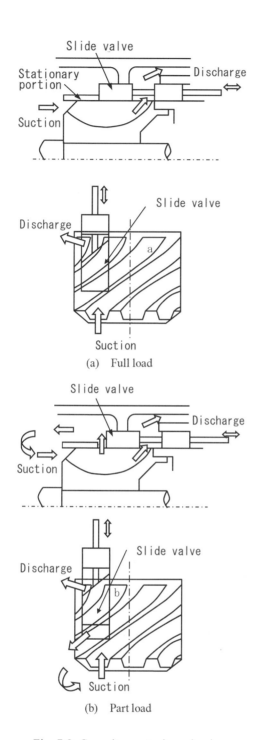

Fig. 7.9 Capacity control mechanism

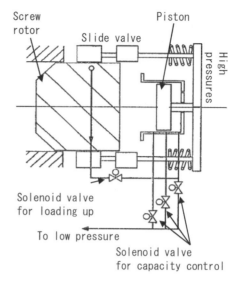

Fig. 7.10 Slide valve drive mechanism

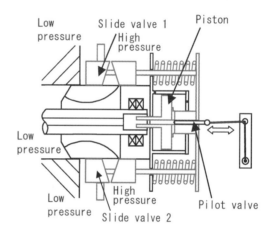

Fig. 7.11 Slide valve drive mechanism (vane motor method)

between the high pressure and the low pressure sides. As a result, when the pilot valve opens slightly to lower the pressure in the right chamber, the piston will stay in a position to balance and cancel the slide valve force. The pilot valve position is controlled by a vane motor. When the pilot valve shifts slightly to the right, the equalization holes will be open to lower the pressure in the right chamber, which in turn causes the piston to move to the right to close the equalization hole. Thus the piston moves following the pilot valve action to provide a continuous capacity control.

As a single screw compressor has two compression spaces, one at the top and the other at the bottom, the capacity control mechanism shown in the figure drives the slide valves for both the top and the bottom compression spaces at the same time. Capacity control for a single screw compressor can be classified in two types depending on how the top and the bottom slide valves are driven:

(a) Simultaneous-drive capacity control method: The top and the bottom slide valves are driven together at the same time.

(b) Individual-drive capacity control method: The top and the bottom slide valves are driven independently from each other so that high efficiency can be maintained even under 50% or lower partial load. In this case, one of the slide valves can be completely pulled off so that no compression takes place on that side while the other slide valve remain in position to continue compression.

7.3.9 Oil separator

In a semi-hermetic single screw compressor, an oil collector device, which serves as both an oil reservoir and an oil separator, is integrated with the compressor unit. While the oil separation system may contain various separation technologies such as collision separation, centrifugal separation and sedimentation separation, the final separation stage is normally achieved by demister-based separation.

The oil separation efficiency requirement differs depending on the product application, but an efficiency of approximately 99.5% is generally considered sufficient for air-conditioning systems. For applications requiring a higher separation efficiency, an additional high-efficiency oil separator needs to be installed separately.

7.4 Rotor Tooth Profile

7.4.1 Rotor tooth profile

The screw rotor tooth profile is determined by how it is machined. Normally, grooves in the screw rotor are machined by a tool that has the same shape as its mating gate rotor[4].

Figure 7.12 illustrates the engagement between a screw rotor and the mating gate rotors, in a cross section made along the broken line and as viewed in the arrowed direction. The contact angle between the screw rotor and a gate rotor can be given as $\tan^{-1}(V_G / V_S)$, with the gate rotor speed V_G and the screw rotor speed V_S respectively, The larger the V_G gets and the smaller the V_S gets, the greater the contact angle will be. On the other hand, the smaller the V_G gets and the larger the V_S gets, the smaller the contact angle will be.

When a radial position on the gate rotor is given, V_G will remain the same during rotation, while V_S will vary depending on the screw rotor radial position. That is, the contact angle changes as the rotors rotate. To avoid interference with the screw rotor, the gate rotors are designed with specific maximum and minimum contact angles.

Due to the gate rotor's maximum contact angle, there will be an opening, or "blowhole", in the space enclosed by the casing, the gate rotor and the screw rotor, as shown in Fig. 7.12. Although this blowhole directly connects the compression chamber to the suction side, the volume of leakage through this blowhole is negligible compared to the chamber volume and therefore can be ignored.

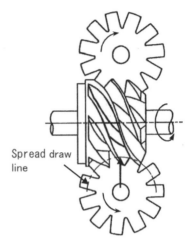

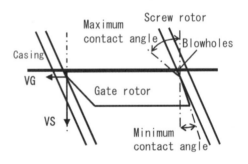

Fig. 7.12 Contact situation

7.4.2 Number of rotor teeth

(1) Number of gate rotor teeth: It is generally believed that a refrigerant compressor needs to have at least three compression chambers formed between its suction and discharge sides to achieve the required compression ratio.

In a typical single screw compressor, each of the two gate rotors with eleven teeth meshes with the screw rotor to a depth that is 0.2 times the gate rotor diameter. With this, three compression chambers are formed from the suction side to the discharge side.

(2) Number of screw rotor grooves: To be able to mesh with two gate rotors at the same time, the screw rotor of a single screw compressor must have an angular groove length of less than 180°.

With eleven teeth on each gate rotor, the angular length of a screw rotor groove would exceed 180° if the number of screw rotor grooves is not more than five. Therefore, the screw rotor needs to have at least six grooves.

With more than six grooves on the screw rotor, the groove length and consequently its leakage time would be shorter, which contributes to leakage loss reduction. However, a shorter groove length will raise the compression speed, which will increase the over-compression loss.

A shorter groove will also force to increase the gate rotor rotation speed, which in turn increases the contact angle between the gate rotor and the screw rotor and may cause issues concerning gate rotor strength. Considering these factors, the screw rotor of a single screw compressor is most commonly designed with six grooves.

7.4.3 Displacement volume

The theoretical volume of a single screw compressor is obtained by integrating the gate rotor surface area that meshes with the screw rotor by the angle of screw rotor rotation from the start to the end of contact between the gate rotor and the screw rotor.

In a simplified manner, the theoretical volume can be obtained from the following equations[2]. As shown in Fig. 7.13, the basic geometric forms of the screw rotor and the gate rotor are determined by the following parameters:

Screw rotor outside diameter: D_S
Gate rotor outside diameter: D_g
Groove width (tooth width): W_r
Distance between the screw rotor and the gate rotor centers: d
Screw rotor engagement angle: θ_s

The basic relations of the parameters are as follows:

$$D_g = D_s \tag{7.1}$$

$$d = 0.8D_s \tag{7.2}$$

$$Wr = 0.15D_s \tag{7.3}$$

In general, the displacement volume of a single screw compressor per rotation can be expressed as follows:

$$V = K_1 \cdot K_2 \cdot D_s^3 \tag{7.4}$$

V: Theoretical displacement volume
K_1: Coefficient based on rotor dimension
K_2: Coefficient based on engagement angle
D_S: Outer diameter of screw rotor

Given the basic geometric forms, K_1 will be 0.364 and K_2 will be determined by the screw rotor engagement angle.

Figure 7.14 shows an example of the relationship between the volume and changes in the compression chamber pressure[3].

7.4.4 New gate rotor tooth profile

The conventional basic geometry of a gate rotor has parallel-sided teeth. However, new compressor models that use gate rotors with fan-shaped teeth, as shown in Figs. 7.15 and 7.16, have been developed in recent years[5-7]. As previously described, the suction volume of a single screw compressor is obtained by integrating the gate rotor surface area

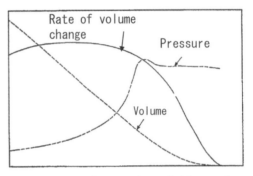

Fig. 7.14 Volume change and pressure change

(a) Conventional shape of the single screw tooth profile

(b) Fan shape of the single screw tooth profile

Fig. 7.15 New shape of the single screw tooth profile

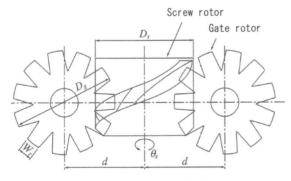

Fig. 7.13 Geometrical shape

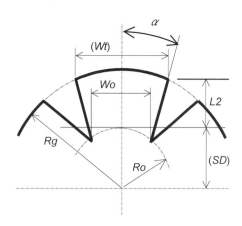

Fig. 7.16 New gate rotor tooth profile

that meshes with the screw rotor by the angle of screw rotor rotation from the start to the end of contact between the gate rotor and the screw rotor. As the tooth width is variable with the fan-shaped profile, a new parameter for the fan angle, α, must be included to obtain the surface area. The tooth tip width W_t can be expressed using the distance L_2 from the basic tooth width W_o as follows:

$$W_t = W_o + L_2 \tan\alpha \tag{7.5}$$

By integrating the gate rotor surface area obtained from the above W_t, the suction volume with the fan-shaped profile can be obtained.

Figure 7.17 shows the relationship between the fan angle α and the volume ratio. It is apparent that the suction volume varies in proportion to the fan angle. By using a new fan-shaped tooth profile that increases the suction volume by more than 20%, the same suction volume will be obtainable with a compressor unit one size smaller. The size reduction will also likely lead to cost reduction.

Figure 7.18 shows the lengths of various leakage sections that are found along the teeth of a single screw compressor with fan-shaped gate rotor teeth, as integrated by the angle of rotation from the start to the end of compression and as compared to those of a compressor with conventional parallel-sided teeth. SL1 to SL7 respectively represent specific leakage sections that are found in the compressor. Compared to the conventional parallel-sided tooth profile, the fan-shaped tooth profile helps shorten the individual leakage sections, which will lead to a higher volumetric efficiency.

However, an excessively large fan angle is likely to cause interference between the gate rotor and the screw rotor, making it difficult to assemble the gate rotors with the screw rotor. Before selecting a gate rotor tooth fan angle in the design stage, potential interference between the rotors must be carefully evaluated by three-dimensional CAD or

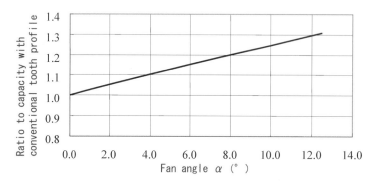

Fig. 7.17 Ratio of fan angle and suction capacity

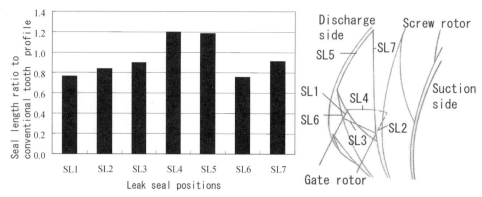

Fig. 7.18 Illustrations of leakage effects & positions

Chapter 7 Single Screw Compressors

other analytical techniques.

The following paragraph explains about the ratio of the numbers of screw rotor and gate rotor teeth. In a single screw compressor, which has two compression spaces formed above and below the gate rotors, the angular length of one screw rotor groove must not be more than 180°. However, the alternative fan-shaped teeth has an effect to extend the effective screw rotor groove length because the gate rotor tooth width mating with the screw rotor groove bottom is larger than that mating with the screw rotor tip. With this, the conventional six-and-eleven tooth configuration will make the angular length of a single screw rotor groove greater than 180°. Therefore, compressors with the new fan-shaped tooth profile has a six-and-ten tooth configuration instead of the conventional six-and-eleven configuration.

The fan-shaped tooth profile has been realized as a result of advances in machining technology. A recently developed new machining technique using a general-purpose machining center, shown Fig. 7.19, has made it possible to produce innovative tooth profiles that previously were not feasible.

7.5 Performance and Noise

7.5.1 Efficiency characteristics

Figure 7.20 shows a breakdown of losses experienced in a single screw compressor[2]. Major loss factors that affect compressor efficiency are motor loss, internal leakage loss in the compression phase and over-compression loss.

Internal leakage loss occurs mainly at gaps between the screw rotor circumference and the casing and gaps at areas of engagement between the screw rotor and the gate rotor. To minimize the gaps between the screw rotor circumference and the casing, it is critical to improve casing machining accuracy and also to reduce the types of deformations mentioned in Section 7.3.7. To minimize engagement gaps,

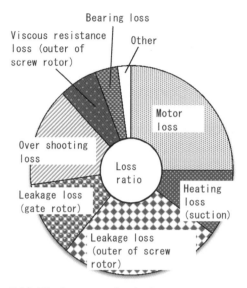

Fig. 7.20 The loss rate of a single screw compressor

not only the dimensional accuracy of the individual rotors but also the relative positioning accuracy between the screw rotor and the gate rotors must be optimized.

Similar to twin screw compressors, each single screw compressor has its own geometrically-determined built-in compression ratio. Depending on the operating conditions, incompatibility between this built-in compression ratio and the operating conditions may cause over-compression and under-compression losses.

Other types of losses include viscosity resistance loss between the screw rotor circumference and the inner casing wall and also bearing losses. Due to the use of rolling bearings, however, the amount of bearing losses experienced in a single screw compressor is relatively small.

7.5.2 Noise characteristics

Figure 7.21 illustrates the noise characteristics of a single screw compressor[2]. The noise level is relatively low compared to that of reciprocating compressors. As the gate rotor is made of engineering plastic material, it will not

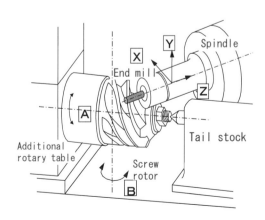

Fig. 7.19 5-axis NC milling

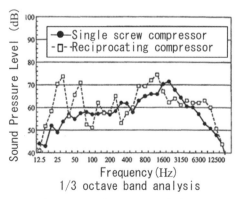

Fig. 7.21 Noise characteristics

generate unpleasant high-frequency noises like the ones generated by contact between metallic parts.

7.6 Other Mechanisms with Single Screw Compressors

7.6.1 Cooling system by oil injection

As shown in Fig. 7.22, an open type single screw compressor has lubrication oil injected into its compression chambers to cool and seal the compression gas. The injected oil is discharged out of the compression chamber together with the refrigerant gas and then is separated from the refrigerant in the oil separator. The oil is cooled by cooling water or refrigerant and again injected into the compression chambers.

In a semi-hermetic model, shown in Fig. 7.23, a mixture of lubrication oil and liquid refrigerant is injected in to the compression chambers to cool the compressed gas. This technique helps simplify the lubrication system by eliminating the need for an oil cooler.

7.6.2 Cooling system by liquid refrigerant injection

Figure 7.24 shows a newer single screw compressor mechanism, which can be classified as an advanced version of the previously described oil-injected, oil-cooled system. This mechanism injects massive amounts of liquid refrigerant instead of oil into compression chambers to obtain compression gas cooling and sealing effects. The reason that this mechanism is applicable to a single screw compressor is that the single screw compressor does not accompany load transmission between the rotors and therefore does not require lubrication between the rotors. While this mechanism offers the advantage of eliminating the need for both an oil cooler and an oil separator, the bearings will have to be lubricated with a smaller amount of oil. Therefore, careful consideration is necessary in designing the lubrication oil feed system.

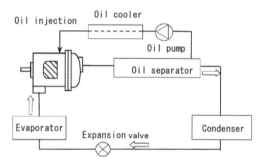

Fig. 7.22 Schematic of oil injection

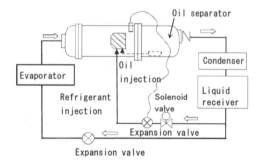

Fig. 7.23 Schematic of oil injection (semi-hermetic)

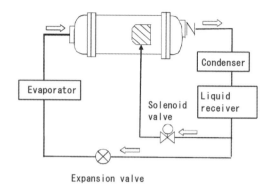

Fig. 7.24 Schematic of liquid injection

7.6.3 Economizer cycle

An economizer is a feature that helps improve the refrigeration cycle efficiency by injecting an amount of intermediate pressure gas into the compression chamber in the middle of the compression phase. As shown in Fig. 7.25, one possible problem related to this feature is that the intermediate-pressure injection port will open immediately after the start of the compression phase. If the capacity-controlling slide valve is activated under this condition, the injection port will connect to the suction side, which will cause the economizer performance to deteriorate significantly. This issue can be solved by using an individual-drive-type capacity control mechanism. Under a 50% or higher load rate condition, the individual-drive capacity control mechanism deactivates the slide valve on the side where the injection port is and operates the slide valve on the remaining side only. With this, efficient operation of the injection can be maintained while functioning the capacity control.

As the single screw compressor has two compression spaces at the top and the bottom of the screw rotor, it is also feasible to configure a three-stage economizer system, which is higher in efficiency than two-stage economizer system, by differentiating the injection port position in the top compression space and that in the bottom compression space.

Chapter 7 Single Screw Compressors

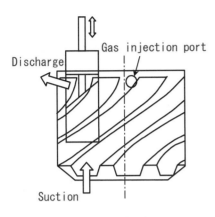

(a) Full load

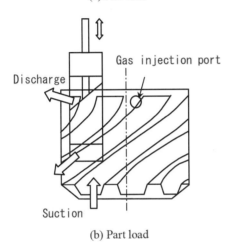

(b) Part load

Fig. 7.25 The subject of the injection in a part load

References

1) E. Shibutani, S. Yoshimura: "Basic Design of Screw Compressors", Turbomachinery, 22 (6), pp. 366-372 (1994). (in Japanese)
2) Japan Society of Refrigerating and Air Conditioning Engineers: "6th EDITION, JSRAE HANDBOOK, II EQUIPMENTS", Japan Society of Refrigerating and Air Conditioning Engineers, Tokyo, pp. 30-33 (2006). (in Japanese)
3) A. Ohtsuki: "Semi-hermetic single screw compressor". Refrigeration, 62 (721), pp. 1175-1181 (1987). (in Japanese)
4) S. Sawai, T. Miyamoto: "Low energy consumption screw chiller", The Journal of Fishing Boat Association of Japan, 241, pp. 394-400 (1982). (in Japanese)
5) K. Ohtsuka: "Miniaturization and efficiency improvement by development of a new tooth form screw rotor", Machine Design, 46 (1), pp. 24-25 (2002). (in Japanese)
6) T. Murono, H. Ueno, K. Ohtsuka, T. Takahashi, T. Susa: "Development of Single Screw Compressor Using New Tooth Profile", Proceedings of 2007 JSRAE Annual Conference, E313, (2007). (in Japanese)
7) T. Murono, H. Ueno, K. Ohtsuka, T. Takahashi, T. Susa: "Development of single screw compressor using new tooth profile", Proceedings of the International Conference on Compressors and Their Systems, London, pp. 183-191 (2007).

Chapter 8
Automotive Air Conditioning Compressors

8.1 History of Automotive Air Conditioning Compressors

Air conditioning in vehicles using vapor compressors first appeared in the United States in the late 1930s. At around the time, cooling systems that used R12 refrigerant was introduced to residential and commercial buildings, and automotive air conditioning systems also used the same refrigerant. In 1938, fundamental requirements for automotive air conditioning system had been listed[1]: 1) cabin cooling and heating, 2) humidification and dehumidification, 3) cabin ventilation, 4) distribution of air and 5) air purification. In 1940, details of a R12-based automotive air conditioning system, which was installed to some luxury automobile models, were reported[2]. In early 1950s, many luxury automobiles that were driven in southern states of the United States were equipped with a cabin cooling system. Starting from 1952, cooling systems were installed to vehicles on production lines at automobile plants[3].

In the early cabin cooling systems, a cooling unit including the evaporator and the blower was housed in the trunk space in the rear of the vehicle, as shown in Fig. 8.1. The current configuration with the air conditioning unit embedded in the dashboard was developed in 1955[4].

Early compressors for automotive air conditioners were reciprocating compressors modified from motels for residential air conditioners. Most of the components were made of iron and the compressor was driven by the engine crank pulley via belt transmission. Compressors used in those days did not have a start/stop control feature like the solenoid clutch used in modern compressors. Air conditioning capacity was adjusted through bypass valve operation.

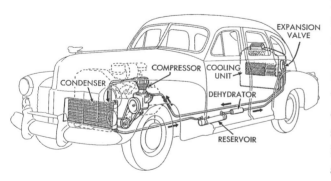

Fig. 8.1 Configuration of automotive air conditioner before 1941 [3]

In winter times when cabin cooing was not needed, it was a common practice to have the compressor driving belt removed from the system.

Reciprocating compressors for automotive air conditioners had a number of improvements including the use of aluminum body and the introduction of high-speed specifications. Also, non-reciprocating types of compressors were developed for automotive air conditioning purposes. General Motors began to produce rotary vane compressors modified from a residential model for their automotive air conditioners in 1953[3]. With the production technology of the time, however, it was difficult to achieve adequate performance and durability. In 1956, a five-cylinder wobble plate compressor was developed[5]. The compressor was equipped with a newly developed solenoid clutch that allowed the user to start or stop the compressor by using control switches. In 1962, a dry single-plate solenoid clutch that could be turned on or off by thermal switch operation was developed, and a swash plate compressor with such clutch was built [6,7]. With this, the fundamentals of modern automotive air conditioner compressor technology were established.

Automotive air conditioners were first brought into Japanese market along with imported automobiles. Air conditioning systems were introduced into domestic luxury automobiles in 1960s and then rapidly spread to compact cars and trucks in 1970s. In the late 1970s, even subcompact (660 cm^3 or lower displacement) models started to be fitted with air conditioners.

Milestones in the compressor technology development in Japan include the development and production of reciprocating compressors in the late 1950s, of swash plate compressors in the mid-1960s and of wobble plate compressors in early 1970s[8]. From the mid 1970s to early 1980s, where market demand for automotive air conditioners increased rapidly, many manufacturers, both Japanese and foreign, developed various types of compressors including rotary vane compressors, radial piston reciprocating (Scotch yoke) compressors, screw compressors, rolling piston compressors[9], Wankel compressors[10] and scroll compressors[11]. Scroll compressors were first implemented for automotive air conditioning before they were used for any other applications. With increasingly greater focus on quietness and energy-saving in automobiles, demand for smaller, lighter, low-noise and yet high speed compressors continued to grow, where models with less fluctuation in driving load/torque were preferred. Rotary vane and scroll compressors grew in market shares, and conventional two-cylinder reciprocating compressors were replaced by five or six-cylinder (later also seven- to ten-cylinder) wobble plate and swash plate compressors. These newer compressors offered less load fluctuation, lower operating noise and faster operation speed. Air conditioning capacity control based on solenoid clutch on-off control had issues of noise and vibration upon compressor start-up and shutdown. To cope with these issues, engineering efforts were made toward a variable discharge volume system in reciprocating piston compressors. In 1985, a swash plate compressor with two-stage discharge volume control was developed[12]. In the United States, a wobble plate compressor with variable volume control was developed. In the latter type of volume control mechanism, the wobble plate inclination angle, which determines the piston stroke, was changeable so that the discharge volume can be changed depending on the cooling capacity demand[13]. Similar variable-volume compressors were developed and sold in Japanese market as well[14]. In 1995, single-sided swash plate compressors with variable volume were developed and sold. Then in 1997, externally controlled variable volume compressors, where the discharge volume could be changed by an external signal, were introduced into the market[15].

Starting from the late 1990s, greater focus on global environment protection brought on more fundamental changes in automotive systems. To reduce the emission of global-warming CO_2, hybrid electric vehicles (HEV) that obtain driving power from both an internal combustion engine and an electric motor for fuel economy improvement, and also fully electric vehicles (EV) that do not emit CO_2 were commercialized. The amount of production of these alternative vehicles has grown rapidly since early 2010s. These vehicles employ newly developed motor-fitted scroll compressors that are driven solely or mostly by a built-in electric motor. Motor-fitted compressor technology, where air conditioning capacity is controlled by adjusting the rotation speed of the built-in electric motor, is being further researched and developed in a completely different engineering approach than that for conventional engine-driven automotive air conditioning compressors for better cabin comfort and higher energy efficiency. Development of heat pump systems is accelerated especially for electric vehicles where engine waste heat is not available for cabin heating. Intensive development efforts are underway to establish such heat pump technology.

While development of these newer, alternative automotive air conditioner compressors continues, this chapter explains about engine-driven compressors, which are still the mainstream of automotive air conditioner compressor technology.

8.2 Automotive Air Conditioning Systems

8.2.1 Characteristics of automotive air conditioning systems

An automotive air conditioner maintains the passenger cabin of a vehicle in comfortable temperature and humidity ranges, and prevents window fogging to provide better visibility for the driver to enhance safety. It is also required to operate stably in a wide cooling capacity range.

Figure 8.2 shows a typical automotive air conditioning system. A compressor and a condenser are fitted inside the engine compartment while a blower, an evaporator and a hot water-based heater core are embedded in the dashboard next to the passenger cabin. A damper to control air flow allows switching between air intake from outside and air circulation in the cabin. Air delivered from the blower passes through the evaporator to be cooled. Part of the air may be diverted from the evaporator to pass through the hot-water heater core so as to control the output air temperature to the target level. The temperature-controlled air then passes through outlet dampers to be distributed to various channels including the defroster, the front outlet and the foot outlet before being fed into the passenger cabin. This temperature control technique is called the air mixing method. Other techniques include the reheating method.

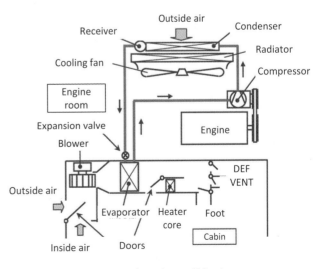

Fig. 8.2 Automotive air conditioning system

Compressors for Air Conditioning and Refrigeration

(1) Configuration of the refrigeration cycle

In the system shown in Fig. 8.2, the compressor is bolted onto the car engine and is driven by the engine crank pulley via belt transmission. Refrigerant gas is compressed in the compressor and enters the condenser located in front of a radiator. In the condenser the refrigerant is cooled and partially liquefied by air flow caused by running of the vehicle and/or the operation of cooling fans (condenser and radiator fans). The refrigerant then passes through the liquid receiver (or, in some systems, through a liquid receiving section of the condenser outlet) to have the liquid and gas contents separated so as to direct the liquid content to the expansion valve located immediately upstream of the evaporator. After decompression and the resulting boiling temperature drop through the expansion valve, the refrigerant evaporates in the evaporator by absorbing heat from the cabin air fed by the blower. The refrigerant then flows through a suction hose and returns to the compressor.

(2) Cooling capacity control

The most commonly used cooling capacity control in an automotive air conditioning system is CCEV, or the clutch cycling expansion valve method. The CCEV method activates or deactivates a power-transmitting solenoid clutch coupled to the compressor so as to run or stop the compressor for cooling capacity control. The CCEV method uses an automatic temperature-controlling expansion valve. This expansion valve operates to keep a constant superheating of refrigerant at the exit of evaporator even if the cooling load or the compressor rotational speed changes so as to sustain system efficiency. Although the CCEV is advantageous in the viewpoint of small power loss, the repeated starting and stopping of the compressor and the resulting shocks and noises are detrimental to comfort of the passengers and drivability of the vehicle. A possible solution to this is VCEV, or the variable displacement compression expansion valve method. The VCEV method uses a variable-discharge-volume compressor to have the discharge volume reduced as required so as to maintain the cooling capacity at an appropriate level without stopping and starting the compressor. With this, the compressor can be operated continuously without on-off clutching. Disadvantages of the VCEV method are complex structure and high cost. Further explanation will be given in Section 8.3.3.

(3) Characteristics of automotive air conditioning systems and compressor requirements

Automotive air conditioning systems have specific characteristics and restrictions that differ from those of general air conditioning systems for buildings. The following paragraphs describe the unique requirements closely relating to the automotive compressor.

a) Cabin temperature of a car parking under the blazing sun may rise significantly (possibly close to 80 °C in some cases), which requires fast cooling. The passenger cabin also absorb massive amount of heat from the hot ambient, and cooling capacity must be secured even in the low speed range of the engine. As a result, an automotive air conditioning system requires a relatively large-capacity compressor for the small cabin space to be cooled. The typical displacement range for compressors that are being used for ordinary passenger vehicles including subcompact models is 80 to 220 cm^3/rev for piston compressors and 40 to 150 cm^3/rev for rotary vane and scroll compressors.

b) When the vehicle is stopped or in slow traffic, the amount of air through the condenser decreases and thus increases the discharge pressure. The pressure will further increase during summertime when the outside air temperature is high. The compressor must be able to operate under those high discharge pressure conditions even in the low engine-speed range. Under that range where effect of internal leakage loss becomes large, it is important to secure good sealing and lubrication at the sliding portion of the compression chambers.

c) Compressors that are directly driven by the engine crank pulley via belt transmission are at risks for seizure and other failures that could significantly affect the safe operation of the vehicle. To decrease these risks, the compressor itself must be designed with high reliability. Fluctuation in the compressor torque also have a negative impact on engine speed control, especially in the low engine speed range. It is necessary to select low-torque-fluctuation type compressors or to employ a multiple-cylinder configuration in the use of piston compressors to level out the torque.

d) As the compressor is directly mounted on the vehicle engine that vibrates with explosive combustion, the compressor must be designed with high resistance against vibration.

e) When a vehicle is moving, rotational speed of the compressor driven by the engine via belt transmission is always changing with rotational speed of the engine. In addition, the compressor in a CCEV system is intermittently stopped and started by solenoid clutch operation. Therefore, not only the solenoid clutch but also the compressor itself must be capable of starting up instantaneously in the all speed ranges. With the typical ratio of the compressor speed to the engine speed, speed range required of the compressor is from 600 to 8,000 rpm and

156

in some case over 10,000 rpm.

f) The compressor is located inside the engine compartment, where it is exposed to sand, dirt, mud and also antifreezing salt on the road in winter. In addition, the compressor may be subjected to various chemicals that drip or splash inside the engine compartment. To withstand these, the compressor must be designed to be dust-, corrosion- and chemical-resistant.

g) Both the compressor and the condenser are exposed to rapid and significant temperature variations inside the engine compartment while the evaporator, installed close to the passenger cabin, experiences only a small and slow temperature variation. Such difference in temperature conditions between these system components may cause the refrigerant to move from one component to another, in some cases repeatedly, even when the air conditioner is turned off. When condensation and evaporation of the refrigerant are thus repeated inside the compressor, lubrication oil in the compressor may be carried out to the condenser and the evaporator and the lubricant film inside the compressor may be washed away by the refrigerant, resulting in poor lubrication in the compressor. Therefore, the compressor must withstand starting up even under an extremely poor lubrication condition.

h) It sometimes happens that liquid refrigerant is locally present inside the suction hose or in the evaporator when the compressor starts up. Therefore, the compressor must also withstand starting up with liquid compression at any rotational speeds.

8.2.2 Types of automotive air conditioning compressors

As discussed in Section 8.1, many types of compressors have been commercialized as a component of the automotive air conditioning systems. These include piston, radial and axial (wobble-plate and swash-plate) type reciprocating compressors as well as rotary vane, screw and scroll type compressors.

Details of compressors that are most commonly used in automotive air conditioners today, i.e. axial (wobble-plate and swash-plate types), scroll, and rotary vane compressors, will be described in Section 8.3 and thereafter.

8.2.3 Power transmission mechanism

To start and stop automotive air conditioning compressors, dry-friction-based single plate solenoid clutches are commonly used.

(1) Functions and characteristics

Most important functions of the solenoid clutch for an automotive air conditioner compressor are to transmit and to cut engine power to the compressor without stopping the engine rotation.

Main features requested of the solenoid clutch:
a) Simply structured, compact and lightweight
b) Capable of fast response with least amount of idling torque
c) Capable of being coupled and decoupled in a wide rotational speed range
d) Easily attachable to the compressor

(2) Structure and operating principle

Figure 8.3 shows an example of solenoid clutch. In principle, a solenoid clutch is comprised of an armature, a rotor and a core.

\<Armature\> The armature section consists of a boss, an armature plate and an elastic component (a set of three leaf springs in this example) connecting them. An approximately 0.5 mm clearance is provided between the armature plate and the rotor end face. The boss is coupled to the compressor shaft with a spline or other torque-transmitting feature, and locked with a nut.

\<Rotor\> The rotor section has a pulley on its circumference to receive power via belt transmission. The rotor is connected to the nose of the compressor housing with a bearing, and locked with a snap ring.

\<Core\> The core section consists of a U-shaped cross-section ring called "stator" and a donut-shaped solenoid coil fitted to the inside of the ring. The core is located in the recess of the rotor with an appropriate clearance between the rotor and itself, and fastened to the compressor housing with a snap ring.

\<Operation\> As shown in Fig. 8.4, when the coil is energized, magnetic fluxes run through the stator, the rotor

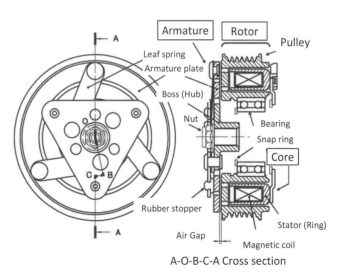

Fig. 8.3 Electromagnetic clutch

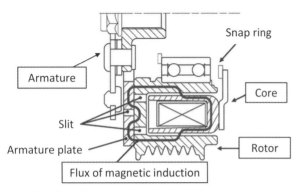

Fig. 8.4 Electromagnetic clutch (in operation)

and the armature so that the resulting magnetic force causes the armature plate to stick fast to the rotor end face. The frictional force between them allows the armature to rotate with the rotor, and torque of the rotor is transmitted to the compressor shaft via the leaf springs and the boss of the armature. When the coil is de-energized, magnetic force is lost and the armature plate is decoupled from the rotor, then it returns to its original position.

To increase the transmission torque, slits with different diameters are formed on the armature plate and rotor surface so as to provide more paths for magnetic flux between the armature plate and the rotor to generate stronger magnetic force.

Another possible structure of the elastic component of the armature is a ring-shaped rubber piece that connects the boss and the armature plate, as shown in Fig. 8.5. The inner periphery of the rubber piece is attached securely to a cylindrical part riveted to the boss while the outer periphery to another cylindrical part riveted to the armature plate. Use of the rubber piece has such advantages as reduction of coupling and decoupling noises, moderation of torsional resonance of the shafts and restriction of vibration transmission from the compressor to the vehicle. Therefore, this solution is preferred for applications where the issues described above need to be addressed.

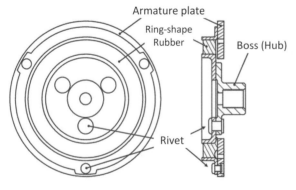

Fig. 8.5 Armature provided with a ring-shaped rubber

8.3 Axial Compressors

Reciprocating piston compressors operate with a circular cylinder and a piston reciprocating inside the cylinder. This type of compressor can be further classified into smaller groups depending on how the shaft rotation is converted into the piston reciprocating motion.

Crank-type piston compressors use a crankshaft and a connecting rod to move the piston back and forth as in a reciprocating internal combustion engine. The cylinder is positioned at the right angle to the shaft axis. Radial type reciprocating compressors (including Scotch yoke models) have cylinders perpendicular to the shaft axis and positioned at the left, right, top and bottom around the shaft axis. The Scotch yoke mechanism converts the shaft rotation into the piston reciprocating motion. Axial type reciprocating compressors have multiple cylinders parallel to the shaft axis, surrounding the shaft axis in a concentric manner. A tilted plate coupled to the shaft is turned to induce a reciprocating motion of the piston.

These piston-type compressors are advantageous in that they operate stably through a wide operation range, but they tend to generate noise and vibration due to the stepwise process of the suction, compression, discharge and re-expansion phases. The excitation force that causes noise and vibration can be dispersed and minimized by having a larger number of cylinders. In the field of automotive air conditioning, five- to ten-cylinder axial compressors are replacing compressors with fewer number of cylinders, such as two-cylinder crank type and four-cylinder radial-type compressors.

Axial reciprocating compressors can be classified into the wobble plate type and the swash plate type.

8.3.1 Wobble plate compressors
(1) Structure

Figure 8.6 shows the cross section of a seven-cylinder wobble plate compressor without the solenoid clutch in place. The wobble plate compressor comprises a front housing with a nose to which a solenoid clutch is coupled, a motion conversion mechanism that converts the shaft rotation into the piston reciprocating motion and is housed in the space called a crank chamber, pistons and cylinders that perform suction and compression of refrigerant gas, a valve assembly that controls suction and discharge of gas, and a cylinder head that separates the suction and discharge chambers and also provides a connecting channel to the refrigerant circuit. In addition, the circumference of the compressor housing has mounting features that allow the

Chapter 8 Automotive Air Conditioning Compressors

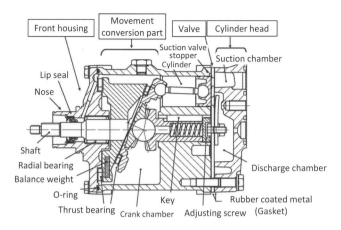

Fig. 8.6 Wobble plate type compressor cross section (without electromagnetic clutch)

compressor to be attached to the vehicle engine (not shown in Fig. 8.6).

\<Shaft sealing mechanism\> An open type compressor for an automotive air conditioning system has a shaft sealing mechanism called "lip seal", fitted to the area where the driving shaft penetrates the compressor housing. Figure 8.7 shows an example of such lip seal.

Outer periphery of a rubber part that is molded together with a rubber lip is pressed into and seals against the inner periphery of the compressor nose. The lip seal provides a double-stage sealing structure consisting of the rubber lip and a PTFE lip. The force pressing the rubber lip on the shaft increases in proportion to the sealing pressure. A metallic retainer partially supports this force, which reduces the force of contact between the rubber lip and the shaft and thereby alleviates friction heat and abrasion. This type of shaft sealing mechanism is mainly employed on variable-capacity compressors (described later) that are designed for high speed rotation and high internal pressure[16].

\<Motion conversion mechanism\> Figures 8.6 and 8.8 show the motion conversion mechanism, which consists of a pair of bevel gears and works with a shaft rotor, a wobble plate and a center ball.

The shaft rotor consists of a shaft and a rotor having a shape like an obliquely cut cylindrical column. The rotor is tightly fixed to the shaft and is rotatably supported by a radial bearing mounted on the front housing and thrust bearings attached to both end faces of the rotor. One of the thrust bearing on the inclined face of the rotor supports the wobble plate and also receives reaction force of gas compression from the pistons via connecting rods.

The pair of bevel gears are shaped to have a spherical socket whose center is at the intersection point of both gear axes. One of the gears is fixed tightly to the wobble plate, and the other with a shank is inserted into the center hole of the cylinder block with an anti-rotational parallel key. The center ball is placed in the spherical socket of the bevel gears, supporting the gear axes and also preventing the wobble pate from slipping down on the rotor slope. Axial clearances between components are adjusted by a spring inserted in the shank center plus an adjusting screw fitted at the rear end of the shank.

The wobble plate and the piston are linked together by the connecting rod with a steel ball welded to each end. The aluminum alloy wobble plate and pistons have a spherical socket with a ring-shaped top over the socket. After fitting each steel ball on the end of the connecting rod into each spherical socket on the wobble plate and the piston, the ring over the socket is plastically deformed to comply with the steel ball, thus constituting a ball joint.

Both the piston and the cylinder are built with aluminum alloy. When parts made of the same metallic material slide against each other, adhesion may occur between them. To avoid this, direct metallic contact is prevented by either using a plastic piston ring or to provide plastic coating on the

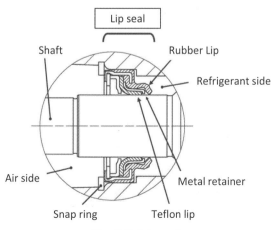

Fig. 8.7 Lip seal

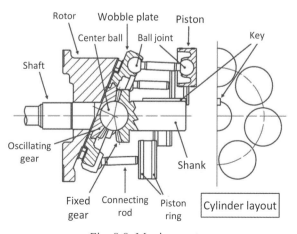

Fig. 8.8 Moving parts

159

piston circumference.

<Valve assembly> As shown in Fig. 8.9, the valve assembly comprises a valve plate, a suction valve on one side of the plate and a discharge valve on the other and a discharge valve retainer that restricts the maximum opening of the discharge valve, and the all parts are fastened together by screws. Like the suction and discharge valves shown in Fig. 8.9, the valve that is fixed at one end of a spring sheet and loosened to open at the other end is called "reed valve".

<Joint sealing> To keep the space inside of the compressor airtight, a rubber coated steel plate is inserted between mating components such as the cylinder block, the valve plate and the cylinder head, and is bolted together. Rubber rings such as square rings and O-rings are also used to seal circular joint areas such as between the front housing and the cylinder block and between piping sections.

(2) Operation

As the shaft rotor rotates, the wobble plate rotates with swing motion around the center ball, whose action is called wobble motion. This is a result of engagement between two bevel gears, one of which is fixed to the cylinder block with the anti-rotational stopper and the other attached to the center of the wobble plate. This wobble movement causes the piston to reciprocate along the inner wall of the cylinder.

As the piston moves down from its top dead center (in Fig. 8.8, the right end represents the top dead center and the piston actually moves from right to left), gas remaining in the dead clearance space begins to expand. When the pressure in the cylinder has dropped below the pressure in the suction chamber of the cylinder head, the suction valve opens to let the gas flow into the cylinder. The maximum lift of the suction valve is restricted by a valve stopper which is a notch dug at the top of the cylinder wall, thus preventing valve damage due to excessive opening and also reducing self-excited vibration that may occur when gas flows in. As the piston moves past its bottom dead center and upward with the suction valve closed, the gas in the cylinder is compressed. When the cylinder pressure has increased above the pressure in the discharge chamber of the cylinder head, the discharge valve opens to let the gas flow out. The discharge valve retainer restricting the maximum of discharge valve lift serves to prevent damage of the discharge valve and to reduce self-excited vibration.

(3) Gas compression work and shaft torque

Referring to Fig. 8.10 in which pistons both at the top dead center and at the bottom dead center are illustrated just for explanation, the piston stroke S_t and the displacement volume V can be expressed as follows:

$$S_t = 2r_1 \sin \alpha \tag{8.1}$$

$$V = \pi n S_t D^2 / 4 \tag{8.2}$$

where
r_1: pitch circle radius of the wobble plate ball joint
α : inclination angle of the rotor slope
n : number of cylinders
D: cylinder diameter

Figure 8.11 shows a status of the compressor whose

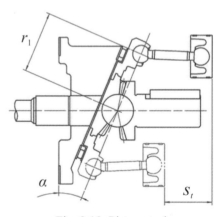

Fig. 8.10 Piston stroke

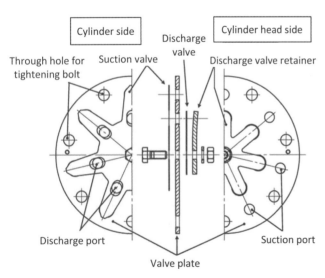

Fig. 8.9 Valve plate assembly

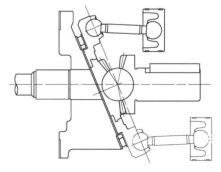

Fig. 8.11 Piston stroke (180 degree rotation)

Chapter 8 Automotive Air Conditioning Compressors

shaft has rotated by 180° from the position shown in Fig. 8.10. Comparing two figures, the bevel gear engagement position has shifted downward, the piston located at the top dead center has shifted to the bottom dead center and the one located at the bottom dead center has shifted to the top dead center.

The relationship between the shaft rotation angle θ and the piston position S can be expressed as follows:

$$S = s_0 + S_t (1 - \cos\theta)/2 \qquad (8.3)$$

where,

reference angle ($\theta = 0$): Top dead center

s_0: top clearance (here, obtained by dividing top clearance volume by cylinder cross section area)

Although a slight tilt in the connecting rod influences the piston position and the friction loss in the sliding areas affects the amount of work, both of these factors are ignored here for the sake of simplification.

Here, let's consider the minutely small work done while the shaft rotates for $\Delta\theta$. With the shaft torque T_1, the shaft rotation work Δw_s is expressed as follows:

$$\triangle w_s = T_1 \triangle \theta \qquad (8.4)$$

And the amount of work Δw_p done by the piston during the time is:

$$\triangle w_p = -(P_2(\theta) - P_c) A \triangle S \qquad (8.5)$$

As the above amounts of work are equal, an equation for the relationship between shaft rotation angle θ and shaft torque T_1 can be obtained. The reason for the negative sign added to the beginning of the right side of Eq. (8.5) is that the pressure direction and the piston movement direction are opposite. Each symbol has the following meaning:

$P_2(\theta)$: pressure in the cylinder at shaft rotation angle θ

P_c: piston's back pressure (= pressure in the crank case)

A: cross-section area of the cylinder (= $\pi D^2/4$)

From Eqs. (8.4) and (8.5), the following is obtained:

$$\begin{aligned} T_1 &= \triangle w_p / \triangle \theta \\ &= -(P_2(\theta) - P_c) A \triangle S / \triangle \theta \end{aligned} \qquad (8.6)$$

On the other hand, the minutely small amount ΔS of the piston stroke is:

$$\begin{aligned} \triangle S &= -S_t (\cos(\theta + \triangle\theta) - \cos\theta)/2 \\ &= S_t (\cos\theta - \cos\theta\cos\triangle\theta + \sin\theta\sin\triangle\theta)/2 \\ &= S_t (\cos\theta(1 - \cos\triangle\theta) + \sin\theta\sin\triangle\theta)/2 \end{aligned} \qquad (8.7)$$

$$\begin{aligned} (1 - \cos\triangle\theta)/\triangle\theta &= 0 \\ \sin\theta\sin\triangle\theta/\triangle\theta &= \sin\theta \end{aligned} \qquad (8.8)$$

By substituting Eqs. (8.7) and (8.8) into Eq. (8.6) to obtain the amount of torque that is generated at each of the individual pistons and summing them up, the shaft torque T can be obtained. With the number of cylinders n, the amount of torque that is generated at i-th piston, T_i, is:

$$T_i = (P_c - P_{2i}(\theta)) A S_t \sin\theta_i/2 \qquad (8.9)$$

where,

$P_{2i}(\theta)$: pressure in the i-th cylinder at shaft rotation angle θ

$\theta_i : \theta + 2\pi (i-1)/n$

Therefore:

$$T = \sum_{i=1}^{n} (P_c - P_{2i}(\theta)) A S_t \sin(\theta + 2\pi(i-1)/n)/2 \qquad (8.10)$$

After obtaining the cylinder pressure $P_2(\theta)$ at the shaft rotation angle θ by theoretical calculation or actual measurement, the shaft driving torque and the compression power can be calculated. The compressor efficiency and loss are estimated based on practical measurement of the shaft driving torque.

(4) Lubrication of wobble plate compressors

Places where lubrication is required in the wobble plate compressor are the motion conversion mechanism where shaft rotation is converted into the piston's reciprocating motion and the portion between the piston ring and the cylinder wall where the sliding motion exists by the reciprocating motion of the piston. The motion conversion mechanism is lubricated by oil that remains in the crank chamber and is stirred by shaft rotor rotation. The piston ring/cylinder portion is lubricated mainly by the oil migrating through the refrigeration cycle with refrigerant, along with the oil in the crank chamber.

There are two different gas flow paths through the compressor. The main one is that refrigerant goes through the suction port, the suction chamber and the suction valve, then compressed in the cylinder and delivered through the discharge valve, the discharge chamber and the

161

discharge port to the refrigeration cycle. The other one is that a portion of refrigerant sucked in to the cylinder leaks through a clearance between the piston and the cylinder into the crank chamber, which is called "blowby gas", and then goes through a balance hole connecting the crank chamber and the suction chamber and returns to the suction chamber.

Part of the lubrication oil that is sucked into the cylinder together with the refrigerant adheres on the cylinder wall and then flows into the crank chamber with the blowby gas. In the crank chamber, the oil is forced to present near the inner wall of the crank chamber by the centrifugal force of the shaft rotor rotation. By locating the balance hole opening near the center axis and away from the inner wall, the amount of oil that escapes from the crank chamber can be reduced, and also it is easier to collect and recover the oil that has migrated into the refrigeration circuit. Note that this solution is most effective from low to medium speed ranges of the compressor where the effect of stirring is relatively low. In higher speed ranges, where the stirring effect becomes greater, lubrication oil in the crank chamber takes the form of floating mist and therefore the effective separation of the mist oil by the balance hole is limited. Under the high speed condition, lubrication in the compressor is entirely relied on the mist oil.

8.3.2 Swash plate compressors
(1) Structure

Figure 8.12 shows a solenoid-clutch-fitted ten-cylinder swash plate compressor. Two compression mechanisms are located across a swash plate in an almost symmetrical structure. Each of the compression mechanisms is comprised of cylinders, a valve assembly and a housing with suction and discharge chambers. A solenoid clutch that transmits and cuts driving power to the shaft is coupled to one of the housings, i.e. front housing. Between the front and rear housings a cylinder block and valve plates are sandwiched, and the whole assembly is fasten tightly with multiple through bolts. Mating surfaces are sealed with a rubber coated steel plate or with other sealing features. The shaft is sealed with a lip seal.

The shaft rotor is fitted with a tilting disk called "swash plate" which is circular when viewed in the shaft axis direction but oval when viewed in the direction normal to itself. The shaft rotor is supported against the cylinder block with the combination of two radial bearings plus the thrust bearing on each end of the swash plate boss. The cylinders are positioned around the shaft at equal intervals and each cylinder is equipped with a double-headed piston which has pistons at both ends. The swash plate and the piston is coupled together with a pair of approximate hemispherical shoes. The spherical side of the shoe contacts a spherical socket of the piston while the flat side contacts the swash plate surface. The shape and material composition of each contact face is designed for high slidability so that it can withstand the high-speed high-load sliding motion. Each piston has a recess at the center to accept the swash plate. This recess is mated with and is guided by the swash plate circumference. This guiding engagement prevents the piston from rotation about its own axis in the cylinder.

Similar to those of a wobble plate compressor, both the pistons and the cylinders are made of aluminum alloy. To prevent adhesion between these components, a plastic piston ring or the plastic coating of piston circumference is used.

(2) Operation

As the shaft rotates, each piston movers back and forth in the cylinder with the function of the swash plate and the shoes. Each shoe slides on the swash plate surface while swinging around the center of the piston socket which is mated with the shoe. When a double-headed piston is in operation, processes in the two cylinders relating to both sides of the piston are different from each other. For example, when a suction stage is underway in one cylinder, compression and discharge stages occurs in the other cylinder.

As the compressor operates, refrigerant gas goes through the suction port into the crank chamber, and then flows into the right- and left-side suction chambers through suction connecting passages provided through the cylinder block. The gas is sucked through the suction valve into the cylinder and compressed in the cylinder, then delivered

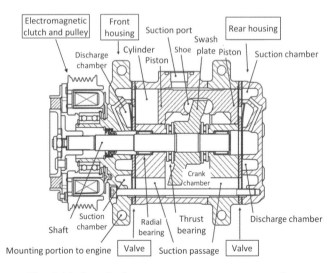

Fig. 8.12 Swash plate type compressor cross section (with electromagnetic clutch)

through the discharge valve into the discharge chamber. The gas in the right- and left-side discharge chambers joins together through discharge connecting passages (not shown) bored through the cylinder block, and flows out through a discharge port (not shown) into the refrigeration cycle.

Although the pistons tend to slide down on the inclined plane of the swash plate by the reaction force of compression via the shoes, they are axially supported by the swash plate and radially by the cylinders.

The clearance between each shoe and the swash plate can be kept constant during shaft rotation. By referring to Fig. 8.13, with the swash plate thickness t, the swash plate tilt angle α to the plane normal to the shaft, and the shoe radius r and height h respectively, the distance L between the shoe tips is expressed as:

$$L = 2r + (t + 2(h-r))/\cos\alpha = \text{Constant} \quad (8.11)$$

With this, the amount of clearance set in the assembly process is maintained.

(3) Gas compression work and shaft torque

By referring to Fig. 8.13, the piston stroke S_t and the displacement volume V can be expressed as follows:

$$S_t = 2r_0 \tan\alpha \quad (8.12)$$

$$V = \pi n S_t D^2/4 \quad (8.13)$$

where,
r_0: cylinder pitch circle radius
α: swash plate tilt angle
n: number of cylinders

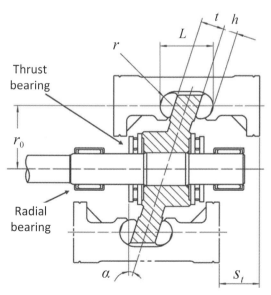

Fig. 8.13 Piston stroke

D: cylinder diameter

The piston stroke for a swash plate compressor is calculated using the product of cylinder pitch circle radius r_0 and $\tan\alpha$ as in Eq. (8.12), while one for a wobble plate compressor using the product of wobble plate ball joint pitch circle radius r_1 and $\sin\alpha$ as in Eq. (8.1). The shaft torque for a swash plate compressor can be obtained in the same way as for a wobble plate compressor.

Each piston head of the double-headed piston works with a pair of cylinders on both sides of the swash plate. Putting the position of each piston head as zero at its top dead center, the position S_f of a given piston head in the cylinder at shaft rotation angle θ is:

$$S_f = s_0 + S_t(1 - \cos\theta)/2 \quad (8.14)$$

where,
Reference angle ($\theta = 0$): top dead center
s_0: top clearance (here, obtained by dividing top clearance volume by cylinder cross section area)

As the other piston head opposing the given piston head works with a phase difference of π, its position is expressed as follows:

$$S_r = s_0 + S_t(1 + \cos\theta)/2 \quad (8.15)$$

With a shaft drive torque T_1 for a double-headed piston, the relationship between a minute shaft work Δw_s and a minute shaft rotation $\Delta\theta$ is expressed as follows:

$$\Delta w_s = T_1 \Delta\theta \quad (8.16)$$

On the other hand, the minute work Δw_p that accompanies the double-headed piston with a minute movement ΔS at one piston head and a minute movement $-\Delta S$ at the other piston head is expressed as follows:

$$\begin{aligned}\Delta w_p &= -(P_2(\theta) - P_c)A\Delta S + (P_3(\theta) - P_c)A\Delta S \\ &= -(P_2(\theta) - P_3(\theta))A\Delta S\end{aligned} \quad (8.17)$$

where,
$P_2(\theta)$: pressure in a given cylinder at shaft rotation angle θ
$P_3(\theta)$: pressure in the opposing cylinder at shaft rotation angle θ
P_c: piston back pressure (= pressure in the crank case)
A: cylinder cross section (= $\pi D^2/4$)

The torque that is required for i-th double-headed piston is:

$$T_i = (P_{3i}(\theta) - P_{2i}(\theta))AS_t \sin\theta_i/2 \quad (8.18)$$

With the total number n_p of the double-headed piston, which is equal to a half of the cylinder number n, the total torque T of the shaft is expressed as follows

$$T = \sum_{i=1}^{n_p}(P_{3i}(\theta) - P_{2i}(\theta))AS_t \sin(\theta + 2\pi(i-1)/n_p)/2 \quad (8.19)$$

In the same way as explained in Section 8.3.1 (3), the above torque is applied to the calculation of compression power requirement, efficiency and loss of the compressor.

(4) Rotational balance

In theory, rotational balance of the swash plate compressor can be achieved perfectly. The center of weight of the swash plate, which is a rotating body, matches its shaft axis and therefore the amount of static unbalance of the compressor is zero. Concerning its dynamic unbalance, by referring to Fig. 8.14, moment M_s that acts to reduce the swash plate inclination due to mass distribution on the swash plate is generated. At the same time, moment M_p that acts to increase the swash plate inclination by the reciprocating inertia of the pistons and the shoes is also generated. As M_s and M_p work in opposite directions, dynamic unbalance can be theoretically canceled out to zero.

In practice, the swash plate compressor is designed to accommodate itself to some unbalance to the extent that the unbalance will not be detrimental in high rotational speed range. The swash plate compressor is generally a machine with superior rotational balance as compared to other reciprocating piston compressors.

Each moment involved can be calculated as follows. The

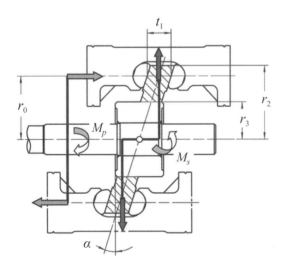

Fig. 8.14 Dynamic balance of swash plate type compressor

coupled moment M_s of the swash plate, with the clockwise direction defined as the positive rotation direction, is expressed as follows:

$$M_s = -U_s \omega^2 \quad (8.20)$$

$$U_s = m_s(r_2^2 + r_3^2)\tan\alpha/4 \quad (8.21)$$

$$m_s = \pi\rho(r_2^2 - r_3^2)t_1 \quad (8.22)$$

where,

U_s: coupled unbalance of the swash plate
m_s: swash plate mass (shaded area in Fig. 8.14)
r_2: swash plate outside diameter (normal to the shaft axis)
r_3: swash plate inside diameter (normal to the shaft axis)
t_1: swash plate thickness (in the shaft axis direction)
ρ: swash plate density

The coupled moment M_p of the pistons is:

$$M_p = U_p \omega^2 \quad (8.23)$$

$$U_p = n_p(m_p + 2m_{sh})r_0^2 \tan\alpha/2 \quad (8.24)$$

where,

U_p: couple unbalance of pistons and shoes
m_p: mass of pistons
m_{sh}: mass of shoes
r_0: radius of cylinder pitch circle
n_p: number of double-headed piston

(5) Torsional resonance of shaft in swash-plate compressors

Due to having cylinders on both sides of the swash plate, the swash plate compressor requires a relatively long shaft, which decreases the torsional rigidity of the shaft. One end of the shaft is coupled with the clutch armature and rotor and the other end is coupled with the swash plate and pistons (Figure 8.15). The natural vibration frequency f_0 of the shaft is determined by inertia moments on the shaft ends and torsional rigidity of the shaft. When this natural frequency coincides with frequency f_c of the shaft torque fluctuation accompanying with compressor operation, torsional resonance occurs.

A torsional resonance has a risk of causing shaft damage and therefore must be avoided. For that purpose, such solutions are used as increasing the natural frequency by reduction of inertia moment with enhancement of the shaft rigidity and/or by reduction of the component weight, or reducing the resonance by applying a ring rubber-type

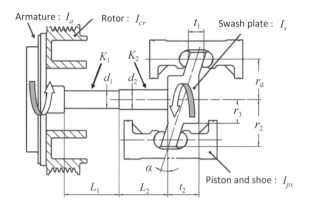

Fig. 8.15 Element related to shaft torsional vibration

armature like the one shown in Fig. 8.5.

With the inertia moments I_1 and I_2 on both ends of the shaft and the torsional spring coefficient K of the shaft, the natural frequency f_0 of torsional vibration of the shaft is expressed as follows under the condition of the shaft mass ignored:

$$f_0 = \frac{1}{2\pi}\sqrt{\left(\frac{1}{I_1}+\frac{1}{I_2}\right)K} \qquad (8.25)$$

where, by referring to Fig. 8.15:

$$I_1 = I_a + I_{cr} \qquad (8.26)$$

I_a: inertia moment of the armature
I_{cr}: inertia moment of the clutch rotor

$$I_2 = I_s + I_{ps} \qquad (8.27)$$

I_s: total inertia moment of the swash plate and the swash plate boss
I_{ps}: reciprocating inertia moment of the pistons and shoes

These values can be calculated by the following equations: Any symbols that have appeared in Fig. 8.14 have the same meaning here.

$$I_s = m_s(r_2^2 + r_3^2)/2 + m_b(r_3^2 + (d_2/2)^2)/2 \qquad (8.28)$$

$$I_{ps} = n_p(m_p + 2m_{sh})r_0^2 \tan^2 \alpha/2 \qquad (8.29)$$

m_b: mass of the swash plate boss

Chapter 8 Automotive Air Conditioning Compressors

$$m_b = \pi\rho(r_3^2 - (d_2/2)^2)t_2 \qquad (8.30)$$

t_2: swash plate boss width (in the shaft axis direction)
The torsional spring coefficient K is:

$$\frac{1}{K} = \frac{1}{K_1} + \frac{1}{K_2} \qquad (8.31)$$

K_1: torsional spring coefficient at diameter d_1

$$K_1 = G\pi d_1^4/32L_1 \qquad (8.32)$$

K_2: torsional spring coefficient at diameter d_2

$$K_2 = G\pi d_2^4/32L_2 \qquad (8.33)$$

where,
G: modulus of rigidity (=shearing modulus), $G = E/\{2(1+v)\}$
E: modulus of longitudinal elasticity (= Young's modulus)
v: Poisson's ratio
d_1, d_2: diameters of the stepped shaft
L_1, L_2: lengths of the stepped shaft

With the compressor rotational speed N and the number n of cylinders, the torque fluctuation frequency f_c of the compressor shaft, which is an excitation source, is expressed by the following equation:

$$f_c = nN \qquad (8.34)$$

(6) Lubrication of swash plate compressors

In the example shown in Fig. 8.12, lubrication oil that circulates together with refrigerant is directly fed to the sliding areas of the compressor. Lubrication oil is directed to the suction port together with the refrigerant and sprayed from the suction port into the center of the crank chamber to be fed to various sliding areas and the bearings.

8.3.3 Variable capacity compressors

One solution to adjust the cooling capacity of an automotive air conditioner is to change the stroke volume of the compressor. The following paragraphs explain about variable capacity compressors where the stroke volume is adjusted by changing the piston stroke.

Such variable capacity compressors can be classified into the wobble plate type and the swash plate type depending on their structure, and also into internally controlled and externally controlled types depending on the control method.

In the basics, a variable capacity compressor adjusts its stroke volume by changing the tilt angle of the swash plate, which results in change in the piston stoke. The angle is changed effectively utilizing the force of the pressure in the crank chamber, which works on the back of the pistons. Posture of the swash plate is controlled in such a manner as the bottom dead center of the piston sifts nearer to the top dead center while the top dead center hardly changes its position. This is effective to reduce the efficiency drop of the compressor when the piston stroke is shortened.

(1) Structure
<Variable capacity wobble plate compressors>
Various types of structures are possible to support the swash plate and also to prevent the rotation of the wobble plate. Figure 8.16 shows the cross section of an internally controlled variable capacity wobble plate compressor as an example.

As shown in Fig. 8.17, a center sleeve is fixed to the inside of a boss of the swash plate with a pair of sleeve pins. The shaft is then inserted into the center sleeve, and a slotted hole on an arm of the swash plate is linked to arms on a shaft rotor with a swash plate support pin. As shown in Figs. 8.16 and 8.18, the wobble plate is rotatably linked to the swash plate boss by the combination of a radial bearing set inside the wobble plate concentrically with the axis of the swash plate, a thrust bearing on the swash plate and a balance ring fixed to the boss end. At the same time, rotation of the wobble plate around its own axis is blocked by the combination of a slider and a monorail

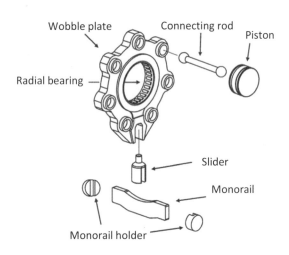

Fig. 8.18 Anti-rotation mechanism of wobble plate

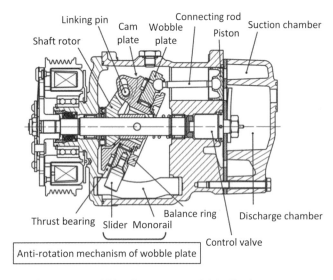

Fig. 8.16 Wobble plate type variable displacement compressor

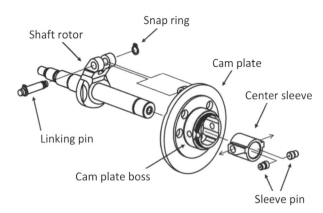

Fig. 8.17 Assembly of shaft rotor and swash plate

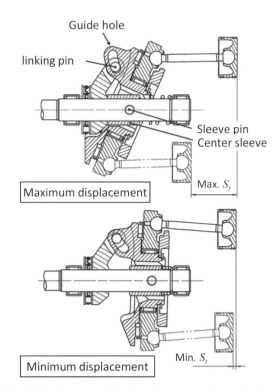

Fig. 8.19 Change of Inclination angle of swash plate

In the rotation block mechanism shown in Fig. 8.18, a cylindrical part of the slider and a slider holding part of the wobble plate form a cylindrical contact, thus allowing a certain degree of angle variation between the wobble plate and the monorail. A slit of the slider is in sliding contact with the monorail, thus allowing the wobble plate to move in the axial direction. A pair of monorail holders coupled to holes on the casing support the monorail, and cylindrical outer surfaces of the monorail holders allow the holders to self-align so as to maintain a surface contact between the monorail and the slider slit.

The pistons are linked to the rim of the wobble plate with approximately equal intervals via connecting rods.

To reduce the stroke volume, as shown in Fig. 8.19, the swash plate support pin, the swash plate arm slotted hole and the center sleeve work together to shift the swash plate toward the cylinder so as to reduce the piston stroke.

<Variable capacity swash plate compressors>

Figure 8.20 shows the cross section of an externally controlled variable capacity swash plate compressor as an example. (This compressor is equipped with a torque limiter and a pulley in place of a solenoid clutch. For further details, refer to Section 8.3.3 (5)).

Unlike the fixed-volume model explained in Section 8.3.2 which has cylinders on both sides, this compressor has cylinders only on one side. As each piston is supported by each cylinder, the compressor has a relatively-long mated connection between the cylinder and the piston.

To accommodate tilt angle variation of the swash plate, a shoe retainer on the piston forms a spherical surface, as shown in Fig. 8.21. The reason for this is, as shown in Fig. 8.13, to keep the shoe-tip length L of Eq. (8.11) constant,

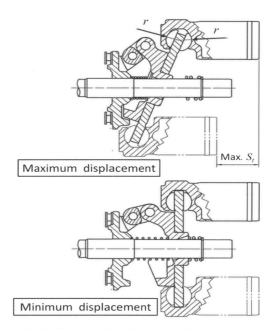

Fig. 8.21 Change of inclination angle of swash plate

regardless of the swash plate tilt angle α. In Eq. (8.11), if the second term $t + 2(h - r)$ is zero, namely $2r = t + 2h$, then L becomes constant equal to $2r$, which means that the shoe retainer must have a spherical surface with radius r.

The piston also has a collar on the other side of the shoe retainer. The collar has approximately the same curvature as an inner radius of the crank chamber, and is guided along the inner wall of the crank chamber to prevent the piston from rotating about its own axis inside the cylinder.

Linkage between a swash plate arm and a rotor arm works to allow the swash plate tilt angle to change and also to allow the swash plate to shift in axial direction. Moreover, a saddle-shaped hole made at the center of the swash plate boss guides the center of the swash plate to remain in line with the shaft axis. With this linkage, the swash plate shifts closer to the cylinder when the swash plate tilt angle becomes smaller. The saddle-shaped hole, which has a narrower center and a wider end on each side, is made by first drilling a pair of intersecting through holes based on the swash plate's maximum and minimum tilt angles and then removing obstructing areas to allow the shaft angle to change freely in the range defined by such intersecting holes.

Similar to the previously explained wobble plate compressor, the above-described linkage can change the stroke volume of the compressor without hardly changing the position of the piston top dead center.

(2) Operating principle

The stroke volume control is achieved by regulating pressure in the crank chamber. As shown in Fig. 8.22, the

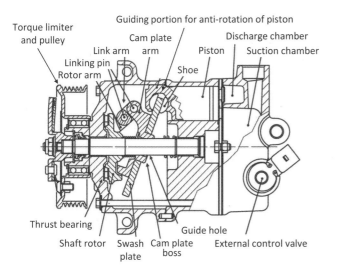

Fig. 8.20 Swash plate type variable displacement compressor

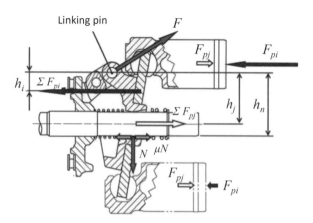

Fig. 8.22 Forces act on the swash plate

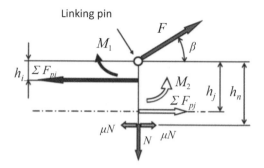

Fig. 8.23 Moment around linking pin

gas compression reaction force works to increase the swash plate tilt angle by pushing down the piston (toward the left in the figure). On the other hand, the pressure in the crank chamber works to reduce the tilt angle by pushing up the piston (toward the right in the figure).

In a fixed capacity compressor, the crank chamber and the suction chamber are connected to each other so that both chambers will be at the same pressure. On the other hand, a variable capacity compressor has a valve or a throttle in the connecting channel between the crank chamber and the suction chamber so as to maintain the crank chamber pressure higher than the suction pressure to reduce the swash plate tilt angle.

The following paragraphs explain how the swash plate tilt angle is controlled in a viewpoint of both forces acting on the swash plate; one caused by the pressure in the cylinder and the other by the pressure in the crank chamber. In a practically operating compressor, the pressure in the cylinder is higher when the piston is near the top dead center, while it is lower when the piston is near the bottom dead center. The resultant force ΣF_{pi} (see Fig. 8.22) that is caused by the cylinder pressure and acting on the swash plate via the pistons acts on the top dead center side of the shaft axis. As shown in Fig. 8.23, a moment M_1 that acts to increase the swash plate tilt angle can be generated by supporting the swash plate at a point outside the acting point of the resultant force.

As the cylinders are positioned at equal intervals surrounding the shaft axis, the resultant force ΣF_{pj} that is caused by the crank chamber pressure and acting on the swash plate via the pistons acts on the center of the swash plate, which arises a moment M_2 that forces to decrease the swash plate tilt angle.

Based on the difference in arm lengths between the resultant forces ΣF_{pi} and ΣF_{pj}, a point to support the swash plate can be selected so as to appropriate for the crank chamber pressure-based capacity control.

The swash plate tilt angle can be controlled as follows: If the crank chamber pressure is maintained at a level of $M_1 = M_2$, the tilt angle remains unchanged. If the crank chamber pressure is increased above the level, the tilt angle decreases. If the crank chamber pressure is decreased below the level, the tilt angle increases. A normal force N generated with link tilt angle β causes frictional resistance μN between the swash plate boss bore and the shaft, which interferes the change of the swash plate tilt angle. After all, $M_1 \geq M_2 + M_r$ is required to increase the tilt angle, and $M_1 \leq M_2 - M_r$ is required to reduce it. Thus the frictional force works to restrict fluctuation in the swash plate tilt angle.

The relationships between forces and moments illustrated in Fig. 8.23 can be expressed by the following equations:

Horizontal axis: $\sum F_{pi} = F \cos \beta + \sum F_{pj}$ (8.35)

Vertical axis: $N = F \tan \beta$ (8.36)

Moment to increase angle: $M_1 = h_i \sum F_{pi}$ (8.37)

Moment to decrease angle: $M_2 = h_j \sum F_{pj}$ (8.38)

Moment by frictional force: $M_r = h_n \mu N$ (8.39)

(In addition to the forces explained above, there are other forces that work on the swash plate, including the reaction force from the cylinder that works to prevent the pistons from sliding down along the swash plate plane and the forces of the springs provided on both sides of the swash plate, but these forces are not explained in any further details here.)

Concerning the rotational balance of the swash plate compressor explained in Section 8.3.2 (4), inertia forces generated by the swash plate rotation and the piston reciprocation bring about moments acting on the swash plate. These moment must be carefully evaluated as they affect the tilt angle control in high rotation speed ranges.

As shown in Fig. 8.24, the momental components that increase the swash plate tilt angle are moment M_{pv} generated by the piston reciprocation and moment M_a generated by the eccentric mass of the swash plate arm and the pin linkage. On the other hand, the momental components that decrease the tilt angle are moment M_{sv} generated by the swash plate itself and moment M_b caused by a balance weight on the swash plate. As these inertia moments are proportional to the square of the angular speed, all related mechanical components must be designed so that these moments does not cause detrimental effects in high rotation speed ranges. To accommodate the swash plate to change its tilt angle, the swash plate needs to be designed relatively thin with smaller M_{sv}. Therefore, it is critical to minimize the piston weight to reduce M_{pv}. In the example shown in Fig. 8.24, a thin-walled hollow piston is used to provide both strength and light weight.

Each moment involved can be calculated as follows:

a) Moment by piston and shoes M_{pv}

$$M_{pv} = U_{pv}\omega^2 \tag{8.40}$$

$$U_{pv} = n_p(m_p + 2m_{sh})r_0^2 \tan\alpha/2 \tag{8.41}$$

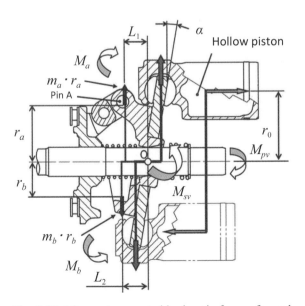

Fig. 8.24 Moment generated by inertia force of swash plate

For the meaning of the symbols, refer to Section 8.3.2 (4).

b) Moment by swash plate itself M_{sv}

$$M_{sv} = -U_{sv}\omega^2 \tag{8.42}$$

$$U_{sv} = m_{sv}(r_4^2 + r_5^2)\sin\alpha\cos\alpha/4 \tag{8.43}$$

m_{sv}: swash plate mass $[= \pi p(r_4^2 - r_5^2)t]$
r_4: swash plate outer diameter (on the swash plate plane)
r_5: swash plate inner diameter (on the swash plate face)
t: swash plate thickness (in the direction along the axis normal to the swash plate plane)

As described in parentheses, r_4, r_5 and t are dimensions on the swash plate plane or along its normal axis. On the other hand, the values used in Fig. 8.14 are dimensions on the plane normal to the shaft or along the shaft axis. Note that the two sets of values relate to different axes. The cylindrically-shaped boss of the swash plate must also be included in the swash plate moment by the same calculation method as for the swash plate.

c) Moment M_a by mass of swash plate arm and pin linkage

$$M_a = U_a\omega^2 \tag{8.44}$$

$$U_a = m_a r_a L_1 \tag{8.45}$$

m_a: mass of the swash plate arm and pin A, plus partial mass of the linkage not allocated to the rotor
r_a: distance from the rotation axis to the center of m_a
L_1: axial distance between point O and the center of m_a

d) Swash plate balance weight M_b

$$M_b = -U_b\omega^2 \tag{8.46}$$

$$U_b = m_b r_b L_2 \tag{8.47}$$

m_b: mass of the swash plate balance weight
r_b: distance from the rotation axis to the center of m_b
L_2: axial distance between point O and the center of m_b

Note: Plus and minus of each moment is given on the basis that the direction to increase the swash plate tilt angle is positive.

(3) Capacity control mechanisms

To induce a larger amount of heat exchange in the evaporator, evaporator temperature needs to be decreased. However, if the evaporator temperature is decreased too low, moisture in the air may cause frosting of the evaporator, which reduces its heat exchange efficiency.

It is needed that the evaporator temperature is maintained around 0°C, which is the lowest possible temperature where frosting can be avoided. As evaporating temperature of refrigerant is in a one-to-one relationship with its pressure, maintaining a constant evaporating pressure leads to keeping the temperature constant. The R134a refrigerant that is commonly used in automotive air conditioners has an evaporation temperature of approximately 0°C at a pressure of 0.3 MPa. Therefore, in a variable-capacity compressor, the suction flow rate into the compressor is controlled so that pressure inside the evaporator does not drop below approximately 0.3 MPa when engine speed is being increased or when cooling load is being decreased.

In situations that the suction pressure is higher than a setting pressure (generally 0.3 MPa but changeable depending on the situation) at the early stage of the cooling or at the high load operation, pressure in the crank chamber is controlled at the same level as the suction pressure so that the compressor runs at its maximum capacity. On the other hand, when the suction pressure drops below the setting pressure due to the cooling capacity being higher than the thermal load, pressure in the crank chamber is increased to reduce the compressor stroke volume. When the suction pressure is approximately the same as the setting pressure, the crank chamber pressure is maintained at a level where the moments to increase and decrease the swash plate tilt angle are balanced and the compressor is operated with an intermediate stroke volume.

(4) Capacity control valve

Even though a vehicle is running with always varying engine speed and thermal load, a capacity control valve serves to maintain an appropriate stroke volume of the compressor by instantaneously controlling the crank chamber pressure in response to the engine speed and the thermal load.

For example, increasing the crank chamber pressure can be achieved by introducing higher pressure gas from the discharge chamber or intermediate pressure gas from the cylinder. On the other hand, decreasing the crank chamber pressure can be achieved by releasing the gas in the crank chamber to the suction chamber. Methods to control the capacity is generally classified into inlet control and outlet control.

As shown in Fig. 8.25, the inlet control employs a control valve set in a gas introduction channel, which runs from the discharge chamber to the crank chamber, to adjust the gas flow through that channel. An orifice with a fixed throttling is provided between the crank chamber and the suction chamber to release the gas in the crank chamber. If

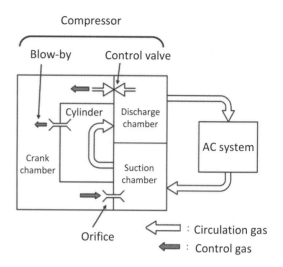

Fig. 8.25 Gas flow of inlet-side control type

the incoming flow from the discharge chamber is stopped, the crank chamber pressure will decrease to be equal to the suction pressure.

On the other hand, as shown in Fig. 8.26, the outlet control employs a control valve set in a gas release channel, which runs from the crank chamber to the suction chamber, to adjust the gas flow through that channel. An orifice with a fixed throttling is provided between the discharge chamber and the crank chamber to introduce the gas in the discharge chamber.

Control valves can also be classified into internally controlled valves and externally controlled valves. The internally controlled valve is stand-alone valve that keeps the suction pressure at approximately constant all the time. The externally controlled valve operates by shifting its action point, which remains fixed in the internally controlled valve, in response to external signals. This method is capable of providing capacity control in a wider range.

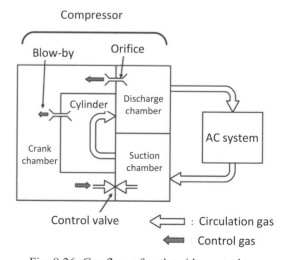

Fig. 8.26 Gas flow of outlet-side control type

‹Internally controlled valve›

Figure 8.27 shows an internally controlled valve being used for the outlet type capacity control, which compensates the suction pressure control point according to variations in the discharge pressure.

The valve comprises a bellows assembly having a support member and a valve element welded on its both ends, a casing, an adapter, and a push rod to sense the discharge pressure. The bellows assembly is welded with its inside kept at almost perfect vacuum so that it can operate without being affected by the ambient temperature. The bellows axially expands and contracts solely in response to variations in the external pressure, thus constituting a pressure sensing unit. The support member on one end of the bellows assembly is attached to the casing. The valve element on the other end constitutes a valve mechanism together with the inner face of the adapter that is in contact with the valve element. The valve element receives a force, via a spring k_2, that is in proportion to the pressure difference between the discharge pressure P_d and the suction pressure P_s acting on both ends of the push rod.

When the bellows assembly contracts, the valve opens to let gas flow from the crank chamber to the suction chamber, causing the crank chamber pressure to decrease. When the bellows assembly expands, the valve closes and the crank chamber pressure increases due to gas flowing in by blowby or through the orifice. (Figure 8.27 shows a valve-closed status.)

In the example shown in Fig. 8.27, the bellows is placed in the crank chamber atmosphere, and it senses both the suction pressure P_s and the crank pressure P_c because a sealing diameter D_2 of the valve is large.

The balance of forces involving the capacity control valve can be expressed by the following equation:

$$(A_1 - A_2)P_c - k\Delta L + A_2 P_s + A_3(P_d - P_s) = 0 \qquad (8.48)$$

where,

A_1: effective bellows area $= \pi D_1^2 / 4$
A_2: valve's sealing diameter area $= \pi D_2^2 / 4$
A_3: push rod cross section area $= \pi D_3^2 / 4$
k: compound spring coefficient ($k = k_1 + k_2$)
ΔL: pre-strain given to the spring

The above equation is converted about crank chamber pressure P_c, as follows:

$$P_c = C_1 - C_2 P_s - C_3 P_d \qquad (8.49)$$

where,

$$\begin{aligned} C_1 &= k\Delta L/(A_1 - A_2) \\ C_2 &= (A_2 - A_3)/(A_1 - A_2) \\ C_3 &= A_3/(A_1 - A_2) \end{aligned} \qquad (8.50)$$

Figure 8.28 shows the relationship between the suction pressure P_s and the crank chamber pressure P_c. With a discharge pressure P_{d1}, when the suction pressure is above P_{s1}, the valve fully opens and the relationship $P_c = P_s$ relationship is formed, as represented by the solid line. When the suction pressure drops below P_{s1}, the relationship expressed by Eq. (8.49) is formed, where the crank chamber pressure increases as the suction pressure decreases. On the other hand, if the discharge pressure increases above P_{d1} to reach P_{d2}, the crank chamber pressure will be controlled to a lower level, as represented by the dotted line. This is effective to prevent cooling-capacity reduction caused by pressure loss in the suction piping between the evaporator outlet and the compressor suction port under the high thermal load

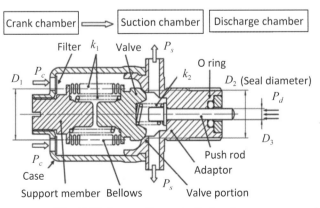

Fig. 8.27 Internally controlled valve (outlet-side control type)

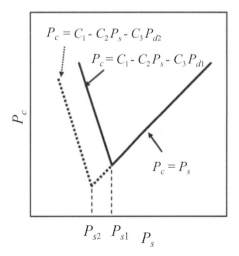

Fig. 8.28 Suction pressure and crank chamber pressure

condition.

When the thermal load is high, the refrigerant circulation rate is usually increase so that pressure loss in the suction piping increases. As the variable capacity compressor controls its capacity in response to variations in the suction pressure, pressure inside the evaporator increases for the amount of pressure loss in the suction piping, causing cooling capacity reduction. Discharge pressure compensation is achieved by lowering the suction pressure control point in response to variations in the discharge pressure that increases with the thermal load increase.

<Externally controlled valve>

Figure 8.29 shows an externally controlled control valve being used for the inlet type capacity control.

A bellows located in the suction pressure region expands when the suction pressure decreases. As the left end of the bellows is fixed to a casing as shown in Fig. 8.29, the expanding bellows moves a push rod and a valve element to the right so as to open the valve. With the valve opening, high pressure gas flows from the discharge chamber into the crank chamber (from P_d to P_c in Fig. 8.29). Due to this increased gas flow into the crank chamber, the crank chamber pressure increases, which in turn decreases the compressor stroke volume.

The electromagnetic force generated by current in a coil acts (in the leftward direction in Fig. 8.29) on the valve element via a moving core to increase the valve closing force. Controlling the intensity of the coil current allows adjusting this valve closing force to lower the action point of the valve. Thus the suction pressure control point can be externally adjusted.

<Externally controlled valve with "off" mode>

As shown in Fig. 8.30, a check valve can be provided at the discharge port of the compressor to regulate the valve-opening differential pressure to approximately 0.05 MPa. Also, an externally control valve with an "off" mode can be combined with a compressor where the compressor piston stroke can be adjusted close to zero. This combination realizes an externally controlled variable capacity compressor where a zero discharge volume can be maintained.

Figure 8.31 shows an externally controlled valve with the "off" mode that is applied to an externally controlled variable capacity compressor. A spring attached on the left side of a bellows moves the bellows to the right together with a push rod and a valve element to keep the valve in open position. With this, gas keeps flowing from the discharge chamber to the crank chamber (P_d to P_c) so that the compressor maintains its minimum capacity condition. If the check-valve-opening differential pressure is set higher than the discharge chamber pressure corresponding the minimum capacity condition, the amount of refrigerant that flows out to the refrigeration circuit can be reduced to zero.

If the solenoid is energized to generate an electromagnetic force that overcomes the spring force, the moving core moves the valve element and the push rod to the left together with the bellows so as to close the valve. With cutoff of the gas flow from the discharge chamber by the closed valve, the crank chamber pressure decreases to approach the suction pressure, which induces the swash plate to tilt. With increasing the discharge gas flow from the cylinder, the discharge chamber pressure increases to open the check valve, which allows refrigerant to flow out into the

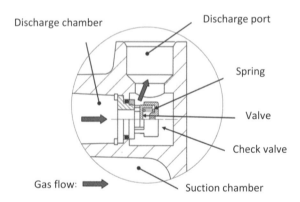

Fig. 8.30 Check valve

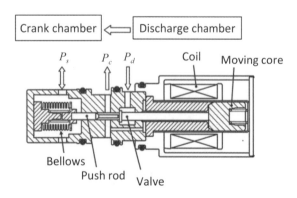

Fig. 8.29 Externally controlled valve (inlet-side control type)

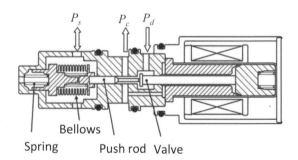

Fig. 8.31 Externally controlled valve with off-mode function (off state)

refrigeration circuit. The operation thereafter is the same as that of an externally controlled valve where the suction pressure control point is changed by adjusting the solenoid current intensity.

If the solenoid current is turned off again, the force of the spring attached to the left side of the bellows moves the valve element to the open position so that the compressor operates in the minimum capacity condition with no refrigerant flowing out into the refrigeration circuit. In this condition, no effective compression takes place although the compressor shaft keeps rotating. (Thus the compressor can be disabled without using a solenoid clutch.)

(5) Externally controlled clutch-less variable capacity compressors and torque limiter

As previously described, an externally controlled variable capacity compressor with an "off" mode is capable of having compression work started or stopped by solenoid current control, thus requiring no solenoid clutch.

However, once such a compressor becomes locked, serious outcomes that are detrimental to the safe operation of the vehicle may occur, such as engine stalling due to overload and compressor drive belt abrasion or breakage. The latter type of failure is particularly problematic as it will also disable other operating components driven by the same belt, including the alternator and the power steering oil pump. A torque limiter can cut off power transmission between the compressor shaft and the pulley, thus protecting the vehicle system from being detrimentally affected by a locked compressor.

There are various methods to achieve such power transmission cut. As an example, Fig. 8.32 shows a mechanism where the power-transmitting joint will be disengaged when an excessive torque acts. A torque transmission disk has three arch-shaped arms, each of which is bolted onto the pulley at its root. The tip of each arm is inserted between a collar of a boss fixed to the compressor shaft and a clamping plate. The collar and the clamping plate are riveted together around the center to form a flat spring. For cutoff torque stabilization, the torque transmission disk has a projection and the clamping plate has a recess that engage together so as to stabilize the pull-out force. When an excessive torque acts, the tips of the torque transmission disk arms will be pulled out from the clamping plate. As each of the torque transmission disk arms has its root bolted to the pulley and its tip lifted to the boss collar as described above, the spring force of the arms will cause the entire disk to stick to the side of the pulley as soon as the tips of the arms are disengaged from the clamping plate. With this, the pulley will not interfere with the boss collar and will

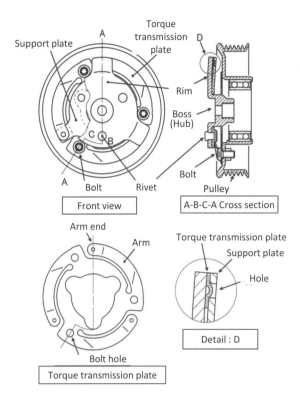

Fig. 8.32 Torque limiter and pulley

continue to rotate smoothly even when the compressor is locked.

Another torque transmission cutoff method involves rupturing of the torque transmission joint when an excessive torque is detected. Due to significant torque pulsation of the engine caused by its explosive combustion cycles, the torque cutoff rupture area will be subject to repetitive loading. Therefore, a torque limiter mechanism with a rupturable joint typically have a load cushioning feature such as a rubber spacer to alleviate component fatigue.

These torque limiters do not have a core essential to a solenoid clutch (see Fig. 8.3), which leads to significant weight and cost reduction.

However, in the torque limiter method, where the compressor shaft keeps turning when the air conditioner does not need to work, a small but constant shaft rotation power is consumed, which constitutes a power loss. When the air conditioner is used frequently, this power loss will not be an issue. On the other hand, the power loss cannot be ignored if the air conditioner is used less frequently. The situation where a solenoid clutch is used will be the other way around: While the compressor is running, it keeps consuming power for its solenoid clutch in order to maintain torque transmission. Therefore, the most appropriate torque transmission on/off control should be selected from a view point of energy efficiency based on how often the air conditioner is actually used.

8.3.4 Noise and vibration of axial compressors

(1) Noise and vibration of automotive air conditioner

It is safe to say that, except for the wind noise from the blower outlet of the passenger cabin unit and the occasional refrigerant flow noise from the expansion valve, a greater part of noise and vibration that is generated when an automotive air conditioner is running comes from its compressor.

Not only the compressor itself is a source of noise and vibration but also it induces suction and discharge pressure pulsations to excite and vibrate other components of the refrigeration circuit. Moreover, the previously described CCEV (cycling clutch expansion valve) mechanism, where a solenoid clutch repeats coupling and decoupling to provide temperature control, generates armature-rotor contact noises and compressor startup jolts. These vibrations and noises may be transmitted through the vehicle structure to reach the driver and passengers and may in some cases be perceived as unpleasant.

The compressor is mounted on the engine, which in turn is mounted on a strongly vibration-attenuating mount for installation in the vehicle. Also, refrigeration circuit components such as the condenser, the evaporator and the pipings are supported by impact-absorbing elements in a floating manner so as to avoid direct contact with the vehicle structure. These solutions are highly effective to reduce vibration transmission from the compressor to the vehicle structure. Nevertheless, excitation from the compressor may still vibrate the vehicle structure when the degree of excitation is strong or if the vehicle structure has a vibration-susceptible element, which can be perceived as audible noise or shaking by human occupants inside the vehicle.

Most of acoustic noise (airborne noise) from the compressor will not normally reach the passenger cabin. However, such acoustic noise may still be perceived in the cabin if the loudness is high or if spectra of the noise are clearly different from frequency bands of the engine noise.

It often happens that noise and vibration from the compressor are perceived more readily when the engine is operating in a specific speed range. This is because both the compressor and the vehicle have vibration-susceptible elements, each having their own natural vibration frequencies. Noise and vibration from the compressor will be amplified when the natural frequency of such element coincides with or approaches the excitation frequency that varies with engine speed.

The compressor noise and vibration pattern differs depending on the compression mechanism.

(2) Noise and vibration of axial compressor

Vibration of an axial compressor is generated in the same way as of a reciprocating piston compressor. Both compressors have a motion conversion mechanism which converts shaft rotational motion into piston reciprocating motion and also have their own moving parts and power transmission members. Action of suction and discharge valves also affects the compressor vibration. In an axial compressor, use of multiple cylinder configuration can help to level out variation of forces to reduce noise and vibration.

Variation of excitation-source forces mostly relates the transition of compression cycle stages including suction, compression, discharge and re-expansion. In a multiple-cylinder axial compressor, where the same number of compressions as the number of cylinders takes place per every shaft rotation, the fundamental frequency of operational vibration is the product of the rotational speed and the number of cylinders. Furthermore, whole number multiplications of this fundamental frequency appears as higher harmonic components.

Figure 8.33 shows vibration analysis of an axial compressor where vibration measurements were taken continuously while the rotational speed was gradually increased. The horizontal axis represents the rotational speed and the vertical axis represents the frequency. The size of each round symbol represents a vibration level. This is a result of continuous measurements of a seven-cylinder compressor operating under high-load condition in a speed range from 1,000 to 6,000 rpm. In Fig. 8.33, each inclined line represents an order of rotation. It is obvious that vibration levels

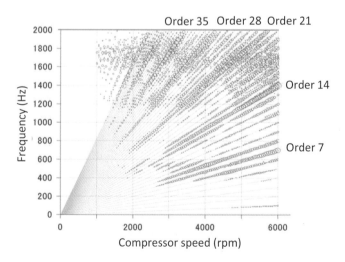

Fig. 8.33 Campbell diagram of compressor vibration

are greater along the seventh-order component line, which is the fundamental frequency for this compressor, and its whole-number-multiplication (14th, 21st and 35th) components. (The relationship between an N rpm rotational speed and the seventh-order frequency f_7 Hz is $f_7 = 7N/60$). It is also apparent that the vibration level becomes greater as the excitation force increases due to rotational speed increase. Even on the same rotational order line, the vibration level becomes greater as the rotational speed increases, indicating that the compressor has its own vibration-susceptible frequency.

In addition to the transition of compression cycle stages, there are many other factors that induce variations in the excitation force intensity, as discussed in the following: Valve-related factors include valve opening delays due to the valve's inertia and, more importantly, oil film adhesion in the contact sealing area between the valve sheet and the valve plate[17]. The valve does not open until an amount of differential pressure across the valve becomes sufficiently strong to overcome the oil film adhesion force. Therefore, pressure in the cylinder drops below the suction pressure when the suction valve opens. On the other hand, it increases above the discharge pressure when the discharge valve opens. These excess differential pressures become suddenly diminished as soon as the valve has opened, where the resulting change in force excites and vibrates the compressor. The amount of force required to overcome viscous oil film adhesion varies depending on the rotational speed. With increase of the compressor rotational speed, the amount of excitation force that is generated by valve opening delay becomes greater.

Valve vibration occurs when gas passes through the suction valve or discharge valve, whose structure is a reed valve. Each valve has a valve stopper to prevent valve damage caused by excessive opening. In some cases the valve may vibrate against its stopper, which superimposes pressure pulsation on the in-cylinder, suction and discharge pressures. The frequency of this vibration depends on the reed rigidity, and therefore is generally lower in suction valves and higher in discharge valves. If the valve lift is restricted too small in an attempt to reduce the valve vibration, resistance of flow through the valve may increase, causing a larger in-cylinder pressure variation during suction and discharge stages and resulting in greater amounts of excitation. Pressure loss through the valve also causes compressor efficiency reduction. Therefore, it is practically impossible to completely eliminate the valve vibration.

When a suction or discharge valve closes, the valve hits the valve plate to generate noise and vibration. A commonly used solution for this is to alleviate impacts by oil film cushioning.

Inadequate accuracy in the machining and assembly of structural elements impairs smooth action of moving parts and induces discontinuous movement of the parts, which can be another factor contributing to greater vibration.

Inadequate rigidity of elements that support force variation may also cause greater displacement of support points, leading to oscillation of mass points, i.e. vibration. Therefore, solutions to avoid resonance must be provided on supporting elements by considering anticipated variations in the forces involved and their frequencies.

Moving parts and connecting elements, around which specified amounts of gaps and clearances are maintained, sometimes can become a source of significant vibration. Under most operating conditions, forces act in the same direction so that components contact each other always in the same position and orientation. However, under some specific conditions or in some chance, the direction of force may be reversed so that parts uncontrollably shift in the clearance area and collide with each other, sometimes repeatedly. Therefore, clearance between components must be designed by evaluating load directions and variations that are expected under all possible conditions.

An automotive air conditioner is required to operate at extremely high rotational speed and therefore needs an appropriate degree of static and dynamic balancing.

Vibrations that are related to specific operating conditions include, among others, noise and vibration caused by liquid compression-induced water hammering, which is an issue that must be solved through system design rather than by compressor-specific solutions.

8.4 Scroll Compressors

8.4.1 History of scroll compressors

The first documented record about the scroll compressor technology in history is said to be that of the patent applied and acquired by French engineer Cruex in 1905. This patent included not only the fundamental principle of the scroll compressor but also an alternative design where an orbiting scroll is provided with wraps on both sides of an end plate so as to cancel out axial loads. And for implementation, the variable-orditing radius mechanism where the orbiting scroll acts in compliance with the fixed scroll profile and the "tip seal" mechanism that has the sealing feature at the tip of scroll wrap were developed in order to decrease leakage flow between the wraps. Another additional feature was the ball coupling mechanism where a ball bearing can act both as a thrust bearing and an anti-rotation device. In 1980s, the world's first ever mass production of scroll compressors started in Japan. Advances in general manufacturing technology such as improved metallic material engineering and numerically controlled machining systems enabled diverse use of scroll compressors including refrigeration and air conditioning systems. For more details about the history of scroll compressors, refer to Chapter 5.

8.4.2 Structure and characteristics of scroll compressors for automotive air conditioners

Figures 8.34 shows a scroll compressor used in an automotive air conditioning system. And Fig. 8.35 shows an internal structure of other scroll compressor for automotive air conditioning system. An automotive scroll compressor receives rotational power from an external drive source such as the vehicle engine. As a result, the compressor has an extended shaft end on the outer side away from the refrigerant circuit, accompanying a pulley and a torque transmission on-off control clutch. As the shaft end needs to be positioned on the outer side, it also must be fitted with a sealing feature to make the internal refrigerant space airtight. In most cases, such shaft sealing is provided by a lip seal. The reason why a lip seal is preferred to other types of seals is that its sealing section is composed of both the shaft and a lip, thus working at a lower circumferential sliding speed than that of a mechanical seal with a larger sealing disk surface. This design is advantageous for use in scroll compressors, which operates more efficiently in high rotational speed ranges than other types of compressors. Other advantages of the lip seal include simple structure, light weight and low cost. Torque fluctuation per rotation and shaft runout of the scroll compressor are smaller than those of single-cylinder and double-cylinder reciprocating compressors. Due to these properties, scroll compressors can be adequately sealed with the lip seal, which is inferior in radial compliance to the mechanical seal.

(a) Scroll wrap geometry

As weight reduction is a high priority for automotive compressors, a commonly used design approach is to make the housing and the scrolls of aluminum alloy and to design the scroll wrap with a relatively large height. The latter technique helps provide larger displacement with a smaller outer diameter of the compressor. Figure 8.36 shows a scroll wrap geometry. To provide the same displacement with the smallest possible outer diameter, the scroll end plate has a radius which is just enough to accommodate the outermost peripheral scroll wrap. On the other hand, in the case of stationary scroll compressors employed for building air conditioners, the scroll end plate radius is larger than the

Fig. 8.34 Scroll compressor for automotive air-conditioner

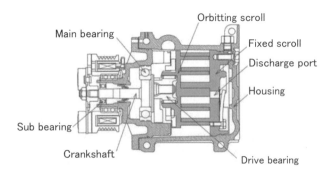

Fig. 8.35 Internal structure of scroll compressor

Fig. 8.36 Scroll for automotive air-conditioner

Fig. 8.37 Scrolls for stationary air-conditioner

outermost peripheral scroll wrap, as shown in Fig. 8.37.

To provide the required displacement volume with the smallest possible size and weight, the scroll must be designed with a high wrap. To support such high wrap with sufficient strength, it is critical to have the proper geometry in the center area of the scroll. As an example of the wrap with greater strength, Fig. 8.38 shows a two-step shape employed at the center of the wrap, where the lower half of the wrap is made thicker than the upper half for greater strength.

Shown in Fig. 8.39 is a scroll wrap design which has evolved under the same concept. The wrap height is changed at a point halfway along the scroll curve. In this wrap height changeover area, R part of the groove and R part of the wrap mesh with each other to form a compression chamber while blocking gas leakage.

(b) Lubrication system

Automotive scroll compressors, which must adapt to significant and frequent changes in vehicle acceleration and deceleration and also in the attitude of running vehicle, are typically lubricated with oil mist that circulates with the refrigerant. The shaft is commonly supported by ball-and-roller bearings. Concerning the means to support the thrust load acting on the orbiting scroll, the compressor shown in Fig. 8.35 uses a slide bearing, while other compressors use a ball coupling bearing or an orbital motion bearing where the bearing elements roll along an orbital track. Such orbital motion-based thrust bearing is explained in Section 8.4.3.

(c) Compression principle

The compression principle of an automotive scroll compressor is little different from that of a stationary scroll compressor. Except for a small number of exceptions, the scroll wrap profile is based on an involute curve.

The rotational speed of an automotive air conditioning compressor changes depending on the vehicle engine rotational speed. During quick acceleration of the vehicle, the refrigerant circulation rate may rapidly increase to cause a steep rise in the discharge pressure. In this situation, the central part of the scroll receives significantly high pressure. This is why a high-strength geometry is required in that part. To cope with that issue, various geometric designs have been proposed for use in the scroll central part. For example, the geometry shown in Fig. 8.40 is a circular arc on which a sine curve is superimposed[18-20].

Figure 8.41 illustrates a compression chamber configuration formed by the fixed and the orbiting scrolls, shown in stages corresponding to different rotation angles. Contact points between the two scrolls exist in the following two areas: (1) between the inside curve of the fixed scroll and the outside curve of the orbiting scroll, and (2) between the outside curve of the fixed scroll and the inside curve of the orbiting scroll. At the stage of rotation angle $\Theta=180°$ shown, there are three contact points, in the order from inside toward the outside, α, β and γ, on the inside curve of the fixed scroll.

Fig. 8.38 Scroll wrap with two-step shape

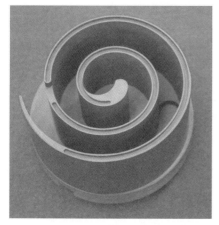

Fig. 8.39 Scroll with multiple wrap height

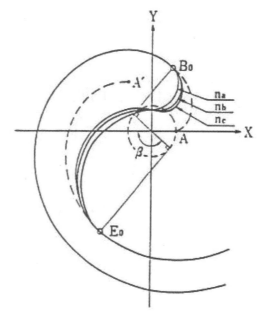

Fig. 8.40 Profile of scroll wrap

Compressors for Air Conditioning and Refrigeration

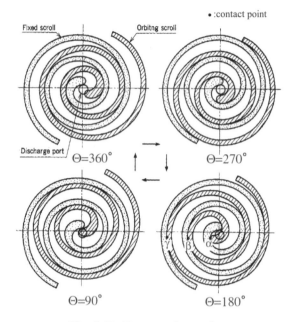

Fig. 8.41 Compression cycle

An important characteristic of a scroll compressor is that the centermost compression chamber volume is reduced down to zero, which helps to minimize re-expansion power loss. Also, the central part of the scroll wrap is made thicker than that of the purely involute-based profile, contributing to greater strength.

8.4.3 Thrust bearing structure

As previously explained, sliding areas of automotive scroll compressors are lubricated by oil mist mixing with the refrigerant. To support rotational load radially acting on the shaft, ball bearings and needle bearings are typically used as radial bearings.

Due to the orbital motion of the compressor, commonly used rolling thrust bearings cannot be applied to supporting the thrust load. Therefore, ball couplings or orbital motion bearings are used as a thrust load-supporting element. Figure 8.42 shows the structure of a ball coupling which is comprised of three elements:

1. Retainer: A ring-shaped disk with three or more pockets to retain balls (sixteen pockets in the example shown). The pocket diameter is determined as the sum of the ball radius at the pocket-ball contact height and the orbiting radius.
2. Race: A ring that support the thrust load of the balls. The race is typically made of bearing steel and is put in grooves on the housing and the orbiting scroll. The race allows balls to rotate on it so as to reduce the slip of balls.
3. Balls: Rolling elements held between the housing and the orbiting scroll. The balls are the same one that is used in a ball bearing.

This ball coupling has advantages of small thrust loss and high reliability with a small amount of lubricating oil, while it has disadvantages of the larger number of parts and increase of weight. In terms of dynamics, the balls that locate in the non-loaded side or in the direction opposite to the orbiting direction are not uniquely positioned in the pockets of the retainer because they are burdened with any thrust load. When the balls enter into the loaded side with the orbiting motion of the scroll, they are suddenly moved to their appropriate position in the pockets, which accompanies a rolling slip between the balls and the race. Critical design requirements include reduction of such slip-induced abrasion and prevention of flaking caused by surface fatigue. To better control the ball position on the non-loaded side with a fewer number of parts, an orbital motion bearing shown in Fig. 8.43 has been developed by a bearing manufacturer. Unlike the ball coupling, the race and the retainer of the orbital motion bearing are in one piece and the ball-retaining pockets are in a grooved shape. Practical orbital motion bearings employ single-piece race-retainers made of a thin plate, which contributes to reduction of weight and cost.

Alternatively, a slide bearing can be used as the thrust load supporting element. Its sliding surface is sometimes coated either with solid lubricant or a tetrafluoroethylene film to assure reliability when the compressor is being started up after a long period stopping or is otherwise in a poorly lubricated condition, for example when the lubricating oil film in the sliding areas is washed away by liquid refrigerant. Such use of a slide bearing as the thrust load supporting element is advantageous to reduction of number

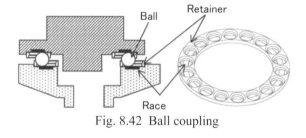

Fig. 8.42 Ball coupling

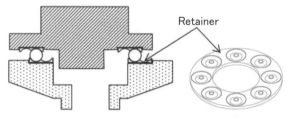

Fig. 8.43 Eccentric motion coupling

Chapter 8 Automotive Air Conditioning Compressors

of parts, size and cost and alleviation of noise.

8.4.4 Capacity control mechanism

The rotational speed of an automotive scroll compressor changes depending on the vehicle engine rotational speed. Therefore, the refrigeration system is designed so that the minimum-required cooling capacity can be obtained at the engine idling speed.

When the vehicle is traveling at higher speed, for example on motorways, where a high engine rotational speed is maintained, capacity control is required for adjustment of cooling capacity. The most commonly used capacity control is to let part of the gas in the compression chamber leak to the suction pressure side, during and immediately after the end of the suction stage.

Figure 8.44 shows a capacity control mechanism using a spool-type valve called "bypass piston". The bypass piston is integrated into the fixed scroll end plate. The bypass position changes its position depending on the level of intermediate pressure generated by a capacity control valve so as to control the amount of leakage to the suction side by changing the amount of port opening. The intermediate pressure that controls the bypass piston position is generated by the difference between suction pressure and discharge pressure.

In order to activate the capacity control mechanism externally, a capacity control valve as shown in Fig. 8.45 is used. The valve can be controlled by electric signal input so as to generate the intermediate pressure for driving the bypass piston. A stepping motor controls a position of the valve center rod to generate the required intermediate pressure. Figure 8.46 shows comparison of *P-V* diagrams of

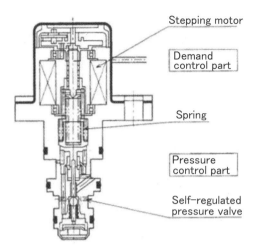

Fig. 8.45 Demand capacity control valve

scroll compressor operating under such capacity control. Under 100% load condition, the pressure starts to rise immediately after the suction is closed off. Under partial load conditions, however, the pressure rise is delayed, indicating that some amount of gas is leaking to the suction side.

Other capacity control mechanisms include one where substantial flow reduction is achieved by high-frequency clutch switching, but this solution has not been actually implemented due to clutch life and noise issues. A similar approach, actually implemented, is a unit control synchronization method where the compressor clutch is deactivated concurrently with closing the expansion valve of the refrigeration system at specific intervals, so as to prevent thermal loss which will occur in case of system pressure equalization. Thus, a compressor not having its own capacity control mechanism can achieve a quasi capacity control practically.

8.4.5 Torque fluctuation and belt life

One of the reasons that scroll compressors are commonly used in air conditioning and refrigeration systems is that a scroll compressor simultaneously forms multiple compression chambers between involute wraps, which

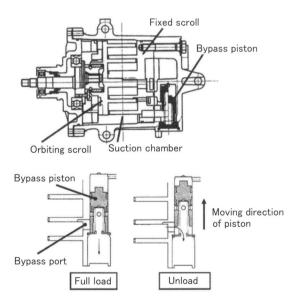

Fig. 8.44 Capacity control mechanism

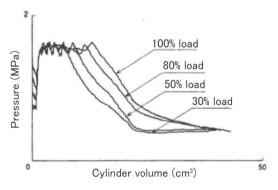

Fig. 8.46 *P-V* diagram at capacity control

entails smaller amount of torque fluctuation than that of a two-cylinder reciprocating compressor or a one-cylinder rotary compressor. Such smaller torque fluctuation leads to less rotational fluctuation, which reduces the amount of slip between the belt and the pulley and extends the belt service life. Other advantages include that the scroll compressor has no parts moving back and forth unlike the reciprocating compressor, which helps in reducing inertia-induced vibration. Although torque fluctuation can be reduced in a reciprocating compressor by using a multiple cylinder configuration, such solution requires a larger number of parts.

A dominant part of vibration and pulsation in a scroll compressor is the primary component of its rotation, whereas that in a multiple-cylinder swash plate compressor is the product of the rotational speed and the number of cylinders. Therefore, in the case of scroll compressors it is relatively easy to avoid the resonance frequency of the mounting bracket, which has many resonance frequencies on the high frequency side, as compared with the case of other types of compressors.

8.5 Rotary Vane Compressors

A rotary vane compressor consists of a cylinder, which may have a circular or non-circular cross section, and a rotor fitted with multiple vanes. As shown in Table 8.1, rotary vane compressors can be classified into the following: (1) York type (consists of a circular cylinder and a rotor eccentrically positioned in the cylinder), (2) Bosch type (consists of an oval cylinder and a rotor positioned at the cylinder center), (3) through-vane type (consists of a non-circular cylinder and vanes that penetrate the rotor)[21,22].

The York type can be further classified according to the direction of the vanes, as shown in Table 8.2, into the "trailing" type, where the vane tips move along the cylinder inner wall in a trailing direction, and the "scooping" type, where the vane tips move along the cylinder inner wall in a scooping direction.

Compared to reciprocating piston compressors or scroll compressors, a rotary vane compressor has a relatively simple structure and therefore can be built with smaller size and lighter weight. As the York type has a cylinder and a rotor that are both circular in shape, it is advantageous in terms of manufacturing cost. Due to the above-described advantages, rotary vane compressors are commonly used in air conditioners for small to medium sized vehicles.

Rotary vane compressors are basically variable-compression-ratio machinery without a specific built-in volume ratio. Therefore, without any extra variable compression ratio mechanism, they can be applied to automotive air conditioners operating under a wide pressure range.

8.5.1 Operating principle

Figure 8.47 shows the compression principle of a typical York type rotary vane compressor. A number of compression chambers are formed being enclosed by walls of the cylinder, the rotor, one or two vanes and the axially positioned housing (not shown). As the rotor and the vanes turn, the compression chamber volume increases and decreases to suck, compress and discharge the working gas.

The suction stage ends and the compression stage starts at the point where the vane tip has moved past the suc-

Table 8.1 Classification of rotary vane compressors

York type	Bosch type	Through-vane type

Table 8.2 Comparison of the York type compressor

York type compressor	
Trailing type	Scooping type

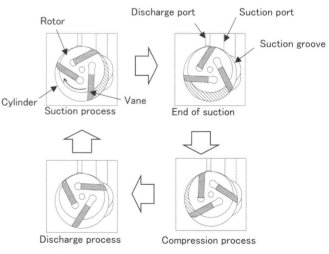

Fig. 8.47 The compression principle of a York type rotary vane compressor

tion groove connected to the suction port, after which the compression progresses as the vane and rotors turn further. When the compression chamber pressure reaches the discharge pressure level, the reed valve mounted on the discharge port is pushed up to discharge the working gas.

8.5.2 Displacement volume

Theoretical displacement volume V of a typical York type rotary vane compressor can be obtained by the following calculation[23].

Figure 8.48 shows the scavenging area in a rotary vane compressor, with the leading side of the vane used as the baseline. The direction of a contact point between the cylinder and the rotor is denoted as x and the direction normal to x is denoted as y. The rotor with radius R_r is inscribed in the cylinder with radius R_c and the centers of the rotor and the cylinder are denoted as O_r and O_c respectively. Point A is the intersection between an extension of the vane's front side line and the cylinder wall. The volume that would exist in a minute gap between the vane tip and the cylinder wall if the vane tip has a curvature is not considered here.

Theoretical volume V is the product of the scavenging area, shown shaded in Fig. 8.48, and the cylinder height H, as expressed by the following equation:

$$V(\theta) = \frac{1}{2}\{R_c^2\psi - R_r^2\theta - ed\sin\phi - R_r d\sin(\phi-\theta)\}H \tag{8.51}$$

where,

$$e = R_c - R_r \tag{8.52}$$

ϕ: Rotary angle of line d connecting the rotor center and point A

Ψ: Rotary angle of the cylinder radius line connecting the cylinder center and point A

The above ϕ and Ψ can be expressed by the following equations:

$$\phi = \alpha + \cos^{-1}\frac{a_l}{d} \tag{8.53}$$

$$\psi = \alpha + \cos^{-1}\frac{a_l + e\cos\alpha}{R_c} \tag{8.54}$$

where,

a_l: Amount of vane offset from the rotor center (with the front side of the vane used as the baseline)

The length of line d is expressed as follows:

$$d = \sqrt{\{R_c\sin(\psi-\alpha)+e\sin\alpha\}^2 + a_l^2} \tag{8.55}$$

As shown in Fig. 8.49, the volume of the scavenging area with the rear side of the vane used as the baseline can be expressed by the following equation (Subscript t means that the value is related to the rear side of the vane):

$$V_t(\theta_t) = \frac{1}{2}\{R_c^2\psi_t - R_r^2\theta_t - ed_t\sin\phi_t - R_r d_t\sin(\phi_t-\theta_t)\}H \tag{8.56}$$

where,

$$\theta_t = \alpha + \cos^{-1}\frac{a_t}{R_r} \tag{8.57}$$

$$\phi_t = \alpha + \cos^{-1}\frac{a_t}{d_t} \tag{8.58}$$

$$\psi_t = \alpha + \cos^{-1}\frac{a_t + e\cos\alpha}{R_c} \tag{8.59}$$

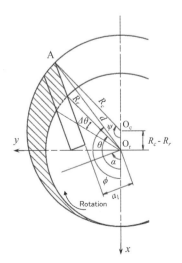

Fig. 8.48 The air scavenging area in a rotary vane compressor with the leading side of the vane

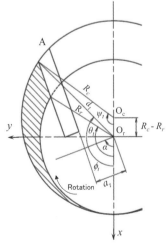

Fig. 8.49 The air scavenging area in a rotary vane compressor with the rear side of the vane

Compressors for Air Conditioning and Refrigeration

$$d_t = \sqrt{\{R_c \sin(\psi_t - \alpha) + e \sin\alpha\}^2 + a_t^2} \qquad (8.60)$$

where

a_t: Amount of vane offset from the rotor center (with the rear side of the vane as the baseline)

A compression chamber in a rotary vane compressors is enclosed either by the cylinder and rotor walls, the inscribed point and a vane, or by the cylinder and rotor walls and two vanes. In a typical three-vane-type compressor as shown in Fig. 8.47, theoretical volume V based on the rotary angle can be expressed by the following equations:

$$V(\theta) = \begin{cases} 0 & (-2\pi/3 \le \theta \le -2\pi/3 + \Delta\theta) \\ V_t(\theta + 2\pi/3) & (-2\pi/3 + \Delta\theta \le \theta \le 0) \\ V_t(\theta + 2\pi/3) - V(\theta) & (0 \le \theta \le 4\pi/3 + \Delta\theta) \\ V(2\pi) - V(\theta) & (4\pi/3 + \Delta\theta \le \theta \le 2\pi) \end{cases} \qquad (8.61)$$

where

$\Delta\theta$: Rotational angle corresponding to the vane slot width (see Fig. 8.48)

8.5.3 Basic structure
(1) Structure

Figure 8.50 shows the cross section of a typical York type rotary vane compressor, and Fig. 8.51 is the cross section of its compression mechanism with three vanes[24]. The compression mechanism is put between front and rear housings. The rotor is supported by needle bearings provided in the two housings.

Three slidable vanes, each inserted into a slit in the rotor, operate to provide three discharges per rotation. A slidable "seal ring" seals a gap between the rotor and the front housing, reducing axial leakage of the working gas.

Figure 8.52 shows the cross section of a seal ring. For better sealing effect, the seal ring has a projecting "lip" on the rotor side so as to contact lightly with the rotor all the time. The compressor is designed so that pressure of fluid leaking from the high pressure side into the rotor gap area acts on the inside of the seal ring to help press the lip more strongly against the rotor to enhance sealability.

The shaft and the rotor are directly driven by the engine crank pulley, by way of a transmission belt and a solenoid clutch attached to the front housing.

Working gas that is sucked into the compressor through the suction port gets compressed in the compression mechanism section before being discharged into the high-pressure

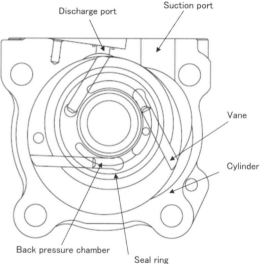

Fig. 8.51 The cross section of the compression mechanism

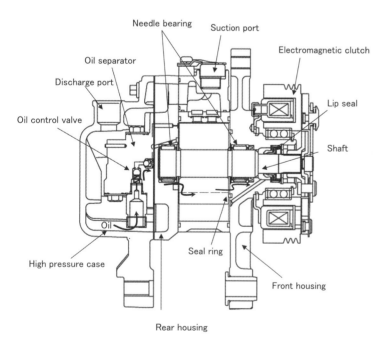

Fig. 8.50 The cross section of a typical York type rotary vane compressor

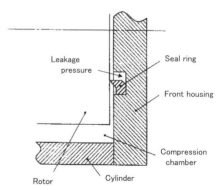

Fig. 8.52 The cross section of a seal ring

casing connected to the rear housing. As shown in Figs. 8.50 and 8.51, the pressure inside the high-pressure casing causes the lubrication oil stored at the bottom of the high-pressure casing to be fed into the vane back pressure chamber to push the vanes out while the compressor is operating. The magnitude of pressure acting on the vanes from the back pressure chamber, i.e. vane back pressure, affects various operating factors such as vane tip slide reliability, the amount of vane-jumping noise and working gas leakage through the vane tip areas. To provide satisfactory performance, noise level and reliability, various engineering solutions are incorporated into the vane back pressure supply mechanism. Figure 8.53 shows a mechanism that supplies lubrication oil (which is the back pressure source) using an oil control valve that adjusts the amount of oil supplied. The oil control valve is a flow control device specially designed to handle pressurized lubrication oil and comprises a throttling, a spring and a ball.

Sliding pairs of parts in the compressor include the shaft and the bearings (needle bearings), the rotor and the side housings, the vanes and the side housings, the vanes and the cylinder, and the rotor and the vanes. Lubrication of these sliding components is done with two lubrication oil feed sources, one of which is the oil control valve and the other is the minute amount of oil mist contained in the suction gas.

Needle bearings are lubricated by positioning the bearings in the oil supply path, where it is exposed to oil flow all the time. For other metallic sliding components, material solutions such as special heat treatment or plastic surface coating are implemented for seizure tolerance and slideability.

Rotary vane compressors for automotive air conditioners are an open type compressor, where sealing elements are fitted to various areas to keep the working gas space airtight. The joint between the cylinder and each side housing is sealed with an O-ring, which is made of refrigerant- and oil-tolerant fluororubber.

As a shaft sealing feature, a lip seal is used to separate the atmospheric pressure space and the discharge pressure space while itself remaining in a rotationally slideable condition (refer to Fig. 8.7).

(2) Operation

Working gas that is ejected from the compression mechanism contains a large amount of lubrication oil. If this gas is discharged into the refrigerant circuit untreated, various detrimental effects, such as heat exchanger efficiency reduction and refrigeration cycle efficiency reduction due to greater pressure loss in the suction piping, will result. To reduce the oil circulation rate by restricting lubrication oil outflow from the compressor, a lubrication oil separator is provided inside the high-pressure casing. The lubrication oil separator in the high-pressure casing utilizes the effects of oil swirling and collision.

As shown in Fig. 8.53, the oil mist-containing gas entering the high-pressure casing is directed to flow back and force through the multiple levels of U-shaped ribs before it reaches the discharge port. Thus the oil mist is repeatedly exposed to collision and centrifugal separation to aggregate into droplets, which flows down to the oil reservoir at the bottom. Lubrication oil in the reservoir is then fed to the vane back space through the oil control valve. Thus the lubrication oil circulates through the entire compressor structure.

As fuel efficiency improvement is a high priority for automotive systems, various engineering efforts are being undertaken to improve the cycle efficiency for automotive air conditioners, which influences vehicle fuel consumption. One of such improvements is the swirl-type lubrication oil separator shown in Fig. 8.54 that provides improved oil separation performance using the effect of swirling flow[25]. With this swirl-based oil separator, the mixture of working gas and lubrication oil that enters the high-pressure casing is directed to the oil separator chamber with a column-and-cone structure in the casing.

With an optimized feed port connecting to the oil separator chamber, an intensive swirling flow is created inside

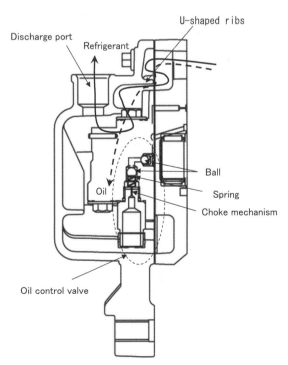

Fig. 8.53 The cross section of the high pressure casing

Compressors for Air Conditioning and Refrigeration

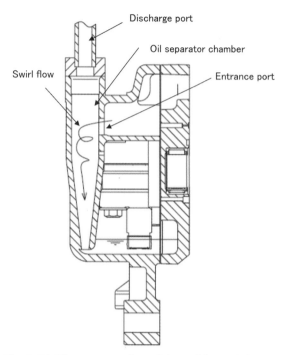

Fig. 8.54 The cross section of the swirl-type lubricant oil separator

the chamber. The oil content separated from the gas by the swirling flow effect is collected in the reservoir at the bottom of the high-pressure casing, and the gas with a reduced oil content is flowing out of the high-pressure casing through the upper discharge port.

8.5.4 Dynamic-mechanical analysis

As previously described, rotary vane compressors can be classified into various types. Of those, the York type, whose cylinder has a circular inner wall and is the easiest to machine, has been in practical use most widely. The following paragraphs provide a dynamic-mechanical analysis of the York type rotary vane rotary compressor[26, 27].

A two-vane trailing type model, shown in Fig. 8.55, where the tips of the vanes are trailing along the cylinder inner wall with the rotor rotation, is discussed as a representative type of the rotary vane compressor. For every one and half rotations of the rotor, a single cycle of suction, compression and discharge is completed.

(1) Definitions of coordinates and variables

Cylinder center O_c and rotor center O_r are defined as shown in Fig. 8.56. The rotor rotates in the clockwise direction. The following analysis is performed focusing on the relative sliding motion between the center of circular arc at the vane tip and the cylinder inner wall. The angle between the perpendicular line drawn from the rotor center O_r to the vane side face and the horizontal line (x axis) is defined as rotor rotation angle θ.

Figure 8.57 shows the cross section of a vane. The relative relationship between the center of vane tip arc, O_b, and the center of vane gravity, G, are identified by dimensions a and b. The vane rotation angle ϕ when the rotor turns is defined as the angular increment of the line segment $\overline{O_c O_b}$ from its original position at the rotor rotation angle $\theta=0$. The angle of the line segment $\overline{O_c O_b}$ against the y axis at $\theta=0$ is denoted as ϕ_0, which can be expressed by Eq. 8.62.

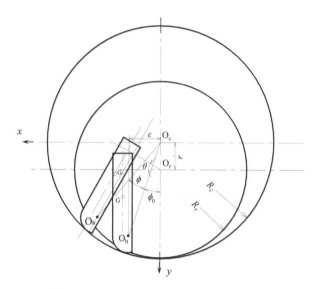

Fig. 8.56 Coordinate system and variables

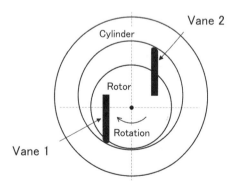

Fig. 8.55 2 vane type compresor

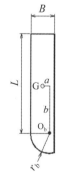

Fig. 8.57 Cross section of vane

184

$$\phi_0 = \sin^{-1}\left(\frac{\varepsilon}{R_c - r_b}\right) \quad (8.62)$$

(2) Relationship between vane rotation angle ϕ and rotor rotation angle θ

The equation of vane motion needs to be described as a function of rotor rotation angle θ. A geometric study can find the relationship between vane rotation angle ϕ and rotor rotation angle θ. By referring to Fig. 8.58, Eqs. (8.63) and (8.64) is obtained.

$$\overline{O_c O_{bh}} = (R_c - r_b)\cos(\phi + \phi_o) \quad (8.63)$$

$$\overline{O_c O_{bh}} = \overline{O_{rh} O_b}\cos\theta - \varepsilon\sin\theta + r \quad (8.64)$$

Where, $\overline{O_{rh} O_b}$ is defined by Eq. (8.65), and then Eq. (8.64) is re-written as Eq. (8.66).

$$\overline{O_{rh} O_b} = (R_c - r_b)\cos(\phi + \phi_o - \theta) - r\cos\theta \quad (8.65)$$

$$\overline{O_c O_{bh}} = \{(R_c - r_b)\cos(\phi + \phi_o - \theta) - r\cos\theta\} \times \cos\theta - \varepsilon\sin\theta + r \quad (8.66)$$

From Eqs. (8.63) and (8.64), a relationship which associates the vane rotation angle ϕ and the rotor rotation angle θ is derived as follows:

$$\{(R_c - r_b)\cos(\phi + \phi_o - \theta) - r\cos\theta\}\cos\theta \\ - \varepsilon\sin\theta + r = (R_c - r_b)\cos(\phi + \phi_o) \quad (8.67)$$

By transforming Eq. (8.67) with substitutions by Eqs. (8.68),

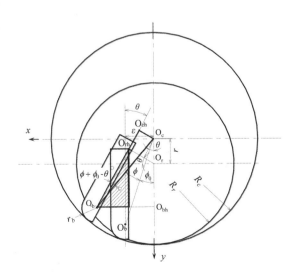

Fig. 8.58 Rotor rotation angle θ and vane rotation angle ϕ

(8.69) and (8.70), Eq. (8.71) is obtained.

$$\alpha = \frac{r}{R_c - r_b}, \quad \beta = \frac{\varepsilon}{R_c - r_b} \quad (8.68)$$

$$A = \sin(\phi + \phi_o) \quad (8.69)$$

$$d = \alpha\sin^2\theta - \beta\sin\theta \quad (8.70)$$

$$A^2\left(\frac{1}{4}\sin^2 2\theta + \sin^4\theta\right) + Ad\sin 2\theta + d^2 - \sin^4\theta = 0 \quad (8.71)$$

Solving Eq. (8.71) for A leads to Eq. (8.72). Finally Eq. (8.72) expresses the relationship between the vane rotation angle and the rotor rotation angle θ.

$$\sin(\phi + \phi_o) = \frac{-2d\sin 2\theta}{\sin^2 2\theta + 4\sin^4\theta} \\ \pm \frac{2\sqrt{d^2\sin^2 2\theta - (\sin^2 2\theta + 4\sin^4\theta)(d^2 - \sin^4\theta)}}{\sin^2 2\theta + 4\sin^4\theta} \quad (8.72)$$

(3) Analysis of vane motion

[Coordinates for the center of vane gravity (x_G, y_G) and its acceleration]

The x and y coordinates for the center of vane gravity, G, is expressed by Eqs. (8.73) and (8.74).

$$x_G = -(R_c - r_b)\sin(\phi + \phi_o) + b\sin\theta - a\cos\theta \quad (8.73)$$

$$y_G = -(R_c - r_b)\cos(\phi + \phi_o) + b\cos\theta + a\sin\theta \quad (8.74)$$

By differentiating each of the two equations twice, the x- and y-direction acceleration can be obtained under the condition that the acceleration of the rotor rotation is not negligible.

$$\ddot{x}_G = -(R_c - r_b)\{\ddot{\phi}\cos(\phi + \phi_o) - \dot{\phi}^2\sin(\phi + \phi_o)\} \\ + \dot{\theta}^2(a\cos\theta - b\sin\theta) + (a\sin\theta + b\cos\theta)\ddot{\theta} \quad (8.75)$$

$$\ddot{y}_G = (R_c - r_b)\{\ddot{\phi}\sin(\phi + \phi_o) + \dot{\phi}^2\cos(\phi + \phi_o)\} \\ - \dot{\theta}^2(a\sin\theta - b\cos\theta) + (a\cos\theta - b\sin\theta)\ddot{\theta} \quad (8.76)$$

Multiplying the above $\ddot{x}_G$ and $\ddot{y}_G$ by vane mass m_b respectively gives inertia forces of the vane.

[Forces acting on the vane]

(a) Gas force

Figure 8.59 shows pressures surrounding the vane and the forces caused by these pressures, and Fig. 8.60 shows

Compressors for Air Conditioning and Refrigeration

the geometric dimensions of and related to the vane. Pressures P_1 and P_2 changer so as to correspond to suction pressure P_s, intermediate compression pressure P_c or discharge pressure P_d depending on the vane rotation angle. Discharge pressure P_d works on the back of the vane to press the vane tip against the cylinder inner wall. As a result of these pressures, the vane receives force W_1 by the pressure difference between P_d and P_1, force W_2 by pressure difference between P_d and P_2, and force W_3 by pressure difference between P_2 and P_1.

$$W_1 = \Delta P_2 \left\{ \frac{B}{2} - r_b \sin(\phi + \phi_o - \theta) + a \right\} H \quad (8.77)$$

$$W_2 = \Delta P_1 \left\{ r_b \sin(\phi + \phi_o - \theta) + \frac{B}{2} - a \right\} H \quad (8.78)$$

$$W_3 = \Delta P_3 \left\{ r_b \cos(\phi + \phi_o - \theta) - \Delta L \right\} H \quad (8.79)$$

Were, H represents the vane height, and ΔL is expressed by the following equations.

$$\Delta L = \sqrt{R_r^2 - \left\{ \varepsilon - \left(\frac{B}{2} - a \right) \right\}^2} - \sqrt{\overline{O_b O_r}^2 - \varepsilon^2} \quad (8.80)$$

$$\overline{O_b O_r}^2 = (R_c - r_b)^2 + r^2 - 2r(R_c - r_b)\cos(\phi + \phi_o) \quad (8.81)$$

(b) Forces of constraint

Forces that work on the rotating vane are shown in Fig. 8.61. In addition to the gas forces, the tip of the vane also receives reaction force F_w from the cylinder inner wall and its frictional force μF_w in the direction opposite to the sliding motion. Inside the rotor slit, the vane is supported at points S and T, causing reaction forces R_1 and R_2. These reaction forces generate frictional forces $\mu_1 R_1$ and $\mu_1 R_2$. These frictional forces change their direction depending on whether the vane is moving out of the slit or into it. When the vane is extending from the slit, the frictional force works in the positive direction ($\delta=1$). On the other hand, when the vane is being retracted into the slit, the frictional force works in the negative direction ($\delta=-1$).

[Equation of vane motion]

By referring to Figs. 8.61 and 8.62, the x- and y-axis force balance equations about the vane and the moment balance equation around the center of vane gravity are expressed as follows:

$$\begin{aligned} & F_W \cos(\phi + \phi_o - \theta) - \mu F_W \sin(\phi + \phi_o - \theta) \\ & + \mu_1 \delta(|R_1| + |R_2|) + (-m_b \ddot{x}_G)\sin\theta \\ & + (-m_b \ddot{y}_G)\cos\theta - W_1 - W_2 = 0 \end{aligned} \quad (8.82)$$

$$\begin{aligned} & -F_W \sin(\phi + \phi_o - \theta) - \mu F_W \cos(\phi + \phi_o - \theta) \\ & + R_1 - R_2 - (-m_b \ddot{x}_G)\cos\theta + (-m_b \ddot{y}_G)\sin\theta - W_3 = 0 \end{aligned} \quad (8.83)$$

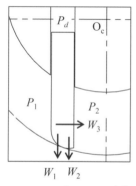

Fig. 8.59 Pressures around vane and forces acting on vane

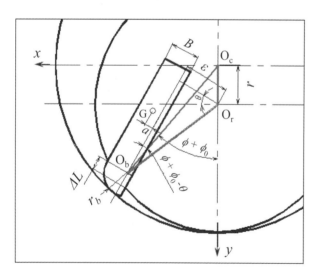

Fig. 8.60 Geometric dimensions around vane

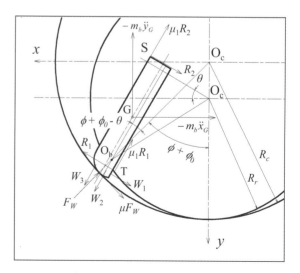

Fig. 8.61 Forces acting on vane

Chapter 8 Automotive Air Conditioning Compressors

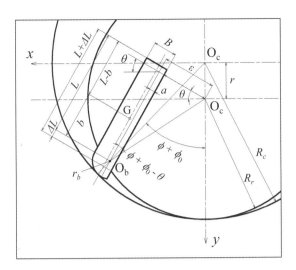

Fig. 8.62 Length of moment

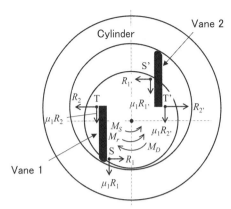

Fig. 8.63 Forces and moments acting on rotor-shaft system

$$-I_b\ddot{\theta} - F_W\{b\sin(\phi+\phi_o-\theta) + a\cos(\phi+\phi_o-\theta)\}$$
$$-\mu F_W\{b\cos(\phi+\phi_o-\theta) - a\sin(\phi+\phi_o-\theta) + r_b\}$$
$$+R_1(b+\Delta L)+R_2(L-b)-\mu_1\delta|R_1|\frac{B}{2}+\mu_1\delta|R_2|\frac{B}{2}$$
$$-W_1\left\{\frac{B}{2}-\frac{\frac{B}{2}-r_b\sin(\phi+\phi_o-\theta)+a}{2}\right\}$$
$$+W_2\left\{\frac{B}{2}-\frac{r_b\sin(\phi+\phi_o-\theta)+\frac{B}{2}-a}{2}\right\}$$
$$-W_3\left\{b+\Delta L+\frac{r_b\cos(\phi+\phi_o-\theta)-\Delta L}{2}\right\}=0$$

(8.84)

By solving Eqs. (8.82) to (8.84) with the behaviors of rotor rotation angle θ and vane rotation angle ϕ clarified, three forces acting on the vane, that are the reaction force F_w from the cylinder inner wall and the reactional forces R_1 and R_2 from the rotor, are obtained.

(4) Equation of motion of rotor-shaft system

In order to determine the shaft torque, loss torques caused by reaction forces and their frictional forces on the vanes and loss torques occurring at the rotor end faces and at the shaft-supporting bearings must be considered.

Moments acting on the rotor-shaft system of a two-vane type rotary compressor are shown in Fig. 8.63. Motor torque M_D drives the rotor-shaft system in clockwise direction. On the other hand, reaction forces and frictional forces from the two vanes (vanes 1 and 2) generate counterclockwise moments.

As the rotor rotation angle of vane 2 is π ahead of the rotor rotation angle θ of vane 1, the reaction forces acting on vane 2 can be calculated using $\theta+\pi$ instead of θ as shown in Eqs. (8.85) to (8.87). Values related to vane 2 are expressed with the same symbols as those for vane 1, with an apostrophe (') added. M_s represents loss torque at the bearings, which can be calculated by multiplying the bearing load by an appropriate friction coefficient. M_r represents viscous loss torque at the rotor end faces, which can be calculated by integrating the shear stress of oil existing in a narrow gap on the rotor end faces.

By referring to Figs. 8.62 and 8.63, the equation of rotational motion of the rotor-shaft system can be expressed by Eq. (8.88).

$$F_W'(\theta) = F_W(\theta+\pi) \tag{8.85}$$

$$R_1'(\theta) = R_1(\theta+\pi) \tag{8.86}$$

$$R_2'(\theta) = R_2(\theta+\pi) \tag{8.87}$$

$$\begin{aligned}I_r\ddot{\theta} &= M_D - R_1\{(R_c-r_b)\cos(\phi+\phi_0-\theta) - r\cos\theta + \Delta L\}\\
&- R_2\{L-(R_c-r_b)\cos(\phi+\phi_0-\theta) + r\cos\theta\}\\
&- \delta\mu_1 R_1(\varepsilon - B/2 + a) - \delta\mu_1 R_2(\varepsilon + B/2 + a)\\
&- R'_1\{(R_c-r_b)\cos(\phi'+\phi_0-\theta') - r\cos\theta' + \Delta L'\}\\
&- R'_2\{L-(R_c-r_b)\cos(\phi'+\phi_0-\theta') + r\cos\theta'\}\\
&- \delta'\mu_1 R'_1(\varepsilon - B/2 + a) - \delta'\mu_1 R'_2(\varepsilon + B/2 + a)\\
&- M_s - M_r\end{aligned} \tag{8.88}$$

Where, distances ΔL and $\Delta L'$ are expressed by the following equations.

187

$$\Delta L = \sqrt{R_r^2 - \left\{ \varepsilon - \left(\frac{B}{2} - a \right) \right\}^2} - \sqrt{\overline{O_b O_r'}^2 - \varepsilon^2} \qquad (8.89)$$

$$\Delta L' = \sqrt{R_r^2 - \left\{ \varepsilon - \left(\frac{B}{2} - a \right) \right\}^2} - \sqrt{\overline{O_b O_r'}^2 - \varepsilon^2} \qquad (8.90)$$

$$\overline{O_b O_r'}^2 = (R_c - r_b)^2 + r^2 - 2r(R_c - r_b)\cos(\phi' + \phi_o) \qquad (8.91)$$

In principle, when values of the motor driving torque M_D and the pressure characteristics during gas compression are given, Eq. (8.88) of rotary motion can be solved with F_w, R_1 and R_2 by the numerical computation method such as Runge-Kutta-Gill method and then rotational behavior of the rotor-shaft system is clarified.

References

1) F. J. Linsenmeyer: "Heating and Air-conditioning of Automobiles ? ", SAE Congress Paper, 380096 (1938).

2) H. Chase: "Heating, Ventilating, and Cooling of Passenger Cars", SAE Journal, 46 (4), pp. 137-146 (1940).

3) W. J. Owen: "Development of the Cadillac Air Conditioner", GM Engineering Journal, 1 (1), pp. 44-49 (1953).

4) J. R. Holmes: "Development of an Automobile Air Conditioning System for Underhood Installation", GM Engineering Journal, 2, pp. 2-9 (1955).

5) J. Dolza, W. K. Steinhagen, P. L. Francis: "A Product Engineering Study: Design of an Axial-Type Refrigeration Compressor", GM Engineering Journal, 3, pp. 40-46 (1956).

6) R. N. Mantey, J. M. Murphy: "Production Design and Development of 1962 Frigidaire Automotive Compressor and Clutch", SAE Congress Paper, 481C (1962).

7) J. Weibel Jr., R. N. Mantey: "The Engineering Development of a Compressor for Automotive Air Conditioning Systems", GM Engineering Journal, 9, pp. 19-24 (1963).

8) H. Hatakeyama, H. Takahashi, S. Kimura, M. Hiraga: "Recent Progress in Technology of the 5-Cylinder Wobbleplate Type Compressor for Automotive Air Conditioning", SAE Congress Paper, 840383 (1984).

9) T. Koda, K. Yoshida, Y. Hoshino, M. Sugihara: "Rolling Piston Type Compressor for Automotive Air Conditioner", SAE Congress Paper, 810502 (1981).

10) I. Ogura: "The Ogura-Wankel Compressor-Application of Wankel Rotary Concept as Automotive Air Conditioning Compressor", SAE Congress Paper, 820159 (1982).

11) M. Hiraga: "The Spiral Compressor – An Innovative Air Conditioning Compressor for the New Generation Automobiles", SAE Congress Paper, 830540 (1983).

12) Y. Nishimura, M. Takagi, H. Kobayashi: "Development of Two-Stage Variable Displacement Compressor for Automotive Air Conditioner", SAE Congress Paper, 850039 (1985).

13) T. J. Skinner, R. L. Swadner: "V-5 Automotive Variable Displacement Air Conditioning Compressor", SAE Congress Paper, 850040 (1985).

14) K. Takai, S. Shimizu, K. Terauchi: "A 7-Cylinder IVD Compressor for Automotive Air Conditioning", SAE Congress Paper, 890309 (1989).

15) A. Kishibuchi, M. Nosaka and T. Fukanuma: "Development of Continuous Running, Externally Controlled Variable Displacement Compressor", SAE Congress Paper, 1999-01-0876 (1999).

16) T. Shimomura, A. Yoshino, H. Ichiyasu, K. Oiyama, K. Kiryu, H. Hirabayashi: "Application of Lip Seals for a High Pressure Type Automotive Air Conditioning Compressor", SAE Congress Paper, 890610 (1989).

17) N. Soda: "The Friction and Lubrication of Solids", Maruzen, Tokyo, pp. 259-261, pp. 288-289 (1975). (in Japanese)

18) T. Hirano, K. Hagimoto: "Study on Scroll Profile for Scroll Compressors (1st Report; Theoretical Analysis)", Proc. of the 23rd Japanese Joint Conference on Air-Conditioning and Refrigeration, 23, Tokyo, pp. 49-52 (1989). (in Japanese)

19) T. Hirano, M. Maeda: "Study on Scroll Profile for Scroll Compressors (2nd Report; Stress, Performance and Noise)", Proc. of the 23rd Japanese Joint Conference on Air-Conditioning and Refrigeration, 23, Tokyo, pp. 53-56 (1989). (in Japanese)

20) T. Hirano, K. Hagimoto, M. Maeda: "Study on Scroll Profile for Scroll Fluid Machines", Trans. of the JAR, 8 (1), pp. 53-64 (1991). (in Japanese)

21) H. Kamiya, T. Yanagisawa, M. Fukuta, T. Shimizu, T. Sakimoto: "Internal Leakage Loss in Through-Vane Compressors: 1st Report, Leakage Flow through Clearance at Vane Tip" Transactions of the Japan Society of Mechanical Engineers (Series B), 60 (573), pp. 1675-1682 (1994). (in Japanese)

22) I. Honda, H. Ohba, Y. Nakashima, M. Yasuda, S. Matsumoto, Y. Sudo: "A Study on a Rotary Sliding Vane Compressor: 1st Report, Numerical Analysis of the Vane's Force", Transactions of the Japan Society of Mechanical Engineers (Series B), 55 (512), pp. 1164-1167 (1989). (in Japanese)

23) M. Fukuta: "Basic Study on Vane Type Refrigerant Compressors", Ph.D. Thesis of Kyoto University, pp. 7-21 (1995). (in Japanese)

24) T. Hirose, T. Kitamura: "High-Efficiency Compact Rotary Compressor Using New Refrigerant HFC134a for Automotive Air Conditioner", National Technical Report, 41 (3), pp. 48-55 (1995). (in Japanese)

25) T. Kitamura, K. Watanabe, T. Kawada, K. Okuzono, N. Tsuchida: "Compressor", Japan Patent, No. 4013554 (2007).

26) K. Sawai, A. Sakuda, T. Nakamoto, N. Iida, T. Tsujimoto, H. Fukuhara, H. Murakami, T. Nagata, N. Ishii: "Development of Small Size Air Compressor for Mobile Fuel Cells", IMechE Conference; Transactions of International Conference on Compressors and Their Systems, London, C639-27, pp.109-117 (2005).

27) T. Nakamoto, A. Sakuda, K. Sawai, N. Iida, N. Ishii: "Development of Air Pump for Fuel Cells", Proceedings of 17th International Compressor Engineering Conference at Purdue, West Lafayette, C158 (2006).

Compressors for Air Conditioning and Refrigeration

Chapter 9 Refrigeration Oil

9.1 Types of Refrigeration Oil

Refrigeration oil refers to lubricant that is used in a vapor-compression type refrigeration and air conditioning equipment.

Refrigeration oil is indispensable to assure adequate lubrication of compressor in refrigeration cycle, but it becomes the factor to reduce its thermal efficiency and affects the cooling capacity. Therefore, it is critical to select the best refrigeration oil for the equipment according to the type of refrigerant and other specifications.

The most commonly used refrigerants are fluorocarbon and natural refrigerants. Most common natural refrigerant are carbon dioxide, hydrocarbons and ammonia.

Refrigeration equipment includes household refrigerators, freezers, showcases, vending machines, room air conditioners, packaged air conditioners, automotive air conditioners, and heat pump water heater systems. Each application requires the best suited refrigeration oil. Table 9.1 shows types of refrigerants and refrigeration oils for various equipment.

Base oils that are used in refrigeration oil is divided into two groups; hydrocarbon compounds including mineral oil (MO) and alkyl benzene (AB); and oxygenated hydrocarbon compounds including polyalkylene glycol (PAG), polyvinyl ether (PVE) and polyol ester (POE). The following

sections describe characteristics of refrigeration oil classified based on the base oil used.

9.1.1 MO refrigeration oil

The base oil that is used in MO refrigeration oil is a lubricating oil distillate obtained from the crude oil refining process that produces naphtha, gasoline and other petroleum products. MOs are classified into paraffinic and naphthenic oils. Naphthenic oils are more commonly used as refrigeration oil. They contain much less wax distillate and therefore are superior in low temperature fluidity compared to paraffinic oils.

For refrigeration equipment that uses capillary tube expansion mechanism such as household refrigerator, it is critical to prevent the capillary blockage by precipitated wax. It is a major reason why naphthenic oil is favored in these types of refrigeration equipment. MO also contains a small amount of sulfur, which has similar effect as lubricity improver which will be described later.

MOs are widely used in isobutane (R600a) household refrigerators as miscible oil and also in industrial ammonia refrigerating machines as immiscible oil.

Reciprocating compressors are most commonly used in R600a household refrigerators. The lower viscosity of MO refrigeration oil is desired for energy saving of R600a household refrigerators. The most typically used refrigera-

Table 9.1 Use example of refrigerant and refrigeration oil to various equipment

Classification		Refrigerant	Refrigeration oil*
Refrigeration equipment	Refrigerator	R600a	MO7/10/15
	Showcase	R404A	POE32, PVE32
		CO_2	PAG68
	Vending machine	R407C	POE68, PVE68
		CO_2	PAG68
	Chiller	R134a, R407C	POE68, PVE68
	Industrial Freezer/Refrigerator	R407C, R410A, R404A	POE68, PVE68
		CO_2	PAG68, POE68
		NH_3	MO56, PAG56
	Freezer for transportation	R134a, R404A	PAG46, POE68, PVE68
Air conditionig equipment	Room air conditioner	R410A	AB22, POE68, PVE68
	Packaged air conditioner	R410A	POE68, PVE68
	Automotive air conditioner	R134a, R1234yf	PAG46/100
Water heater	Heat pump water heater	CO_2	PAG46/100

*Addition number indicates the ISO viscosity grade

190

tion oil in R600a household refrigerators is low-viscosity naphthenic oil in the viscosity grade range of VG7 to VG15.

A possible concern about using naphthenic oil for R600a is that both R600a and naphthenic oil are hydrocarbon compounds and are extremely miscible with each other. Since solubility is too good, R600a is endlessly dissolved in naphthenic oil. It becomes a cause of viscosity reduction and possibly leads to defective lubrication.

However, from the viewpoint of flammability concerns, the amount of R600a used in household refrigerator is limited to 150 g or less by IEC (International Electrotechnical Commission) standard. In fact, R600a of only 65 g has been used in the household refrigerator of 320 L class[1]. On the other hand, approximately 200 g of refrigeration oil is charged into a household refrigerator of this class. Therefore, there is less R600a than refrigeration oil. As a result, there is no significant risk of viscosity reduction of refrigeration oil by R600a dissolution.

Naphthenic oil is also used in refrigerating machinery with ammonia as immiscible oil. Such use of naphthenic oil has a long history, more than 130 years since the machine using ammonia was first developed by Linde and Boyle in 1875. A major characteristic of MO refrigeration oil that is used in industrial ammonia refrigerating machines is that they have a viscosity of VG32 or higher. This is approximately as high as that of the lubricant for general industrial machines. As ammonia is a highly reactive refrigerant, naphthenic oil which is chemically stable with little trace of impurities of nitrogen and sulfur is favorably used for ammonia refrigerating machines.

9.1.2 AB refrigeration oil

The base oil that is used in AB refrigeration oil is a petrochemical product made from naphtha and AB is classified in a synthetic hydrocarbon compound. AB can be classified into two types; the soft AB type, which has a normal alkyl chain; and the hard AB type, which has a branching alkyl chain.

AB is the most chemically stable base oil that can be used for refrigeration oil based on chemical branching. As the hard AB type is more stable than the soft AB type, hard AB is much more commonly used in comparison with soft AB. AB is immiscible with HFC, ammonia and carbon dioxide refrigerants but are fully miscible with HC refrigerants.

AB has been traditionally used with older refrigerants such as R22 and R502. However, AB is now being used more in special low temperature refrigerating machines and

also in some R410A room air conditioners as immiscible oil. Since AB is immiscible with R410A, use of AB with R410A has a possibility for oil return concern. In most cases this concern has been resolved by the use of low viscosity AB or installation of oil separator in the system[2, 3].

9.1.3 PAG refrigeration oil

The base oil that is used in PAG refrigeration oil is synthetic oil made from propylene oxide by polymerization. PAG is an oxygenated hydrocarbon compound that has an ether structure on the main chain[4]. One of important characteristics of PAG is that it has the highest viscosity index of all the base oils that can be used for refrigeration oil, having superior viscosity-temperature relationship.

This characteristic is advantageous for lubrication as it will reduce the load on a compressor that is caused by temperature variation. In addition, the ether structure of PAG does not allow hydrolysis. However, one disadvantage is that PAG has low oxidation stability and becomes oxidized in air. PAG also has low volume resistivity and therefore is inferior for electric driven hermetic compressors.

PAG is commonly used in R134a automotive air conditioners[5], and also in R1234yf automotive air conditioners starting from 2011[6]. For global warming control, CO_2 is suitable as a low-GWP (global warming potential) natural refrigerant. Some manufactures have begun use of CO_2. As part of this trend, CO_2 heat pump water heater system has been newly developed[7], using PAG[8]. CO_2 has high volume resistivity and its mixture with PAG does not detrimentally affect its electrical characteristics for electric driven hermetic compressor. PAG is also used in CO_2 vending machine[9] and showcase[10].

In addition, under some circumstances, PAG can now be used in R134a electric automotive air conditioners. This is achieved by the improvement of dielectric strength of electric motor design[11]. PAG is also used in canned motor type ammonia refrigeration system for industrial freezer[12].

9.1.4 PVE refrigeration oil

The base oil that is used for PVE refrigeration oil is synthetic oil made from vinyl ether by polymerization. PVE refrigeration oil is an oxygenated hydrocarbon compound that has ether structure at side chain[13].

PVE has good miscibility with many different refrigerants. Further, PVE has excellent electrical property because of high volume resistivity unlike PAG. PVE is oxidized in air by having ether structure as same as PAG, and it does not suffer from hydrolysis. PVE is thermally stable in the presence of refrigerant.

A difference from PAG despite having ether structure, is that PVE has reduced viscosity-temperature property because of its lower viscosity index[14].

PVE is commonly used in R410A room air conditioner and packaged air conditioner, R404A showcase, R407C vending machine, chiller and industrial freezer and refrigerator, and R410A gas engine heat pump system.

9.1.5 POE refrigeration oil

The base oil that is used in POE refrigeration oil is synthetic oil made by joining polyol and carboxylic acid through dehydration synthesis. POE is an oxygenated hydrocarbon compound that has an ester structure[15].

Characteristics of POE include superior miscibility with refrigerant, and good oxidation stability and thermal stability even in air. POE also has favorable electrical property. On the other hand, POE reacts with water by hydrolysis[16] for having ester structure and it also is known to undergo tribochemical reaction on friction surface[17, 18].

POE is commonly used in R410A room air conditioner and packaged air conditioner, R404A showcase, R407C

vending machine, chiller, industrial freezer and refrigerator as same as PVE. More recently, POE has started to be used in R134a electric automotive air conditioner. Figure 9.1 shows the chemical structures of various refrigeration oils. And Tables 9.2 and 9.3 show the general specification for refrigeration oils of several viscosity grades.

9.2 Interaction with Refrigerant

Refrigeration oil and refrigerant circulate in refrigeration cycle together as mixture. Therefore, appropriate refrigeration oil must be selected by considering type of compressor and type of refrigerant as well as the operating condition of the refrigeration cycle.

In particular, it is important to understand about various characteristics such as physical properties, lubricity and stability and performance properties of refrigeration oil in refrigerant. The main physical properties include miscibility, solution properties (solubility and mixture viscosity) and electrical insulatability. Performance properties of refrigeration oil include stability, lubricity and compatibility with

Fig, 9.1 Chemical structure of various refrigeration oil

Table 9.2 General specification of refrigeration oil (1)

Item	MO					AB			
ISO viscosity grade	VG7	VG10	VG15	VG32	VG56	VG15	VG22	VG32	VG56
Viscosity (@40℃; mm²/s)	7.448	10.2	15.12	29.15	54.59	15.1	22.7	37.99	50.45
Viscosity (@100℃; mm²/s)	2.025	2.383	3.008	4.317	5.959	3.0	3.7	4.794	5.528
Viscosity index	43	26	13	3	11	12	-12	-12	-5
Density (@15℃; g/cm³)	0.8798	0.9082	0.9079	0.9185	0.9151	0.867	0.867	0.8706	0.8714
Acid number (mgKOH/g)	0.00	0.00	0.00	0.01	0.01	0.01	0.01	0.00	0.00
Flash point (℃)	148	156	164	178	188	168	178	188	194
Pour point (℃)	-50>	-50>	-50>	-50	-35	-50>	-50>	-42.5	-37.5
Sulfur (ppm)	60	253	337	250	350	-	-	-	-
Specific volume resistance (Ωm)	2.0×10^{11}	2.9×10^{12}	5.0×10^{12}	3.7×10^{12}	8.0×10^{12}	2.4×10^{12}	2.5×10^{12}	2.8×10^{12}	3.1×10^{12}

Table 9.3 General specification of refrigeration oil (2)

Item	PAG			PVE			POE		
ISO viscosity grade	VG46	VG68	VG100	VG32	VG68	VG100	VG32	VG68	VG100
Viscosity (@40℃; mm²/s)	48.26	68.73	104.9	32.4	66.57	105.4	30.91	66.39	97.49
Viscosity (@100℃; mm²/s)	10.24	13.93	20.1	5.12	8.037	10.99	5.201	8.187	10.91
Viscosity index	207	212	217	78	84	87	96	89	96
Density (@15℃; g/cm³)	0.9954	0.9973	1.015	0.9315	0.9369	0.9403	0.9815	0.9595	0.9689
Acid number (mgKOH/g)	0.00	0.00	0.00	0.00	0.00	0.00	0.00	0.00	0.00
Flash point (℃)	212	216	226	178	206	207	265	250	268
Pour point (℃)	-50	-50	-42.5	-47.5	-37.5	-32.5	-50	-40	-37.5
Specific volume resistance (Ωm)	1.2×10^7	1.5×10^7	2.3×10^7	1.9×10^{11}	1.8×10^{11}	1.5×10^{11}	1.5×10^{12}	9.4×10^{11}	3.4×10^{10}

organic materials. Table 9.4 shows a summary of properties of refrigeration oil.

9.2.1 Miscibility

In a refrigeration cycle, it is ideal if refrigeration oil and refrigerant are uniformly mixed with each other. However, in many temperature conditions, refrigeration oil and refrigerant separate into two phases. Evaluating about miscibility between refrigeration oil and refrigerant provides clues to in what temperature ranges their dissolution and separation occur.

Miscibility test method is to measure two-phase separation temperature at various oil content ratios. Two phase separation curve obtained through such measurement is specific physical property to each mixture of refrigeration oil and refrigerant. Such data package is useful and necessary for the designing of a refrigeration cycle.

The following paragraphs further explain about two-phase separation temperature curves by referring to Fig. 9.2. The curve on high temperature side in the figure is called upper two-phase separation curve, and the curve on low temperature side is called lower two-phase separation curve. Some refrigeration oil and refrigerant mixtures may exhibit two-phase separation curve on both upper and lower sides. Points K and K' are called upper critical solution temperature and lower critical solution temperature, respectively[19].

There are two liquid phases in two-phase separation region, oil-rich phase and refrigerant-rich phase. It is separated into upper and lower phases by magnitude of density at each phase. Here, it takes the point A in two-phase separation region as an example, and it shows how to determine the oil content ratio in each of oil-rich phase and refrigerant-rich phase at point A. Oil content ratio in oil-rich phase at point A is same as oil content ratio at point C. Similarly, oil content ratio in refrigerant-rich phase at point A is same as oil content ratio at point B. The weight ratio of oil-rich phase and refrigerant-rich phase becomes $\overline{AB}/\overline{AC}$.

If the temperature is linearly lowered from point A, which is in upper two-phase separation region, at constant oil content ratio, the line enters miscible region and it will pass through point D at 15°C. The line will then enter lower side two-phase separation region and reach point E at -15°C.

Table 9.5 shows a summary of miscibility for various

Table 9.4 Requirement of refrigeration oil

Requirement		Evaluation item
Physical properties	Miscibility	Phase separation
	Solubility	Ref. solubility
		Mixture viscosity
	Electric insulation	Volume resistivity
Performance	Lubricity	Friction
		Wear
		Seizure
		Fatigue
	Stability	Thermolysis
		Oxidation
		Hydrolyisis
	Material compatibility	Chemical stability

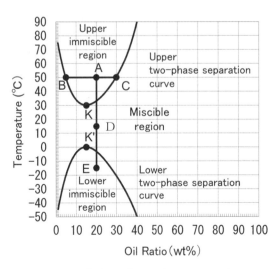

Fig. 9.2 Description of two-phase separation curve

Table 9.5 Miscibility of refrigeration oil

Refrigerant		MO	AB	PAG	PVE	POE
HFC	R134a	×	×	○	○~◎	◎
	R1234yf	×	×	○	○~◎	◎
	R410A	×	×	○	○	○
	R407C	×	×	○	◎	○
	R404A	×	×	○	○	○
Natural	CO₂	×	×	○	○	○
	R600a	◎	◎	○~◎	◎	◎
	R290	◎	◎	○~◎	◎	◎
	NH₃	×	×	○	○	Disabled

× : Immiscible, ○ : Two-phase separation, ◎ : Completely miscible

refrigeration oil and refrigerant mixtures.

MO and AB are immiscible with each HFC, CO_2 and NH_3 refrigerants. PAG, PVE and POE are either two-phase separation or completely miscible with these refrigerants. In addition, R600a and R290 are completely miscible with all refrigeration oils.

As typical examples, Fig. 9.3 shows the two-phase separation temperature curves for R134a with various viscosity grades of PAG. Figure 9.4 shows the similar curves for R410A with PVE and POE of VG68. PAG and PVE exhibit upper side two-phase separation curves. On the other hand, POE exhibits both upper and lower two-phase separation curves.

9.2.2 Solution properties

The solubility and mixture viscosity of refrigeration oil and refrigerant combination at temperature and pressure conditions at each part of refrigeration cycle are important in the design of the cycle. These properties affect lubrication of compressor's sliding part and also thermal efficiency of the condenser and evaporator.

In the field of refrigeration engineering, solubility refers to the proportion of refrigerant that will be dissolved into the refrigeration oil. To measure solubility in actual refrigeration cycle, the typical method is to obtain a sample of refrigeration oil and refrigerant mixture into sampling container. Provided in the testing equipment, measurement tools are integrated to measure the amounts of refrigeration oil and refrigerant in the container by gravimetric method to calculate solubility. Or, alternative methods are to measure the solubility by refractive index, electrostatic capacity and sonic speed[20-23].

To measure solubility in laboratory, first thing is to charge each amount of refrigeration oil and refrigerant into pressure vessel at each temperature. Next, amount of liquid phase refrigerant is calculated by subtracting amount of gas phase refrigerant from total refrigerant in the vessel. Solu-

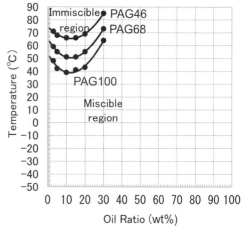

Fig. 9.3 Two-phase separation curve for R134a with PAG

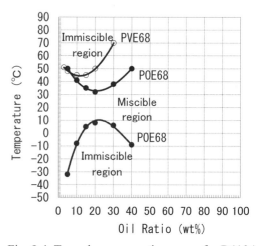

Fig. 9.4 Two-phase separation curve for R410A with PVE & POE

bility can be calculated from amount of liquid phase refrigerant and refrigeration oil. In this way, the solubility chart shows the relationship between pressure and solubility at various temperature[24].

The equation that represents the relationship between pressure, temperature and solubility is called solubility equation. Hildebrand has conducted a theoretical study to

Chapter 9 Refrigeration Oil

establish the solubility equation. It can now be estimated based on solubility parameters shown in the following formula[25, 26]:

$$P_2 = P_2^0 X_2 \exp\left\{\frac{V_2 \phi_1^2 (\delta_1 - \delta_2)^2}{RT}\right\} \quad (9.1)$$

(P_2: Pressure of constituent 2, P_2^0: Saturation vapor pressure of constituent 2, X_2: Mole fraction of constituent 2, V_2: Volume of constituent 2, ϕ_1: Volume fraction of constituent 1, δ_1: Solubility parameter of constituent 1, δ_2: Solubility parameter of constituent 2, R: Gas constant, T: Absolute temperature)

In recent years, other estimation methods have been developed based on polynomial approximation or state equations[27-30]. It is known that solubility in saturation vapor pressure region and supercritical region exhibit different behaviors[31, 32].

On the other hand, mixture viscosity refers to the viscosity of refrigeration oil and refrigerant combined. To measure mixture viscosity in actual refrigeration cycle, the typical method is to use vibration viscometer in the measurement part of cycle[33].

To measure mixture viscosity in laboratory, predetermined amounts of refrigeration oil and refrigerant are charged in a pressure vessel which was placed inside a viscometer. Next, it measures the pressure and mixture viscosity at various temperatures.

At this time, solubility corresponding to the measured solution viscosity can be obtained from temperature and pressure on the solubility chart. Alternatively, there is a method to measure four factors of pressure, solubility, mixture viscosity and temperature at the same time[34].

Based on the measured solubility and mixture viscosity, it is possible to draw mixture viscosity lines on the viscosity temperature chart. Daniel chart is obtained by drawing the mixture viscosity lines and the isobaric lines on viscosity temperature chart, for example as shown in Fig. 9.6.

The following paragraphs explain about the viscosity theory of mixture liquid, which provides a basis for theoretical analysis of mixture viscosity. According to the viscosity theory of the ideal solution, the viscosity of a two-constituent mixture can be expressed by the following Arrhenius equation:

$$\log \eta = X_1 \log \eta_1 + X_2 \log \eta_2 \quad (9.2)$$

(η: Viscosity of the mixture, η_1: Viscosity of constituent 1, η_2: Viscosity of constituent 2, X_1: Volume fraction of constituent 1, X_2: Volume fraction of constituent 2, $X_1 + X_2 = 1$)

Alternatively, Kendall equation uses mole fraction instead of volume fraction. For more details, see referential document 35[35].

In the field of refrigeration engineering, other mixture viscosity equations by Grunberg-Nissan, Frenkel, Yokozeki, McAllister and Michaels-Sienel have been developed based on Kendall equation[36]. Various estimating equations based on polynomial approximation have also been proposed[37, 38].

A problem with these estimating equations is that their parameters are based on measurements taken with CFC refrigerants of the past. Care should be taken when applying these estimating equations to hydrocarbon and CO_2 refrigerants. To obtain a comprehensive estimating equation, it is necessary to measure many actual data with a wide range. For this reason, it is important to establish a rapid method of measurement.

The above-described solubility chart, mixture viscosity chart and Daniel chart show the mixture properties which is unique to each specific combination of refrigeration oil and refrigerant. These are important details to understand how refrigeration oil and refrigerant will behave in a refrigeration cycle.

As typical examples, solubility chart and Daniel chart for PAG46, PVE68 and POE68 are shown in Figs. 9.5 to 9.10. PAG46 is used in R134a automotive air conditioners, PVE68 and POE68 are both used in R410A stationary air conditioning systems. These charts are all obtained from polynomial approximation based on actual measurements.

9.2.3 Electrical insulation

Electrical components for stationary air conditioning system must be selected so that the leak current will not be more than 1 mA. There is no covering to the motor terminal in electric compressor unit, leak current may be an issue. Particularly when the compressor is started up with a large amount of liquid refrigerant in the sump, the electrical terminals would be immersed in refrigeration oil dissolving into refrigerant. In such situation, the potential level of leak current will be the highest.

Therefore, refrigeration oil is required to have characteristics of insulating oil and is desired to have superior insulation property. Tables 9.2 and 9.3 show volume resistivity of various refrigeration oils.

MO, AB, POE and PVE have high volume resistivity and therefore will provide sufficient electrical insulation. Due to this characteristic, these types of oils are used in hermetic type compressors. In contrast, PAG is easy to conduct electricity because oxygen content is high on the chemical structure. Therefore, it provides lower volume resistivity.

Compressors for Air Conditioning and Refrigeration

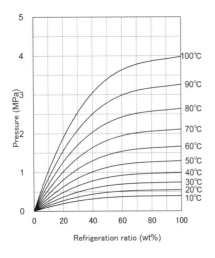

Fig. 9.5 Solbility for R134a with PAG46

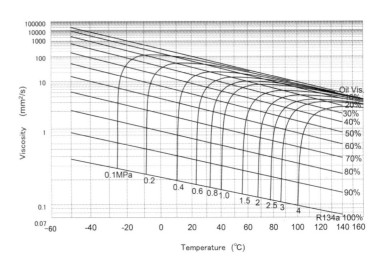

Fig. 9.6 Daniel chart for R134a with PAG46

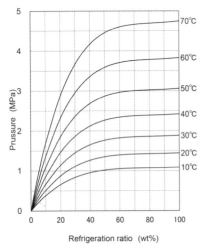

Fig. 9.7 Solbility for R410A with PVE68

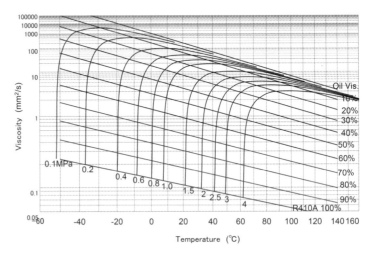

Fig. 9.8 Daniel chart for R410A with PVE68

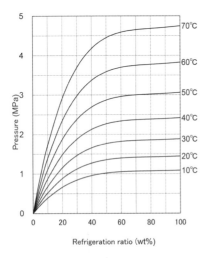

Fig. 9.9 Solbility for R410A with POE68

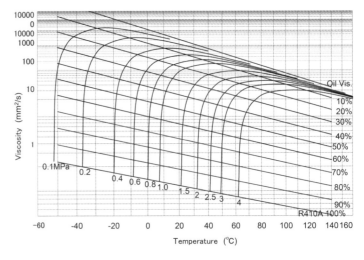

Fig. 9.10 Daniel chart for R410A with POE68

Table 9.6 shows volume resistivity of various refrigerants. CO_2 and R290 of HC refrigerant have higher volume resistivity. On the other hand, ammonia and HFC refrigerants have lower volume resistivity. Refrigeration oil and refrigerant circulate in the compressor together as a mixture, the use of PAG, creates a situation in which electric insulation is insufficient by a combination of refrigerant. It is critical to examine the electrical insulation of refrigeration oil and refrigerant mixture.

For this purpose, hermetic type volume resistivity measurement apparatus can be used to measure volume resistivity of the mixture in a similar way as measuring the volume resistivity of insulating oil[39].

Figure 9.11 shows PAG-CO_2 and PAG-R134a mixtures and the relationship between refrigerant ratio and volume resistivity. As CO_2 has a high volume resistivity by itself, its volume resistivity when mixed with PAG is not reduced to less than 10^7 Ωm, which is considered a sufficient electrical insulation. Therefore, the PAG-CO_2 combination is commonly used in heat pump water heater systems, vending machines and showcases.

On the other hand, the volume resistivity curve of PAG-R134a combination has a bottom at refrigerant ratio of 90 wt%. It shows that this combination is inferior in electrical insulation. For this reason, PAG is not used with R134a and other HFC refrigerants using hermetic type compressors.

Table 9.6 Volume resistivity of refrigerant (@25℃)

Refrigerant	R134a	R410A	R407C	R404A
Volume resistivity (Ωm)	1.1x10⁶	3.9x10⁵	7.4x10⁵	8.5x10⁵

Refrigerant	R1234yf	CO₂	R290	NH₃
Volume resistivity (Ωm)	1.3x10⁶	7.5x10¹¹	7.3x10⁹	1.0x10⁵

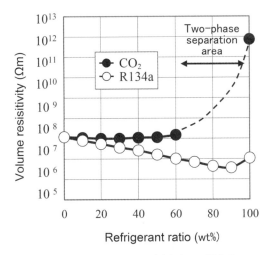

Fig. 9.11 Volume resisitivity of PAG

9.2.4 Lubricity

Table 9.7 shows a list of frictional surfaces of various compressors. There are many types of compressors that are being used, and each has its specific frictional surface on sliding portions. Radial bearing and thrust bearing are portions of sliding that are common to all type of compressors. Roller bearing and sliding bearing are used for sliding portions. While sliding parts are commonly built with steel, aluminum is widely used in automotive air conditioning compressors for weight reduction[40]. Depending on the application, these sliding parts may have PTFE or DLC coating[41] or may be heat treated[42].

Although these sliding portions are designed to be fluid film lubrication, they enter into boundary lubrication at start-up or overload. In situations such as oil film breakage caused by high temperature or being washed with liquid refrigerant, these sliding portions will operate without refrigeration oil and experience wear.

The main purpose of refrigeration oil is to maintain lubrication of the sliding portions in the compressor. Therefore, lubricity is the primary requirement for refrigeration oil.

Lubricity of refrigeration oil depends on the base oil and the type and quantity of lubricity improvers; such as extreme pressure agent and oiliness agent. The most appropriate package of base oil and lubricity improvers must be selected according to such factors as operating environment of compressor and material composition of sliding portions.

The base oils for refrigeration oil are described in Sections 9.1.1 to 9.1.5. Although dependent on the lubrication condition, most of base oils are low in reactivity to sliding part materials. Therefore deterioration of the friction surfaces does not occur by tribochemical reaction. However, it is known that friction surface could locally get as hot as over 250°C. At that time, deterioration of base oil is likely to occur by tribochemical reaction. It is therefore desirable to select a thermally stable base oil[43].

Table 9.8 shows additives that are commonly used in refrigeration oil. Additives such as lubricity improver, stabilizer and defoaming agent are typically used for refrigeration oil[44, 45]. Extreme pressure agent is used under severe lubrication state of high speed or high surface pressure. Extreme pressure agent forms a reaction film on the friction surface to prevent seizure. Examples of extreme pressure agents are phosphate and phosphite. On the other hand, oiliness agent forms physical adsorption film onto friction surface in order to reduce friction. Examples of oiliness agents are alcohol and ester. Phosphorus additives are known to be effective on sliding part made of steel such as SKH and

Compressors for Air Conditioning and Refrigeration

Table 9.7 Frictional surface of various compressor

Unit Type	Comp. Type	Frictional surface (F.S.)	Material	Common F.S.	Material
RAC PAC HPWH	Rotary	Cylinder/Vene/Rolling piston	Steel SKH SCM FCD	Radial bearing Thrust bearing	Steel SUJ FCD FC
	Scroll	Fixed scroll/Turning scroll/Oldham ring			
	Swing	Piston/Swing busch/Cylinder			
Refrigerator	Reciprocating	Piston/Cylinder			
Industrial freezer	Reciprocating	Cylinder/Piston ring/Piston pin			
	Screw	Male screw rotor/Female screw rotor			
Automotive air conditioner	Swash plate	Swash plate/Shoe/Piston/Cylinder	Aluminum ADC AC		
	Rotary	Cylinder/Sliding vene/Rotor			
	Scroll	Fixed scroll/Turning scroll/Oldham ring			

Table 9.8 Refrigeration oil additive

Lubricity improver	Extreme pressure agent	Phosphate Phosphite
	Oiliness agent	Alcohol Ester Fatty acid
Stabilizer	Antioxidant	Hindered phenol
	Acid catcher	Epoxy
Defoaming agent		Polysiloxane

SCM. They form an iron phosphate film onto steel surface to prevent seizure. Also, alcohol is an effective oiliness agent for aluminum alloy such as ADC.

Where sliding part is surface-coated, there is a case that extreme pressure agents detrimentally affect the coating film. Therefore it is important to select the best appropriate additive as lubricity improver for different sliding components.

The following paragraphs explain about testing methods that are used for evaluating lubricity of refrigeration oils and also about test data that can be obtained from these methods.

Approximately twenty years ago, seizure and wear test using Falex tester had been done for evaluation of refrigeration oil lubricity. Falex tester was widely used because Falex test methods are incorporated into ASTM standards and also because test pieces were easily obtainable.

This basic Falex tester is still being used in some cases to obtain basic data of refrigeration oils. In an attempt to evaluate lubricity in the actual refrigerant oil environment, a chamber is installed around the Falex tester[46].

In recent years after CFC regulations of 1987, many new refrigerants such as R134a and R410A have been developed. Due to the increasing needs of lubricity evaluation in the refrigerant atmosphere, various hermetic pressure type friction tester have been developed.

As shown in Table 9.9, various hermetic pressure type

friction testers are used. They include ones for basic evaluation and also ones that incorporate the sliding part of actual compressor[47]. As an example, Figure 9.12 shows a vane on disk tester. Fatigue life tester for rolling bearing in EHL, viscosity-pressure coefficient measurement apparatus and oil film thickness measurement apparatus have been implemented.

These testers can be used to measure friction coefficient, seizure load and wear. In addition, various surface analyzers such as X-ray photoelectron spectroscopy (XPS) are used to establish the relationship between lubricity and friction surface material composition.

Figure 9.13 shows the effect of moisture content in POE related to the increased wear of vane and disk which is obtained using vane on disk tester. The figure shows that wear of vane steeply increases as moisture content increases above 500 ppm[48].

Figure 9.14 shows the result of evaluation using block on ring tester about how block wear differs by refrigerant type. The refrigeration oil used in this evaluation was PAG. This data provides that the block wear width is smaller in CO_2 atmosphere than in R134a atmosphere.

Figure 9.15 shows PAG and POE friction coefficients obtained using swash-plate shoe tester operated in CO_2 atmosphere. This data shows that friction coefficients with PAG are lower than those with POE[49].

Table 9.10 shows comparison of seizure load and wear with several chemical structures of PAG which obtained using pin V-block tester. It is apparent that the type of PAG that is capped with methyl ether on both ends offers superior resistance at seizure and wear[5].

Figure 9.16 shows the result of comparative fatigue life tests with two kinds of viscosity grades of PAG in N_2 and CO_2 atmospheres. This data provides that the fatigue life in CO_2 atmosphere is shorter than that in N_2 atmosphere. This is because that CO_2 being soluble in PAG will more reduce the viscosity of mixture than N_2 being almost insoluble in

Chapter 9 Refrigeration Oil

Table 9.9 Hermetic type friction tester for oil/ref. mixtire

Tester	Contact geometry	Example simulation point	Evaluation item
Ball on disk tester	Point contact - sliding	–	Friction coefficient Wear Seizuer load Surface roughness Oil film analysis XPS EPMA TOF-SIMS FIB-TEM Nanoindenter
Vane on disk tester	Line contact - sliding	Vene/Rolling piston Radial sliding bearing	
Block on ring tester			
Pin/V-block tester			
Ring on disk tester	Plane contact - sliding	Oldham ring	
Pin on disk tester		Swash plate/Shoe	
Swash plate/Shoe tester			
Block on plate reciprocating tester		Piston/Swing busch	
Bearing fatigue life tester	Line contact - rolling/sliding	Rolling bearing	Fatigue life
Oil-film thickness measuring instrument	Point contact - rolling/sliding	Rolling bearing	Oil film thickness

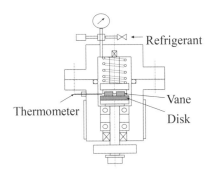

Fig. 9.12 Vane on disk tester

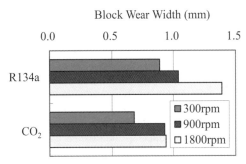

Fig. 9.14 Block on ring test result

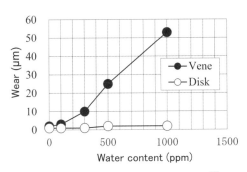

Fig. 9.13 Vane on disk test result [48]

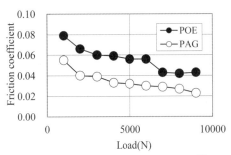

Fig. 9.15 Pin on disk test result [49]

Table 9.10 Lubricity of PAG [5]

Chemical structure of PAG	Seizuer load (N)	Wear (mg)	
		Pin	V-block
HO-(PO)$_n$-H	1780	11.0	18.9
CH$_3$O-(PO)$_n$-H	2230	9.7	24.6
CH$_3$O-(PO)$_n$-CH$_3$	5340	0.1	0.1
CH$_3$COO-(PO)$_n$-CH$_3$	1780	8.9	20.1
CH$_3$COO-(PO)$_n$-OCOCH$_3$	1780	8.5	261
CH$_3$O-(EO/PO=5/5)$_n$-CH$_3$	5340	0.3	0.1
CH$_3$O-(EO/PO=3/7)$_n$-CH$_3$	5340	0.6	0.3

PAG[49]. Table 9.11 shows viscosity-pressure coefficients of PAG46 and PAG100. The table shows that PAG100, which exhibits longer fatigue life in Fig. 9.16, has higher viscosity-pressure coefficients. This indicates that there is a correlation between viscosity-pressure coefficient and fatigue life.

Provided above is an overview of typical lubricity test results. For more details, see referential documents [50, 51].

9.2.5 Stability

The mixture of refrigeration oil and refrigerant that circulates in a refrigeration cycle is exposed to wide range of temperatures between high to low. The temperature of friction portions and refrigerant discharge port rise the most in a compressor, which may be possibly leading to thermal deterioration of refrigeration oil. Also, refrigeration cycle contains residual oxygen at installation work and also contains moisture in refrigeration oil. Therefore, refrigeration oil deteriorates though oxidation by residual oxygen. And also

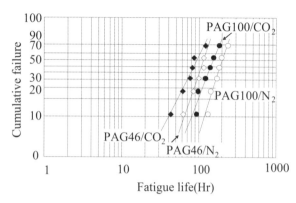

Fig. 9.16 Fatigue life test result

Table 9.11 Viscosity-pressure coefficient of PAG

Refrigeration oil	Viscosity-pressure coefficient (GPa^{-1})		
	40℃	100℃	150℃
PAG46	17.7	13.9	11.4
PAG100	19.6	15.1	12.8

Table 9.12 Stability of refrigeration oil

	MO	AB	PAG	PVE	POE
Thermolysis	○	○	○	○	○
Oxidation	○	○	△	△	◎
Hydrolyisis	○	○	○	○	△

△ is solvable by additive or desiccant

Table 9.13 Bonding energy of refrigeration oil

Bond type		C-H	C-C	C=C	C-O	C=O
Bonding energy (kJ/mol)	σ bond	411	366	366	378	378
	π bond	–	–	260	–	237
Bond distance (Å)		1.09	1.54	1.34	1.43	1.23
Refrigeration oil		MO	MO	–	–	–
		AB	AB	AB	–	–
		PAG	PAG	–	PAG	–
		PVE	PVE	–	PVE	–
		POE	POE	–	POE	POE
Characteristic		Reducing ←――――→ Oxidizing				

refrigeration oil deteriorates though hydrolysis by moisture. These deterioration products are deposited as sludge in the cycle. It may be causing discharge valve operation failures, terminal insulation failures, capillary tube clogging and ultimately reduction in thermal efficiency.

To address these issues, refrigeration oils include additives such as antioxidants and acid catchers like the ones listed in Table 9.8 which may be provide better stability.

Refrigeration oil stability can be evaluated with sealed tube test and autoclave test. These are the accelerated aging tests. Usually these tests are conducted in the temperature range of 175°C to 200°C and for a holding period of 10 to 30 days. This testing is simulates 10 to 30 years of actual compressor use in a short amount of time.

Table 9.12 shows a stability comparison of several refrigeration oils against thermolysis, oxidation and hydrolysis[52]. Such difference in stability between refrigeration oils is due to their different chemical structures. Table 9.13 provides chemical structure information about refrigeration oils shown in Fig. 9.1. It shows the type of chemical bond, the bonding energy, the bonding distance and the characteristics about functional group[53, 54].

MO and AB are hydrocarbon compounds with C–C bonds and C–H bonds, and AB consists of C=C bonds further plus. Both of them is excellent in stability.

On the other hand, PAG and PVE are ether compounds with C–O bonds. They are inferior in oxidation stability. As shown in the following formula, ether compound reacts with oxygen. The oxidative degradation accelerates by hydroperoxide formed through α hydrogen abstraction[55].

Therefore, it is imperative that an antioxidant agent is added to PAG and PVE.

$$RCH_2-O-CHR_2 + O_2$$
$$\rightarrow RCH_2-O-C(-OOH)R_2$$
$$\rightarrow \text{Oxidative degradation products}$$

POE is ester compound and has C=O bonds. It produces carboxylic acid and alcohol by hydrolysis. This hydrolysis occurs in the following scheme[56].

$$RCOOR + H_2O \rightarrow RCOOH + HOR$$

To prevent this hydrolysis, it is critical to remove moisture in the refrigeration system. A typical moisture removal method is to install molecular sieves inside the refrigeration system.

POE used for refrigeration oil is a specially-structured hindered ester, which has better oxidation stability than MO and AB.

As explained in the above paragraphs, stability of refrigeration oil is closely related to its chemical structure and also is affected by external factors such as heat, oxygen and water.

Other factors that affect stability of refrigeration oil are metalworking fluids. These fluids are used in component manufacturing processes. These include press oil, cutting oil, grinding fluid, copper tube drawing oil, rust preventive oil and so on.

In the working of sintered alloy, press oil will enter inside the sintered alloy component. Though most washes away, some residual oil remains inside the sintered alloy. Therefore, residual oil can affect the stability of refrigeration oil.

Table 9.14 shows the result of a contamination effect evalu-

Table 9.14 Influence of contaminaton

Contamination	Evaluation item	PAG	PVE	POE
Sulfur-Based cutting oil (1%)	Oil appearance	Yellow	Yellow	Yellow
	Catalyst appearance	Cu-black	Cu-black	Cu-black
	Acid number (mgKOH/g)	0.01	0.01	0.01
Grinding liquid (1%)	Oil appearance	Good	Good	Good
	Catalyst appearance	Good	Good	Good
	Acid number (mgKOH/g)	0.34	0.31	0.47
Rust preventive oil (1%)	Oil appearance	Good	Good	Good
	Catalyst appearance	Good	Good	Good
	Acid number (mgKOH/g)	0.03	0.02	0.04
Copper Tube drawing oil (1%)	Oil appearance	Good	Good	Good
	Catalyst appearance	Good	Good	Good
	Acid number (mgKOH/g)	0.01	0.01	0.01

Test condition : 175°C, 14days, oil/R410A=20g/20g, catalyst (Fe,Cu,Al)

ation by autoclave testing. Sulfurous cutting oil is found to discolor copper catalyst. Also, grinding fluid, which is typically alkaline, accelerates hydrolysis of POE and phosphoric additives. Also it has raised the acid value of refrigeration oil. It is critical to wash the components before assembly to remove these metalworking fluids and rust preventive oil.

9.2.6 Compatibility with organic materials

Organic materials that are used in hermetic compressor include enamel coating on motor winding wire, stator lacing cord, cluster socket and insulation film and tube. These components use polyamideimide heat resistant plastic, PET and PPS. Also, connection hose, lip seal, O-ring, gasket and piston coating for automotive air-conditioner use organic materials such as rubber, nylon and PTFE.

If any of these organic materials are not fully compatible with refrigeration oil or refrigerant, the part may become flaked, harden or swollen. For example, massive elution of oligomer from PET may cause capillary tube clogging. Many types of rubbers are used as sealing. They must be selectively used depending on the purpose. If a rubber part shrinks, its sealing effect is lowered to cause leakage. Therefore, the most appropriate rubber must be used.

For that reason, organic material must be checked for compatibility with refrigeration oil and refrigerant by autoclave test in the same way as for stability evaluation. Compatibility test is performed by immersing organic material into the refrigeration oil and refrigerant mixture inside autoclave. After the test, it is necessary to check if the refrigeration oil has deteriorated or if properties of organic material has changed.

As a test method to evaluate the elution of an organic material such as PET oligomer, there is a measurement of the floc point.

Floc point is technique originally invented as a means to evaluate wax precipitation from MO.

Table 9.15 shows the test result of organic material compatibility evaluation. Compared to MO, PVE and POE are less prone to cause swelling of HNBR (hydrogenated nitrile butadiene rubber) or elution of oligomer from PET.

9.3 Refrigeration Oil Selection Method and Considerations for Use

9.3.1 Refrigeration oil selection method

The fundamental role of refrigeration oil is to assure lubrication of compressor by providing oil film between the sliding parts. For this reason, the appropriate viscosity grade of refrigeration oil for compressor must be determined by selecting adequate mixture viscosity in the high-temperature, high-pressure region of sliding part on Daniel chart that is described in Section 9.2.2.

Further, for heat transfer characteristics and oil return property, it is common to select refrigeration oil with good miscibility with refrigerant. Specifically, low temperature equipment such as freezer, needs to select refrigeration oil with critical solubility temperature of not more than -20°C. If provided with the oil separator in refrigeration system, a refrigeration oil that is immiscible with refrigerant can be selected.

Hermetic compressors with a built-in motor need to select refrigeration oil of good electrical resistance.

The above-described refrigeration oil properties; including viscosity and mixture property, miscibility and electrical resistance; primarily depend on the type of base oil and are most important factors to be considered when selecting refrigeration oil.

However, in a compressor that entails high discharge pressure and heavy load, it is difficult to maintain oil film on the sliding parts. In this case, a refrigeration oil that has appropriate type and amount of lubricity improver should be selected and validated by various lubricity and durability tests.

Also, if the refrigeration equipment is operated for long hours

Table 9.15 Material compatibility test result

Material	Refrigeration oil	MO	PVE	POE
	Refrigerant	R22	R134a	R134a
HNBR(*1)	Oil appearance	Good	Good	Good
	Acid number (mgKOH/g)	0.02	0.01	0.01
	Weight change rate (%)	7	2	3
PET(*2)	Oil appearance	Cloud	Good	Good
	Acid number (mgKOH/g)	0.15	0.06	0.09
	Amount of oligomer (mg)	43	12	15

(*1)Test condition : 150°C , 30days, oil/ref.=40g/20g
(*2)Test condition : 150°C , 30days, oil/ref./PET=50g/5g/20g

under high temperature condition, it is especially important to provide a chemically stable refrigeration oil. In this kind of situation, a refrigeration oil that include appropriate type and amount of stabilizer is required. As a note, these lubricity improver and stabilizer must not affect the organic material used in refrigeration equipment.

In this way, the reliability of compressor has been assured by selecting refrigeration oil whose base oil satisfies physical properties such as viscosity, and with the best type and amount of additives that satisfies chemical properties such as stability.

9.3.2 Considerations for use of refrigeration oil

Some refrigeration equipment such as household refrigerators are produced to be finished in a complete operating status at manufacturing plants and then shipped and sold to consumers. Other equipment, such as packaged air conditioner for commercial buildings, are assembled and connected on site.

In the case of equipment that is produced at manufacturing plants, compressor and other components such as condenser and evaporator used for refrigeration system are washed and assembled under strict contamination control. Refrigeration oil is charged into the compressor after being dehydrated to the specified moisture control limit. The amount of residual air in the cycle is also controlled to the specified limit by adequate evacuation.

On the other hand, equipment that is assembled and connected on site require measures to be taken to in order to prevent absorption of moisture in refrigeration oil and also to sufficiently remove residual air from inside the system.

PAG, PVE and POE absorb moisture more readily than MO and AB as shown in Tables 9.2 and 9.3. This is because that PAG, PVE and POE are oxygenated hydrocarbon compounds. Oxygen in the molecule attracts moisture in the air by hydrogen bond.

As a result, if the container of refrigeration oil is left open during one day, PAG, PVE or POE would likely absorb more than 1000 ppm of water. Therefore, after filling refrigeration oil from the container into equipment, the container must be immediately capped so as to prevent remaining oil in the container from contacting the outside air. POE requires particularly careful moisture control as it reacts with absorbed moisture by hydrolysis.

As explained in Section 9.2.5, PAG and PVE are inferior in oxidation stability. Therefore removal of residual air in equipment is crucial. Insufficient air removal, which is likely to happen to commercial building air conditioner having long piping sections, may lead to discoloration of refrigeration oil. Installation work must be carefully controlled so as not to leave air in completed refrigeration equipment.

Care must also be taken to prevent generation of copper oxide during brazing process of copper piping and contamination by copper swarf during assembling process of the piping.

When a refrigerant system is switched over from R22 to R410A, the refrigeration oil must also be switched to PVE or POE for R410A. In this situation, the common procedure is to flush the refrigeration circuit so as to remove MO for R22[57]. Alternatively, the washing-less technology for the refrigerant switchover has also been studied in order to ensure that there is no problem of remaining a small amount of MO in existing piping[58].

References

1) K. Takaichi, S. Yamada: "The Development of Freezer-Refrigerator using Non-flon (CFC-free, HCFC-free, HFC-free) Refrigerant", Refrigeration, 77 (896), pp. 494-498 (2002). (in Japanese)

2) M. Nakayama, Y. Sumida, S. Suzuki, H. Makino, K. Segawa: "Development of R410A room air conditioner using alkylbenzene oil", Proceedings of the International Symposium on HCFC Alternative Refrigerants 1998, Kobe, pp. 6-11 (1998). (in Japanese)

3) T. Sakamoto, Y. Shimomura, S. Suda, Y. Yamamoto, U. Sasaki, M. Saito, M. Sunami: "HFC Refrigeration Lubricant for HFCs", Proceedings of the International Symposium on HCFC Alternative Refrigerants 1998, Kobe, pp.89-93 (1998). (in Japanese)

4) H. Ikeda: "Technical Trend of Synthetic Oil using Refrigeration Lubricant", The Tribology, 20 (12), pp. 12-14 (2006). (in Japanese)

5) M. Kaneko, T. Konishi, Y. Kawaguchi, M. Takagi: "The Development of PAG Refrigeration Lubricants for Air Conditioner with HFC134a", SAE Paper, No. 951052 (1995).

6) M. Kaneko, T. Matsumoto, T. Tokiai, H. Suto, H. Shimosaki: "Evaluation of characteristics of PAG Lubricant for Air Conditioner with HFO1234yf", Proceedings of the 2011 JSRAE Annual Conference, Tokyo, pp. 667-670 (2011). (in Japanese)

7) K. Hashimoto: "Development of the CO_2 heat pump water heater for residential use and its prevention effect on global warming", PETROTECH, 8 (32), pp. 592-597 (2009). (in Japanese)

8) M. Kaneko, T. Tokiai, H. Ikeda: "The Development and Practical Use of Refrigeration Lubricants for Heat Pump Water Heater System with CO_2", PETROTECH, 8 (32), pp. 561-565 (2009). (in Japanese)

9) Y. Kimura, N. Ishita: "Transition and Future Prospect of Refrigerant for Vending Machines", Refrigeration, 82 (959), pp. 767-770 (2007). (in Japanese)

10) I. Tomochika: "Numerical Simulation for Heat Exchanger Design of the Refrigerated Display using Natural Working Fluid", Refrigeration, 84 (979), pp. 401-406 (2009). (in Japanese)

11) K. Matsunaga, K. Inui, O. Ishida, H. Kikuchi, K. Iijima, J. Hara: "Refrigerant Trend of Vehicle Air- conditioning System", Refrigeration, 82 (959), pp. 771-775 (2007). (in Japanese)

12) A. Matsui: "Ammonia Refrigeration Machinery", Refrigeration, 73 (853), pp. 985-988 (1998). (in Japanese)

13) J. Yagi: "Market and Technology Trends of Synthetic Base Oils using Polyvinylether", The Tribology, 18 (12), pp. 22-24 (2004).

14) M. Kaneko, S. Sakanoue, T. Tazaki, S. Tominaga, M. Takagi, M. Goodin: "Determination of Properties of PVE Lubricants with HFC Refrigerants", ASHRAE Trans. 1999, 105 (2), pp. 5-12 (1999).

15) M. Muraki: "Refrigeration lubricant based on polyolester for alternative refrigerants", Proceedings of the International Symposium on R22 & R502 Alternative Refrigerants '94, pp. 101-106 (1994).

16) M. Sunami: "Refrigeration Oil for HFCs (Hygroscopic and Hydrolytic Reaction", Refrigeration, 78 (905), pp. 177-182 (2003). (in Japanese)

17) M. Muraki, D. Dong, T. Sano: "Friction and Wear Characteristics of Polyolester Base Lubricant in Refrigerants Environment", Journal of Japanese Society of Tribologists, 43 (1), pp. 43-49 (1998). (in Japanese)

18) M. Kaneko: "Deterioration of Refrigeration Oils on Friction Surface with HFC134a", Proceedings of JAST Tribology Conference, Osaka, pp.275-277 (1997). (in Japanese)

19) Japanese Association of Refrigeration: "Refrigerating and Air Conditioning Technology", Japanese Association of Refrigeration, Tokyo, pp.73-75 (1998). (in Japanese)

20) M. Ito, M. Fukuta, T. Yanagisawa: "Characteristics of Critical Angle Sensor for Concentration Measurement of Refrigerant and Oil Mixture", Proceedings of the 43rd Japanese Joint Conference on Air-conditioning and Refrigeration, 43, Tokyo, pp. 77-80 (2009). (in Japanese)

21) M. Fukuta, T. Yanagisawa, Y. Ogi, J. Tanaka: "Measurement of Concentration of Refrigerant in Refrigeration Oil by Capacitance Sensor", Transactions of JSRAE, 16 (3), pp. 239-248 (1999). (in Japanese)

22) Y. Ozaki, T. Hotta, T. Hirata: "Oil Circulation Ratio On-line Measurement in CO_2 Cycle", Denso Technical Review, 8 (2), pp. 87-93 (2003). (in Japanese)

23) M. Fukuta, T. Suzuki, T. Yanagisawa: "Concentration Measurement of Refrigerant/ Refrigeration Oil Mixture by Sound Speed", Transactions of JSRAE, 25 (4), pp. 391-400 (2008). (in Japanese)

24) H. Takahashi: "Characteristics of Refrigerant-Oil Mixture; R290-Mineral Oil, POE and PAG", Proceedings of the International Symposium on HCFC Alternative Refrigerants and Environmental Technology 2002, Kobe, pp.160-164 (2002). (in Japanese)

25) J. H. Hildebrand, R. L. Scoot: "The Solubility of Nonelectrolytes", Reinhold (1950).

26) K. Shinoda: "Solution and Solubility", Maruzen, Tokyo, pp.95-130 (1974). (in Japanese)

27) K. Takigawa: "Measurement of Refrigerant Solubility of Refrigeration Oil and the Predictive Calculation", Nisseki Mitsubishi Technical Review 2000, 42 (4), pp. 18-23 (2000). (in Japanese)

28) H. M. Lee Kang, S. C. Zoz, M. B. Pate: "Solubility of HFC32, HFC125 and HFC134a with Three Potential Lubricants", Proc. 1994 Int. Ref. Conf. Purdue, West Lafayette, pp. 437-442 (1994).

29) A. M. Yokozeki: "Solubility and Viscosity of Refrigerant-Oil Mixtures", Proc. Int. Comp. Eng. Conf. Purdue, West Lafayette, pp. 335-340 (1994).

30) J. S. Fleming, Y. Yan: "The prediction of vapour -liquid equilibrium behavior of HFC blend-oil mixtures from commonly available data", Int. J. Refrigeration, 26, pp. 266-274 (2003).

31) A. Hauk, E. Weidner: "Carbon Dioxide as an environmentally benign Coolant for Climatisation of Automobiles", Proc. Int. Symp. Supercritical Fluids, Atlanta, pp. 1-10 (2000).

32) K. Nagahama, K. Yamato: "Phase Equilibrium and Density Data of CO_2+Lubricant Oil Mixtures for CO_2 Heat Pump Cycle", Proceedings of the 2001 JSRAE Annual Conference, pp.185-188 (2001). (in Japanese)

33) K. Takigawa: "Measurement of Refrigerant Dissolved Viscosity of Refrigeration Oil and the Predictive Calculation", Nisseki Mitsubishi Technical Review, 44 (1), pp. 13-17 (2002). (in Japanese)

34) M. Kaneko, H. Ikeda, T. Tokiai, H. Suto, M. Tamano: "The Development of Ether Lubricant for All kinds of Refrigeration System with CO_2", Proceedings of the International Symposium on New Refrigerants and Environmental Technology 2006, Kobe, pp.65-68 (2006).

(in Japanese)

35) T. Ishikawa: "Theory of the Mixed Solution Viscosity", Maruzen, Tokyo, pp. 9-29, (1968). (in Japanese)

36) H. H. Michels, T H. Sienel: "Viscosity Modeling of Refrigerant/Lubricant Mixture", Proc. Int. Conf. Ozone Protection Tech. Baltimore, pp. 96-105 (1997).

37) V. Z. Geller, M. E. Paulaitis, D. B. Bivens, A. Yokozeki, "Viscosity for R22 alternatives and their mixtures with a lubricant oil", Proc. 1994 Int. Ref. Conf. Purdue, West Lafayette, pp. 49-55 (1994).

38) S. Sakanoue: "Characteristics of Solubility for Synthetic Refrigeration Lubricant with New Refrigerants", Idemitsu Technical Report, 40 (6), pp. 22-25 (1998). (in Japanese)

39) M. Kaneko, H. Ikeda, T. Tokiai, H. Suto: "Current Status of Lubricant for CO_2 Refrigerant", Proceedings of 2006 JSRAE Annual Conference, Fukuoka, pp. 425-428 (2006). (in Japanese)

40) S. Nakayama, M. Kurahashi, K. Takenaka: "Development of forged aluminum alloy piston for automotive air conditioning", Journal of the Japan Institute of Light Metals, 40 (4), pp. 312-316 (1990). (in Japanese)

41) H. Nakao, H. Maeyama, N. Hattori, T. Takayama: "Wear-Reducing Technologies for Rotary Compressor Using CO_2 Refrigerant", Transactions of the JSRAE, 25 (4), pp. 365-374 (2008). (in Japanese)

42) S. Nakajima: "Present and Future of Wear Resistant Materials for Alternative Refrigerant Compressors", Journal of Japanese Society of Tribologists, 43 (3), pp. 195-198 (1998). (in Japanese)

43) H. Takahashi: "Consideration about Base Oils and Additives for Refrigeration Oil", Proceedings of the International Symposium on New Refrigerants and Environmental Technology 2006, Kobe, pp.77-82 (2006). (in Japanese)

44) K. Tagawa: "Trends of Additive Technology for Refrigeration Oils", The Tribology, 17 (196), pp. 41-43 (2003). (in Japanese)

45) K. Tagawa, K. Sawada: "Effect of Refrigeration Lubes on the Efficiency of Refrigeration Systems (Part2) -Improvement of Frictional Property-", Proceedings of the International Symposium on New Refrigerants and Environmental Technology 2004, Kobe, pp. 93-96 (2004). (in Japanese)

46) K. S. Sanvordenker, W. J. Gram (Authors), M. Yamamoto (Translation): "Laboratory Testing under Controlled Environment using a Falex Machine", Refrigeration, 50 (571), pp. 408-412 (1975). (in Japanese)

47) Y. Yamamo, J. Kim, S. Gondo: "Friction and Wear Characteristics of polyvinylether (PVE) as a Lubricant for Alternative Refrigerant", Transaction of the Japan Society of Mechanical Engineers (Series C), 64 (624), pp. 395-400 (1998). (in Japanese)

48) H. Hirano: "Influence of Water in Oil on Wear and Friction", Proceedings of the International Symposium on HCFC Alternative Refrigerants and Environmental Technology 2002, Kobe, pp.58-62 (2002). (in Japanese)

49) M. Kaneko, H. Ikeda, T. Tokiai, H. Soto: "The Evaluation of PAG Refrigeration Lubricants for Automotive A/C with CO_2", Proc. ASIATRIB 2006, Kanazawa, pp. 275-276 (2006).

50) H. Nakao, T. Matsugi, K. Yano: "Influence of Critical Solubility Temperature on Scuffing Characteristics of Hydrofluorocarbon Refrigerant and Ester Oil Mixture", Journal of Japanese Society of Tribologists, 52 (12), pp. 880-887 (2007). (in Japanese)

51) M. Akei, K. Mizuhara, T. Taki: "Development of Oil Film Thickness Measuring Apparatus in Pressurized Refrigerant Atmosphere and Evaluation of Refrigeration Lubricants", Journal of Mechanical Engineering Laboratory, 48 (4), pp. 197-206 (1994). (in Japanese)

52) M. Kaneko: "The Status of Development of Refrigeration Oil with Refrigerant Change", Refrigeration, 82 (959), pp. 741-745 (2007). (in Japanese)

53) Chemical Society of Japan: "Kagaku Binran Kisohen II (Handbook of Chemistry), 3rd Ed.", Maruzen, Tokyo, pp.322-325 (1984) (in Japanese)

54) Kyoritsu Shuppan: "Encyclopedia of Chemicals Volume 7", Kyoritsu Shuppan, Tokyo, p. 12 (1989). (in Japanese)

55) Y. Kamiya: "Organic Oxidation Reactions", Tokyo Print Center, Tokyo, pp. 360-368, (1973). (in Japanese)

56) K. Iizuka, A. Ishiyama: "Tribological Technology Trend of Ecologically Conscious Design for Domestic-Use Compressor", Journal of Japanese Society of Tribologists, 48 (7), pp. 564-570 (2003). (in Japanese)

57) T. Kawanishi, O. Morimoto: "Cleaning System on Multi Air Conditioner with Existing Piping (Principle and Installation Method)", Refrigeration, 78 (905), pp. 192-196 (2003). (in Japanese)

58) Y. Kushiro, Y. Watanabe: "The Feature of the System that is able to Divert the Refrigerant Piping without Washing", Refrigeration, 78 (905), pp. 197-203 (2003) (in Japanese)

Chapter 10 Motors and Inverters

10.1 Motor Structure and Performance

The history of motor technology goes back many years. As shown in Fig. 10.1, more than 170 years have elapsed since the most primitive form of electric motor was built for the first time in the world[1]. These earliest motors were produced approximately in the same period when the basics of electromagnetism was first understood by mankind. The fundamentals of the electric motor mechanism were established during this period. The first type of electric motor ever developed was wound-field DC motors. Wound-field DC motors have since been used for many decades for its simplicity and easy controllability. The induction motor was created much later, in 1880s. Induction motors are still being used in many applications due to their structural rigidity and good maintainability. At around the same time when the Alnico (aluminum-nickel-cobalt) magnet was first created in 1930, research and development of the permanent magnet-based synchronous motors has started. With non inverter technology developed yet, the synchronous motors of this period were directly connected to an AC power source and needed a damper winding, or squirrel cage winding, to start itself. These early synchronous motors were only applied to limited uses as they were too crude to offer the full advantages of synchronous technology. The implementation of inverter technology after 1950 brought on variable speed control capability, which drastically improved synchronous motor performance. Now synchronous motors are widely used in diverse applications.

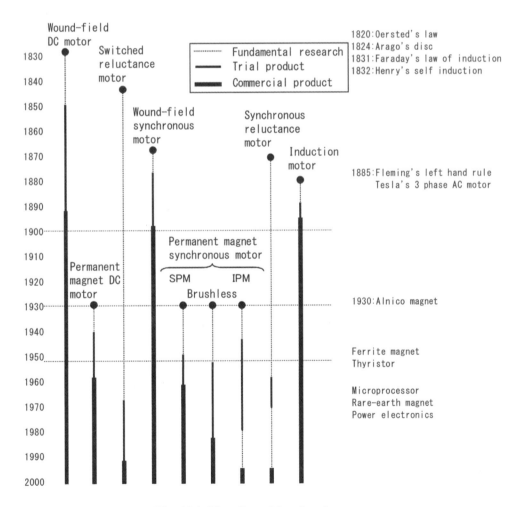

Fig. 10.1 Time line of the electric motor

10.1.1 Motor classifications by drive power waveform

Electric motors can be classified in many ways, for example by the type of power supply used, torque generation mechanism or other structural, applicational or operating characteristics. Figure 10.2 shows motor classifications by the type of drive power source used[2]. Drive power sources for electric motors can be largely classified into direct current sources, alternating current sources and non-sinusoidal (pulsed) power sources.

Synchronous motors (DC motors) powered by a direct current source can be divided, according to how their magnetic field is created, into wound field types and permanent magnet field types. The armature coil, connected to a direct current drive source by the combination of brushes and a commutator, generates torque by interacting with the magnetic field created inside the motor. The brush-and-commutator connection method allows torque to be generated continuously even while the motor is turning. Because of the use of brushes, these types of motors are sometimes referred to as "brushed permanent magnet synchronous motors".

Motors driven by sinusoidal power source are largely divided into induction and synchronous motors. An induction motor rotates with a delay from the rotation of the magnetic field created by the sinusoidal power source. On the other hand, synchronous motors run fully synchronized with the rotating magnetic field. Synchronous motors are classified into multiple types depending on how the magnetic field is created. Most commonly used synchronous motors are permanent magnet field motors and reluctance motors. Permanent magnet field motors are further divided into two types depending on how the permanent magnets are positioned inside the motor. One type is SPMSM (surface permanent magnet synchronous motor), where the permanent magnets are mounted on the surface of the rotor, and the other type is IPMSM (interior permanent magnet synchronous motor), which has permanent magnets embedded inside the rotor.

Motors that are powered by non-sinusoidal, or pulsed, drive source can also be technically considered a synchronous motor. They are structured generally the same way as standard synchronous motors. Commonly used non-sinusoidally driven motors include permanent magnet field stepping motors. Characteristics of the most commonly used electric motors are shown in Table 10.1.[3] Of the motor types listed, the induction-type motors offer the widest range of capacity and can work either with the commercial power supply or with an inverter. Also, the rigid structure makes induction motors suitable for high speed operation and easily adaptable to various applications. Due to these advantages, induction motors are one of the most widely used electric motors today. One disadvantage of the induction motor is that it incurs greater amounts of rotor loss compared to other types of motors, which causes heating in the rotor, shaft and bearings. Therefore, bearing service life can be a difficult issue in the use of induction motors. Of

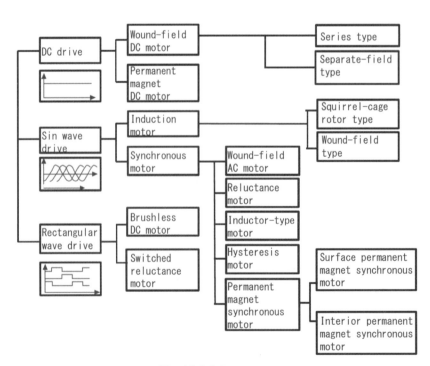

Fig. 10.2 Motor types

Table 10.1. Comparison of various motors[3]

	DC motor	Induction motor	SMPSM	IPMSM	SynRM
Structure		Bar	Magnet	Magnet	Flux barrier
Maximum output	~several MW	~several tens of MW	~several kW	~several kW	~several kW
Typical drive	DC drive chopper drive	AC PWM drive	PWM drive	PWM drive	PWM drive
Size	△	○	◎	◎	○
Weight	△	○	◎	◎	○
Inertia	△	○	◎	◎	◎
Efficiency	△	○	◎	◎	○
Power factor	○	○	◎	◎	○
Rotor speed	△	◎	△	○	◎
Electric time constant	○	△	◎	○	○
Bearing lifetime	○	○	◎	◎	◎
Noise & vibration	○	○	◎	◎	△
Cost	◎	◎	○	○	◎
Features	· Simple control · Low life-span · Low speed	· Low cost · Continuous speed · Easy PWM drive	· Frequent START/STOP · Small-size motors	· Frequent START/STOP · medium-size motors	· Specific applications · Low cost

the types of permanent-magnet-based synchronous motors, surface permanent magnet synchronous motors (SPMSMs) are compact and high-performance and therefore well cater to today's greater energy saving demands. As a result, use of SPMSMs are rapidly increasing despite their relatively high cost. SPMSMs are often used in servo-controlled applications, where the motor is required to be started and stopped at short intervals. IPMSM (interior permanent magnet synchronous motor), another type of permanent-magnet-based synchronous motor, is designed to be able to use reluctance torque in addition to the permanent magnet torque. This characteristic contributes to a high-torque, high efficiency performance. In addition, the positioning of permanent magnets away from the rotor surface makes the motor readily adaptable to high speed applications. Due to these advantages, IPMSMs are being increasingly adopted in compressors and fan motors. The reluctance motor, a type of non-sinusoidally driven motor, has a rotor composed of an iron core without any winding. Due to this simple structure, reluctance motors can be produced at low cost and are also suitable for high speed applications that are subject to hard acceleration and deceleration. However, the performance of reluctance motors is inferior to that of permanent magnet-based synchronous motors as they oper-

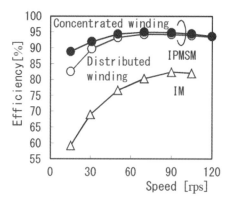

Fig. 10.3 Efficiency comparison of motor types

ate solely on reluctance torque.

Figure 10.3 provides a comparison of efficiency between induction motor (IM) and IPMSM (interior permanent magnet synchronous motor), both commonly used in air conditioning and refrigeration applications. The rating output of both motors is 750 W. Comparison was made at the torque level of 2 Nm. The diagram shows that the concentrated winding type of IPMSM offers the highest peak efficiency while the induction motor exhibits the lowest efficiency. This is due to a larger loss occurring in the induction motor, which is caused by the secondary copper loss in the rotor's conductor bar, as illustrated in Fig. 10.4. This type of copper loss does not occur in a permanent magnet-based synchronous motor because it does not have a rotor conductor bar. Other types of motor losses, including iron loss that occur in the electrical steel sheets, primary copper loss in the stator's magnet wire, mechanical loss at bearings by rotor rotation and windage loss in the motor air gap, occur in the same way both in induction motors and in permanent magnet-based synchronous motors.

One reason for the high efficiency of IPMSM is that, in addition to using the magnet-generated torque in the same way as SPMSM, it can also utilize the reluctance torque that works on the rotor iron core surface. This contributes to reduction in the primary copper loss in stator winding and thus provides superior efficiency. Especially in low speed rotation ranges where the rotating magnet field frequency is also low, the proportion of iron loss (iron core loss) to the total motor loss is smaller and therefore reduction in copper loss has a greater effect. This is the reason why efficiency improvement is apparently more significant in low speed ranges. Furthermore, compared to the distributed winding-type stator, the concentrated winding-type stator provides greater reduction in winding resistance and thus significantly reduces primary copper loss. With this, the concentrated winding IPMSM exhibits particularly good efficiency in low speed ranges.

10.1.2 Motor structure and components
(1) Rotor

Figure 10.5 shows the rotor and its iron core of an induction motor (IM). An induction motor comprises an iron core made of punched out and laminated electrical steel sheets, with aluminum secondary-conductor die-cast into it.

Figure 10.6 shows the rotor of a surface permanent magnet synchronous motor (SPMSM). Figure 10.7 shows the rotor of an interior permanent magnet synchronous motor (IPMSM). A major difference between SPMSM and IPMSM is the positioning of their permanent magnets. Major elements of a motor rotor include an iron core, permanent magnets, end plates and rivets, each of which is explained in the following paragraphs.

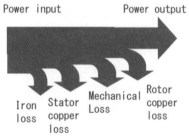

Fig. 10.4 Depiction of motor losses

Fig. 10.5 Rotor(left) and rotor core(right) of IM

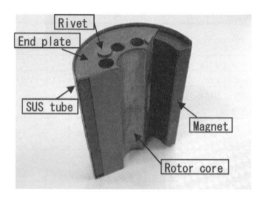

Fig. 10.6 Rotor of SPMSM

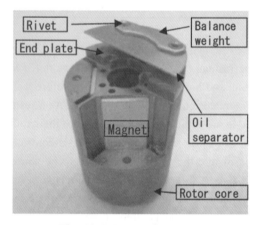

Fig. 10.7 Rotor of IPMSM

(a) Rotor core

The rotor core in an electric motor is typically built by punching out a number of 0.5 mm non-oriented electrical steel sheets to the required form and laminating them to a specific height. The laminated sheets are usually fastened together by dowel joint or other mechanical jointing techniques. For performance optimization, high grade materials such as electrical steel sheets that are as thin as 0.35 mm or are 3 W/kg or lower in the JIS W15/50 scale may be used. Also, the laminated iron core may be annealed to relieve internal stress from the punching process for magnetic performance improvement.

(b) Secondary conductor

The rotor of an induction motor has aluminum secondary conductors which are die-cast into the laminated iron core. The conventional secondary conductor part used to be made of copper, which was fitted into grooves in the core and had both ends brazed to a copper ring for short-circuiting. Currently, most secondary conductors, even those in several tens of horsepower class, are die-cast into the iron core. Use of aluminum as the secondary conductor is advantageous as it is more easily workable than copper and therefore better suited for mass production. Aluminum is also lower in cost than copper. JIS-standard aluminum ingots (99.7% or higher purity) are commonly used for secondary conductor production.

(c) Permanent magnet

The major types of permanent magnets used in a motor rotor are ferrite magnets and rare earth magnets. For use in compressors, both types of magnets are produced by metallic powder sintering. Ferrite magnets, made by press-forming and sintering iron oxide powder, can be produced at low cost but their magnetic force and coercive forces are relatively weak. One caution about the use of ferrite magnet is that its coercive force is directly proportional to temperature, so that it gets stronger when temperature increases and weaker when temperature decreases. Due to the weaker magnetic force, the ferrite magnets are mostly used in SPMSM where the magnets are mounted on the rotor surface.

Rare earth magnets are classified into two types, neodymium magnets and samarium magnets, and neodymium magnets are dominantly used as motor rotor magnets. The strong magnetic energy of neodymium magnet, which can be as high as over 400 kJ/m^3 in some cases, greatly contributes to motor performance improvement. Due to their inherently strong magnetic force, neodymium magnets can be made more compact and thinner to provide the same magnetism. As a result, they are dominantly used in IPMSMs where the magnets are embedded inside the rotor. Contrary to the case of ferrite magnets, the magnetic force of neodymium is inversely proportional to temperature, so that it gets weaker when temperature increases and stronger when temperature decreases.

(d) End plate

End plates are fitted to each end of the rotor iron core mainly to prevent the permanent magnets from coming out of the rotor body. End plates must be a non-magnetic material to avoid short-circuiting the magnetic fluxes. SUS and aluminum are typically used.

(e) Rivet

Rivets are used for fastening the rotor iron core together with the end plates and the balance weights. Rivets are normally made of iron but may sometimes use SUS.

(f) SUS pipe

SPMSMs, which have permanent magnets mounted on the rotor surface, often have a SUS pipe fitted over the rotor as an additional protection to prevent magnet fragments from scattering in case they are shattered inside. Like rivets, the protection pipe over the rotor must be a non-magnetic material to avoid short-circuiting the permanent magnets, which is one of the reasons why SUS is used. The SUS pipe also needs to be made as thin as practical so as to reduce the vortex current that occurs as a result of magnetic flux pulsing induced by the rotor's rotation, but it also needs to be thick enough to withstand the centrifugal force from rotation.

(g) Oil separator

An oil separator separates the oil content from the discharge refrigerant so that oil can be returned to and circulate inside the compressor again. The oil separator may in some cases be attached to the motor rotor. The oil separator is most commonly built with iron.

(h) Balance weight

Balance weights are provided to balance the centrifugal forces and moments that are generated as a result of the rotational movement of the compressor piston and the eccentric crank. In general, a pair of balance weights are fitted to the rotor, one on each end of the rotor in 180° opposed positions. Balance weight may be either brass for the high specific gravity or iron for the low cost.

(2) Stator

Figures 10.8 and 10.9 show a distributed-winding and a concentrated-winding stators, respectively. In a distributed-winding stator the windings are distributed over multiple slots, while in a concentrated-winding stator each coil is wounded directly onto a stator tooth. Due to the widely spread winding configuration, the distributed-winding stator can provide a sinusoidal or nearly sinusoidal rotational magnetic field relatively easily and thus generates less noise, while the longer winding length of the distributed-winding stator incurs more copper loss. On the other hand, the concentrated-winding stator has a shorter winding length and thus generates less copper loss, therefore achieving high efficiency at lower cost. However, the greater harmonic components of the rotating magnetic field tend to increase noise and vibration.

(a) Stator core

Similar to the rotor core, the stator core is built by punching out a number of non-oriented electrical steel

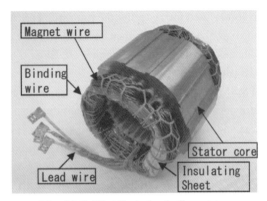

Fig. 10.8 Distributed-winding stator

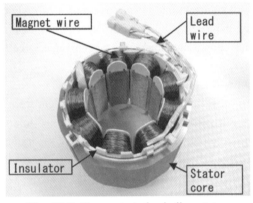

Fig. 10.9 Concentrated-winding stator

sheets to the required form and laminating them to a specific height. The material composition of a stator core is similar to that of the rotor core. The laminated steel sheets are usually fastened together by dowel joint or other mechanical jointing techniques, but in some cases they may be joined by circumferential welding.

(b) Magnet wire

Magnet wire, which is wound to construct coils, is usually JIS C 3102 - compliant electrical soft copper wire. While coils of a distributed-winding stator are constructed in the same way as those in an induction motor, a concentrated-winding stator has the magnet wire directly wound onto the stator teeth at high speed using a wire feed nozzle. Therefore, the magnet wire covering must be highly abrasion resistant to withstand nozzle contact during winding operation.

(c) Slot insulation

Slot insulation comprises an insulation paper, which was traditionally produced from plant-derived fiber but is now dominantly made of polyester film due to moisture absorption consideration.

(d) Insulator

The concentrated-winding stator, which has the magnet wire directly wound around the stator teeth, requires additional insulation to electrically separate the stator core and the winding.

The insulation is also required to help physically support and maintain the winding form, for which the polyester film slot insulator often used in induction motors is not strong enough. A commonly used approach is to mold a PPS plastic layer over the stator teeth, over which the magnet wire is wound. Other than PPS, LCP or PBT plastic may also be used as stator insulation. In addition to this plastic layer insulation, the stator may also have a polyester film insulation inside the slots, similar to the one used for conventional slot insulation.

(e) Lead

The stator lead comprises a plastic terminal and the lead wiring. Small "cluster"-type terminal is usually made of molded phenolic or PBT plastic and houses a number of female contacts ("receptacle" terminal). The lead wiring comprises a conductive core and an insulation coating. The core is a number of fine copper wires twisted together and formed into shape with polyester film and fiber or with PTFE film and polyester braiding. More recently, instead of using the lead wiring, the lead terminal is constructed by putting an insulation coating directly onto the magnet wire and swaging a receptacle terminal over it.

Motors with larger outputs, that entails high amperage and requires tabs to be fitted to both receptacle side and hermetic lead-pin side, do not use the cluster-type terminal. While the lead wiring is to make electrical connection between the motor wire and the casing-side lead-pin, it needs to be flexible enough and also mechanically strong enough to withstand vibration that occurs during compressor transportation and when the compressor is starting or stopping.

10.1.3 Motor component fabrication

Figure 10.10 shows the motor component fabrication process, from iron core sheet punching to stator and rotor completion.

(1) Punching out and laminating iron core sheets

The electrical steel sheets are first slit to the punching die width and coiled into a hoop, before punched out into the shape of stator and rotor by a high speed continuous punching press.

For high volume production models, progressive punching dies are used to produce iron core sheets. For smaller models, two or three-row dies are used so that more than one piece of iron core can be punched out at the same time.

Not only punching out the steel sheet into shape, many punching dies being used for high volume iron core production also form raised tabs into the sheet, as shown in

Chapter 10 Motors and Inverters

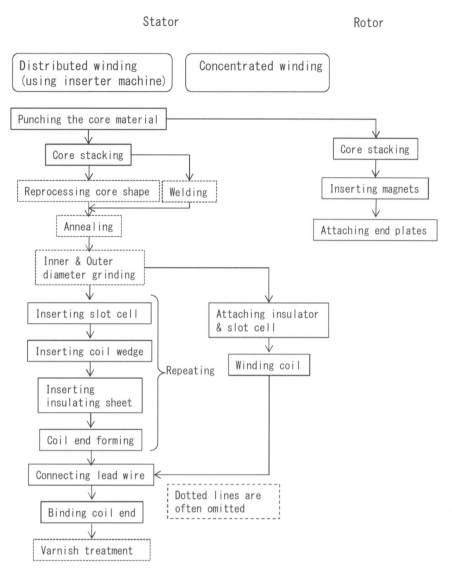

Fig. 10.10 Stator assembly flow chart

Fig. 10.11, so that the required numbers of core sheets are already nested and laminated when they are taken out from the die. For small volume models, non-progressive, single-impression dies or notching presses may be used to punch stator and rotor core sheets separately. Iron core sheets that come out of the die in non-laminated state must be grouped into the required quantity and be exactly stacked on a fixture, with the inside diameter, slots and other features accurately matched, before they are laminated together by clamping or other mechanical jointing method.

(2) Inside diameter and outside diameter machining

Stator dimension variation, which may occur when the iron core sheet punching accuracy is not good enough or when welding stress causes deformation in the laminated core, causes inferior iron core characteristics or makes the iron core unable to be shrunk-fit. Such deformed or incor-

Fig. 10.11 Laminated core connected by caulking

rectly dimensioned iron core can be corrected by cutting or grinding of its inside and outside.

(3) Heat treatment

Mechanical deformation in the iron core sheet that occurs during the punching process may detrimentally affect the magnetic characteristics of the punched core sheet. Annealing may be done to restore such degraded core sheet characteristics.

Especially when a low grade electrical steel sheets is used, annealing helps to enlarge the crystal grain size of the metallic microstructure to reduce vortex current. In addition, annealing is effective to reduce residual metalworking fluid that will otherwise be brought into the refrigerant circuit.

When the amount of carbon contained in the electrical steel sheets increases, iron loss increases as a result. Therefore, annealing must be done in nitrogen or other decarburizing gas. The heat treatment furnace may be either continuous or stationary.

Besides annealing, other heat treatment processes that could be implemented include oil burning, which is done at a lower temperature than annealing and works on metalworking fluid reduction only, and Fe_3O_4 film formation process to protect sheared steel surface from rusting.

Fig. 10.12 Coil wedge inserter

Fig. 10.13 Nozzle drive winding machine

(4) Coil insertion

In the case of a distributed-winding stator, the coils may be inserted into the slots by hand or by an automatic inserter where the coils wound by a high speed winder is mechanically inserted together with wedges, as shown in Fig. 10.12. The wedges are fitted with a specified width of hoop material and will be cut, formed and inserted automatically.

In the case of a concentrated-winding stator shown in Fig. 10.13, the wire is fed through the nozzle, passed through the slot opening and wound directly onto the stator core teeth.

(5) Coil end lacing

Coil ends may be laced manually or automatically by a lacing machine. Coil end lacing is usually not required on a concentrated-winding stator where the coils do not need to be secured after winding.

(6) Varnishing

In the case of large capacity motors, which vibrate more when operating or starting up, the coils may need to be additionally secured by applying a refrigerant-tolerant varnish in order to prevent insulation breakdown caused by vibration.

(7) Die casting

An AC induction motor requires aluminum die casting to form the secondary conductor. The required quantity of punched rotor sheets are stacked together and set in the die. To minimize voids around the slots and end ring porosity, such die casting parameters as configurations of gate and degassing, temperatures of die and melt, injection pressure and injection speed must be carefully designed and controlled.

(8) Riveting

A permanent magnet-based synchronous motor has permanent magnets fitted to the rotor. The example shown in Fig. 10.14 has permanent magnets inserted into the rotor core before the top and the bottom of the core is covered by

Fig. 10.14 Rotor riveting process

non-magnetic end plates and secured by riveting.

10.1.4 Alternative stator fabrication method

The core of a motor stator are conventionally constructed in one piece. A newer, alternative method shown in Fig. 10.15 has an articulated stator core, where the core is fabricated in multiple segments and then connected through articulated joints. In the case of a conventional one-piece core, where the coil wire needs to be fed from inside of the core through a wire feed nozzle, the wire cannot be wound very densely inside the slot due to the nozzle's dead space. On the other hand, with an articulated core, the core segments can be spread the other way around the joints as shown in Fig. 10.15 so that the wire can be wound much more densely. It is also possible to stack the coil wires into a heaped bale structure inside the slot, enabling the coils to occupy a much larger proportion of space inside the slot.

10.2 Motor Design

The following sections explain about induction and permanent magnet synchronous motors, which are most commonly used for air conditioning and refrigeration, especially about their torque generation mechanism, operating principle and how the motor design specifications should be determined according to the anticipated compressor load.

10.2.1 Operating principle of induction motors

Torque generation by an electrical motor is a conversion of electric work into rotational kinetic work which is achieved by the electromagnetic induction caused by interaction between magnetic flux and electric current. An induction motor operates by utilizing the electromagnetic force that occurs between the rotating magnetic field and the induced current. As explained in Section 10.1.1, the induction motor is the most commonly used type of alternating current motor, being applied to general-purpose mechanical operations as well as air conditioning and refrigeration uses, as it can work either with the commercial power supply or with an inverter and is also easier to handle with its simple and rigid structure.

Induction motors can be classified into two types depending on its secondary rotor conductor design, that are the squirrel-cage type and the winding type. Of the two, the squirrel-cage type is much more commonly used. The following paragraphs explain specifically about the squirrel-cage type induction motor. When the squirrel-cage type rotor conductor shown in Fig. 10.16 is placed inside a rotating magnetic field like the one shown in Fig. 10.17, voltage will be induced in such a way that electric current runs through each conductor, in the ⊙ and ⊗ directions shown, as governed by the Fleming's right hand rule[4]. As the ends of the rotor are short-circuited by end rings, the induced electromotive force generates electrical current through the conductors. Assuming that both ⊙ and ⊗ in the Fig. 10.17 represent electrical current, a ⊙-direction current passes through the bars facing the N pole, which,

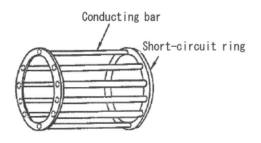

Fig. 10.16 Squirrel-cage rotor

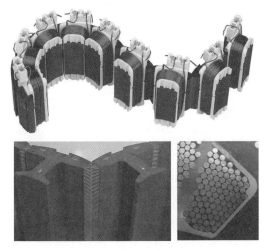

Fig. 10.15 Joint-lapped core and perfectly aligned coil

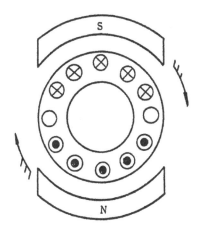

Fig. 10.17 Squirrel-cage rotor under rotating magnetic field

under Fleming's left hand rule, generates a rotational force in the same direction as the magnetic field rotation. Bars facing the S pole also generates a rotational force working in the same direction, causing the rotor to turn in the same rotational direction as the magnetic field. However, the rotor rotation speed is slower than that of the magnetic field. This difference in speed is called "slip". If the rotor rotation speed and the magnetic field rotation speed are exactly the same, the rotor conductor bars would not pass across the magnetic field and no electromagnetic force would be induced, hence no rotational force.

Induction motors may be either three-phase or single-phase. In a three-phase induction motor, the rotating magnetic field is generated by applying a three-phase alternating current to the winding around the stator core. In a single-phase induction motor, a rotating magnetic field cannot be produced only by applying a single-phase alternating current to the pair of stator wirings. Thus the rotor cannot by itself start to rotate from a stationary state. Therefore, a single-phase induction motor requires some kind of phase offset, for example providing a start-up, or "auxiliary", winding in addition to the main winding, to obtain a rotating magnetic field to start the rotor turning. As long as the rotor does not stop turning, the rotational movement will be maintained by the action of the magnetic field generated by the induced current through the rotor. The phase offset can be achieved by providing a capacitor on the auxiliary winding located at electrical angle of 90° from the main winding to run a leading current, which will generate and accelerate a rotating magnetic field. Once a steady-state operation is established, the auxiliary winding will be switched off. Or, the auxiliary winding may remain connected in some cases where the capacitor size is small.

10.2.2 Operating principle of permanent-magnet synchronous motors

A permanent-magnet synchronous motor has a permanent magnet-fitted rotor inside the stator, as shown in Fig. 10.18[2]. Similarly to an induction motor, the rotor turns following the rotational movement of the magnetic field created by the stator windings. The rotor rotation speed of a permanent-magnet synchronous motor is exactly the same as that of the magnetic field and is synchronized with it. The ⊙-direction current that passes through the stator windings facing the N pole of the rotor's permanent magnet generates, as governed by the Fleming's left hand rule, a rotational force in the same direction as the magnetic field rotation. Windings facing the S pole also generate a rotational force working in the same direction, causing the rotor to turn in the same rotational direction as the magnetic field.

Permanent magnet synchronous motors can be classified, as shown in Fig. 10.19, into two types according to the way that permanent magnets are fitted to the rotor; the surface permanent magnet synchronous motor, or "SPMSM", where permanent magnets are mounted on the rotor surface, and the interior permanent magnet synchronous motor, or "IPMSM", where permanent magnets are embedded inside the rotor[1].

An SPMSM generates torque solely by the principle illustrated in Fig. 10.18, while an IPMSM generates another type of torque in addition to the torque generated by the Fig. 10.18 principle. That other torque generation mecha-

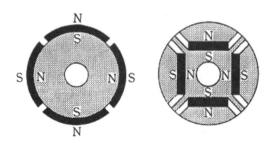

(a) SPMSM   (b) IPMSM

Fig. 10.19 Rotor structure of SPMSM(left) and IPMSM(right)

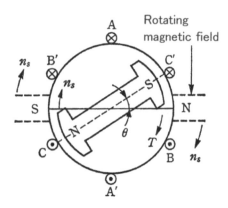

Fig. 10.18 Synchronous motor working principle[2]

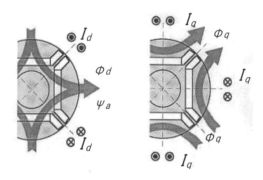

Fig. 10.20 Principal IPM magnetic flux paths. d axis(left). q axis(right) [5]

nism is illustrated in Fig. 10.20[5].

When a three-phase alternating current is represented as a combination of direct-current components projected onto a d- and q-axis orthogonal coordinate system, the torque generated by the combination of the permanent magnet's magnetic flux Ψ_a and the q-axis electric current I_q orthogonal to that is the "magnet torque", or the torque created by the magnetic field of the permanent magnet. On the other hand, the torque generated as a result of difference between the following two electromagnetic forces, one is the product of magnetic flux Φ_d ($=L_d I_d$) generated by the combination of the d-axis current I_d and the d-axis inductance L_d and the q-axis current I_q and the other is the product of the magnetic flux Φ_q ($=L_q I_q$) generated by the combination of the q-axis current I_q and the q-axis inductance L_q and the d-axis current I_d, is "reluctance torque". In short, a reluctance torque is the torque generated as a result that the iron core in contact with the permanent magnet embedded inside the rotor follows the rotating magnetic field of the stator. An IPMSM is designed to be able to use both the magnetic torque and the reluctance torque, and therefore can obtain a larger total torque than a SPMSM. Both the magnetic torque/reluctance torque ratio and the general torque peak value change depending on the current phase running through the stator, as shown in Fig. 10.21. Therefore, an appropriate current phase control must be implemented to operate the motor. Note that magnetic flux Φ_q runs through the iron core section only and therefore is prone to cause magnetic saturation of the iron core. When magnetic saturation occurs, only the reluctance torque decreases, as represented by the dotted line in Fig. 10.21.

10.2.3 Motor design specifications (specification requirements, output-torque relationship)

The following paragraphs explain the motor designing procedure taking the example of a compressor motor. The design work starts based on design specifications, which would be presented as shown in Table 10.2. The remaining part of this section focuses on the two types of motors that are most commonly used as a compressor motor, those are the three-phase induction motor and the permanent magnet synchronous motor[4].

At first, the design specification document should fully clarify the mechanical design specifications of the compressor for which the motor will be employed. The maximum allowable temperature rise is an especially important parameter as it affects what type of motor insulation should be selected and, in the case of a permanent magnet motor, how much magnetic force reduction is expected.

To plan the motor design specifications, the first thing to do is to select the type of motor. The characteristics of each type of motor is as explained in Section 10.1.1. The current trend is to select a permanent magnet motor for applications where energy saving is of greater importance and to select an induction motor where low cost and reliability are more important. The number of phases of the induction motors are dominantly three phases, but single-phase motors may be selected for some small size applications. For details about single-phase induction motors, refer to Reference 3).

The number of poles, output power, voltage, frequency, the maximum torque and the maximum output are the parameters that should be determined depending on the compressor capacity needed. In the case of an inverter-driven motor, where the frequency will be variable, the relationship between output power, the frequency range and the number of poles must be determined based on the capacity range required. The following paragraphs give further details.

In electric motor technology, "starting current" refers to the amount of inrush current that runs through the motor immediately after it is powered, and "starting torque" refers to the amount of torque that is generated when the motor starts to turn from a stationary state. With a non-inverter-driven induction motor, its starting current may be several times higher than the rated current and may affect other functions. On the other hand, only a fraction of the rated torque can be obtained as the starting torque, and the motor may be unable to start up if its starting torque is lower than the compressor starting load. Therefore, the starting current and torque must be determined considering these factors. Inverter-driven motors are capable of much smoother start-up as long as the starting voltage and the frequency are appropriately selected according to the compressor starting load condition and the current capacity.

"Standard operating output characteristics" refers to motor performance parameters. Each standard operating output

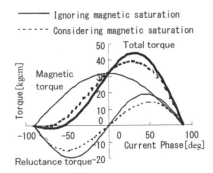

Fig. 10.21 Torque-Phase characteristics

Compressors for Air Conditioning and Refrigeration

Table 10.2 Motor design properties

		Notes
Intended use		Compressor
Refrigerating capacity	kW	
Refrigerant	R-	
Operating temperature	℃	
Design		Compressor type
		Stator outer diameter
		Rotor inner diameter (shaft diameter)
		Number of cylinder
		Balance weight size
Motor type		IM or PMSM
Drive		Single, dual or PWM
Phase		
Pole number	P	
Rated output	W	
Rated voltage	V	
Rated frequency	Hz	
Maximum output	W	
Maximum torque	Nm	
Inrush current	A	JIS or other standard
Input	W	
Current	A	
Efficiency	%	
Power factor		
Motor resistance	Ω	
Revolution	rpm	
Applicable temperature	℃	
Insulation grade		
Maximum instantaneous current	A	Demagnetizing current (PMSM)
Sound level	dBA	
Size	mm	Rotor & stator diameter and height
Net mass	g	
Due date		

characteristic is both a design value and a requirement. Of the characteristics listed, efficiency is particularly important as it also affects the compressor performance. Power factor refers to what percentage of the power given to the motor is effectively used. If the power factor is too small, an exces-

sively large power capacity and, in the case of an inverter-driven motor, a needlessly large inverter capacity are required, which would lead to significant cost increase.

The higher the temperature tolerance class is, the higher the reliability will be, but the material cost will be higher.

The commonly selected temperature tolerance classes are Classes E (120°C) to F (155°C). Ultimately, the temperature tolerance class of a motor must be selected based on the level of anticipated temperature increase estimated from the related equipment design specifications.

The following paragraphs explain about the basic relationship between motor dimensions and its output power and torque characteristics. Motor output power can be expressed by Eq. (10.1)[6].

$$P = KD^2Ln \tag{10.1}$$

$$n = \frac{f}{p} \tag{10.2}$$

Here, P is output power, D is rotor diameter, L is rotor length, n is rotational speed, f is power supply frequency, and p is the number of pole pairs (number of poles/2). K is called "power coefficient", of which approximate value can be calculated from the motor's specific magnetic and electric loadings. For further details, refer to Reference 5). As previously explained, the induction motor has a "slip" where the rotor rotation frequency is 2 to 3 % slower than the drive power frequency. On the other hand, permanent magnet motors are fully synchronous. Hence, the rotor rotational frequency is equal to the drive power frequency. The stator's outside diameter varies depending on the number of poles. With two to six-pole motors, which are the most commonly used configurations for compressors, the stator diameter needs to be approximately 1.5 to 2 times larger than D in order to accommodate the magnetic wire. As indicated by Eq. (10.1), the motor output power can be changed by changing the rotor diameter, the rotor length or the rotational speed. What is important is to determine the motor specifications based on the compressor specifications and limitations.

Motor torque T can be expressed by the following equation:

$$T = \frac{P}{2\pi n} \tag{10.3}$$

As indicated by Eq. (10.4), the amount of power available for the motor is fixed according to the inverter or other driving power supply. Thus, Eq. (10.5) is obtained by reorganizing the torque equation using motor voltage and current.

$$P = \sqrt{3}VI\eta\rho \tag{10.4}$$

$$T = \sqrt{3}I\frac{V\eta\rho}{2\pi n} \tag{10.5}$$

Here, V is motor voltage, I is motor current, η is motor efficiency and ρ is motor power factor.

Assuming that the motor efficiency and power factor are approximately constant, $V\eta\rho/(2\pi n)$ represents the motor voltage depending on the rotor rotational frequency, which can be determined based on the maximum voltage and maximum rotational speed. As I represents motor current, the equation shows that the maximum torque will be determined by the maximum current that can be supported by the driving power supply.

The above indicates that the dimensions, output power and torque of a motor can be determined more or less singularly, while the actual motor design reflects various engineering efforts for efficiency and power factor improvement, such as maximum rotor speed increase through flux-weakening control and motor structure and material refinements, toward higher capacity, better energy saving and motor size reduction.

10.3 Motor Materials and Evaluation Methods

Motors, like the one shown in Fig. 10.22, that drive a fully enclosed (hermetic) compressor used for air conditioning and refrigeration purposes are commonly referred to as "hermetic motors". Hermetic motors operate immersed in refrigerant or refrigeration oil.

Figure 10.23 shows the stator of a hermetic motor and its main components.

During compressor operation, its motor windings are exposed to high-temperature and high-pressure gas refrigerant and also to refrigeration oil with massive amounts of refrig-

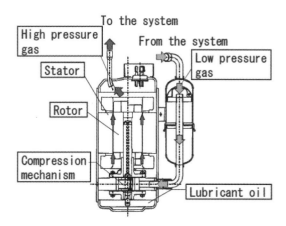

Fig. 10.22 Cross-sectional model of air conditioner compressor

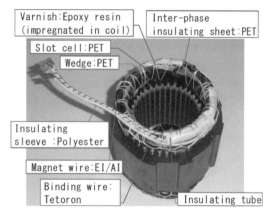

(a) Distributed winding

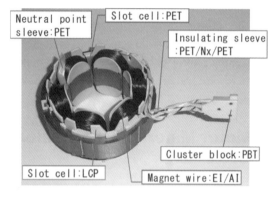

(b) Concentrated winding

Fig. 10.23 Hermetic motor stator and its main components

erant dissolved into it. When the compressor is not running, liquid refrigerant collects inside the compressor. Insulation materials that will be exposed to the above-described conditions for long periods of time must be such that offers a semi-permanent service life under repeated temperature cycles from very hot to very cold and in the gas-liquid mixture atmosphere where the gas and liquid phase ratio keeps changing. To be capable of operating under these conditions, insulation materials must satisfy the following requirements:
- Not soften, swell, melt or foam under high-temperature and high-pressure or low-temperature and low-pressure conditions in a gas-phase or liquid-phase atmosphere.
- The dielectric strength and the insulation resistance do not significantly deteriorate in the atmosphere of refrigerant and refrigeration oil mixture.
- No ingredient material dissolves in the refrigerant or refrigeration oil.

Excessive elution of the insulation material (by dissolution into the refrigerant) will cause the eluted substance to deposit in a capillary tube or valves in the refrigeration circuit, incapacitating the refrigeration cycle. However, use in a refrigerant-filled atmosphere may provide the following advantages where certain deterioration factors can be avoided or reduced, which may in some cases make motor operation easier than general mechanical applications or make the same capacity available with a smaller sized motor.
- Less interaction with oxygen or water, therefore reduced probability of deterioration through oxidation or hydrolysis.
- Cooling of the motor windings by the refrigerant helps mitigate coil heating, delaying thermal deterioration.

10.3.1 Magnet wires

Enameled magnet wire that constitutes stator windings is always exposed to the atmosphere of refrigerant and refrigeration oil mixture. Therefore, the magnet wire winding is subject to various dynamic and static actions such as massive heat and the compression and expansion of strongly dissolvent or elution-causing refrigerant. Therefore, enameled magnet wire must satisfy the following requirements:
- Not soften, swell or elute after being exposed to the refrigerant for a long period of time.
- Not exhibit blisters (bubbles) under pressure changes caused by compressor operation.
- Not prone to reduction in the dielectric strength or insulation resistance.
- Not deteriorate, or accelerate the deterioration of, insulation materials and the lubricant oil.
- Easily workable during the manufacturing process and also highly resistant to mechanical stress that occurs during the motor winding process.

Table 10.3 shows the performance requirements for an enameled magnet wire and the corresponding product characteristics.

Conventional coating materials for enameled wires, such as polyvinyl formal resin, were not very refrigerant-tolerant. In 1970s and thereafter, wires coated with a combination of two different resins, either of which will not by itself provide sufficient performance, began to be used. One such coating formula is a polyamide-imide topcoat, which is both strong and flexible but very expensive, added to the strong but not very flexible polyester imide undercoat. During the mechanized high-speed winding operation, the magnet wire may get stretched by being pulled by the winding guide or fixture or in some cases part of the coating may be scraped off.

Table 10.3 Magnet wire properties and requirements

Requirements	Properties	Tests
Workability (scratch resistance)	Abrasion strength Friction	Unidirectional scrape Coefficient of static friction Coefficient of kinetic friction
Winding workability	Flexibility	Flexibility test
Refrigerant resistance	Refrigerant resistance at overload	Blistering test Dielectric breakdown test after blistering
	Refrigerant exraction	Extraction test
Compatibility of wire and varnish	Surface wettability	Fixing strength test
Varnish adhesion	Film adherence	Adhesion test

After coil insertion, forming operation is required to terminate the coil ends to the specified dimension, where the wire will be subject to severe stress including guide and forming tool impacts as well as friction, which puts the wire coating at greater risk of damage. As long as the coating film's tensile strength and elasticity are sufficiently high, the wire will be strong enough to withstand these conditions. To summarize, an enameled magnet wire must satisfy the following mechanical characteristics:
- Superior abrasion resistance of the coating film
- Strong adhesion between the coating film and the conductor.
- Good surface slip

In 1980s, wires with alternative coatings, where the topcoat contains particulate lubricant to acquire self-lubricity, were developed. With this, the magnet wires became more slippery, which helped reduce insertion pressure, vibration and abrasion during coil insertion. Then in 1990s, scratch-proof wires, which offer improved film-wire adhesion and abrasion resistance, started to be used.

Motor coil winding operations are being increasingly mechanized and made faster. Also, more motor models adopt concentrated winding instead of distributed winding so that coils occupy a larger proportion of space inside the stator. With these changes, tougher, more advanced performance than ever is being sought of magnet wires. Table 10.4 provides a comparison between different types of magnet wires. Figure 10.24 shows temperature tolerance service life curves of various magnet wires.[7]

10.3.2 Varnish

Hermetic motors that are used in hermetic compressors suffer electromagnetic and mechanical vibrations in the coils during starting, which gets greater as the motor capac-

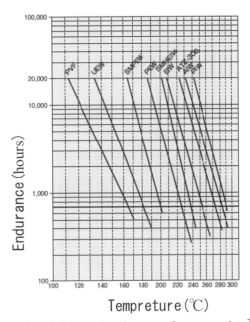

Fig. 10.24 Thermal endurance of magnet wires[7]

ity increases. These vibrations have damaging effect on the coils, potentially shortening the motor service life.

As motor coils are always enveloped in the refrigerant-refrigeration oil mixed atmosphere, the coil varnish must, in addition to providing general coil varnish characteristics such as adhesion to the magnet wire, adaptability and mechanical strength, be oil- and refrigerant-tolerant. Cases where varnish ingredients dissolved into the refrigerant have precipitated to clog refrigerator capillary tubes and others where, due to the use of strongly moisture-absorbent lubrication oils, hydrolysis was caused by moisture contaminated in varnish, have been reported.

Refrigerant-tolerant impregnating varnishes started to be used in 1970s. These varnishes are typically produced by

Compressors for Air Conditioning and Refrigeration

Table 10.4 Properties of magnet wires

Type	Thermal class (°C)	Features	Operational precautions	Applications
Polyvinyl formal enameled wire (PVF)	105	· Mechanically strong coating and good flexibility · Good thermal shock resistance · Strong in hydrolytic degradation	· Crazing prone (Preheating prevents crazing from developing.)	· Transformer
Polyurethane enameled wire (UEW)	130	· Soldering is possible without stripping off coating · Excellent electrical characteristics with high frequency	· Coating is mechanically weak. · Vulnerable to aromatic solvents. · Crazing prone. (Preheating prevents crazing from developing.)	· Coils for electronic equipment · Coils for communication equipment · Coils for electrical meters · Micromotors · Magnet coils
Polyester enameled wire (PEW)	130-155	· Good electrical characteristics · Good heat resistance · Good solvent resistance	· Mediocre resistance to thermal shock. · Poor resistance to hydrolytic degradation; care must be taken when used in sealed equipment.	· General purpose motors · Magnet coils
Polyester-imide enameled wire (EIW)	155-200	· Good heat resistance · Good thermal shock resistance · Mechanically strong coating · Excellent resistance to hydrolytic degradation · Excellent resistance to refrigerants	· Film detachment is difficult.	· Class-F motors · Freon motors · Microwave oven transformers · Magnet coils for heat-resistant components · Motors for electrical equipment
Polyamid-imide enameled wire (AIW)	220	· Mechanically strong coating · Good heat resistance · Good overload characteristics just below IMW	· Coating flexibility is slightly inferior to PEW.	· Transformers for heat-resistance equipment · Motors for electric tools · Hermetic motors · Motors for electrical equipment
Polyimide enameled wire (PIW)	240 (280)	· Top-level heat resistance among enamelled wire. · Excellent overload characteristic · Good resistance to chemical solvents	· Coating is somewhat weak mechanically.	· Motors for heat-resistant equipment · Equipment for airplanes

combining and modifying different types of resins, such as phenolic, epoxy, urethane and acrylic resins, so as to be better suited to the respective use conditions.

As part of hermetic motor manufacturing productivity improvement efforts, varnishing operations are being increasingly mechanized and made faster. In addition, varnishes are now subject to volatile organic compounds (VOC) restrictions. To better cope with these new situations, non-solvent-based and water-based varnishes have been developed.

Varnish evaluation methods include the followings:
- Adhesion tests (helical coil test, Stracker test)
- Refrigerant tolerance (sealed glass tube test etc.)
- Polyester film adhesion
- Hydrolysis stability
- Solvent elution

10.3.3 Insulating papers (films)

Thin-film insulation material, or "insulating paper", is used for insulation in and around a compressor motor, including slot insulation, inter-phase insulation, wedge, pole and phase connection insulation as well as external coil end insulation. Insulating paper may be made from cellulose-based (craft paper, fish paper, rag paper and other specially processed paper), cellulose-and-plastic based or plastic based materials, processed into a paper or film form and specially treated.

The most commonly used type of insulating paper is polyethylene terephthalate (PET, a type of polyester) film. PET film is least water absorbent and can be dehydrated and dried relatively easily through a dryer process, contributing to significantly shorter production time than other materials.

(a) Polyester film

Polyester-based insulation material is provided in various forms such as films, fibrous tapes, cords, sleeves and braidings. The chemical composition of polyester (poyethylene terephthalate, PET) film is expressed by the following equation.

$$HO-CH_2-CH_2-\left[O-CO-\langle\bigcirc\rangle-CO-O-CH_2-CH_2\right]_n-OH$$

PET is a polymer created by the condensation polymerization of terephthalic acid and ethylene glycol.

Polyester has a very low moisture absorptivity (approximately 0.4%). As moisture should be reduced as much as practical in a refrigeration system, low moisture absorption is one of the major reasons that makes polyester suitable as electrical insulation material in refrigeration systems.

Potential factors that deteriorate polyester films include moisture, oxygen, heat and light. When considered specifically as a motor insulation material, the primary deterioration effects on polyester film are those caused by moisture and heat. Of those deterioration effects, hydrolysis caused by moisture is the most damaging. Hydrolysis breaks polyester film molecules apart to embrittle the film.

To minimize hydrolysis of the polyester film, equilibrium moisture content and temperature must be considered when including the film in the product design,

(b) Other types of films

One type of insulation film that is being increasingly used in recent years is temperature tolerance class B (130°C) polyester, which offers not only the inherent advantages of PET, such as mechanical strength, electrical insulation and chemical resistance, but also improved temperature tolerance life and hydrolysis resistance. Its content of low-molecular impurities such as oligomer is also significantly reduced compared to conventional products.

In cases where better temperature tolerance life is desired, Class F (155°C) polyethylene naphthalate (PEN) or wholly aromatic polyamide ("aramid paper", temperature tolerance 220°C) may be used. Table 10.5 shows a comparison between different types of insulation materials. Figure 10.25 shows temperature tolerance life curves of commonly used insulation films.

(c) Insulator materials

For energy saving, almost all the hermetic motors be-

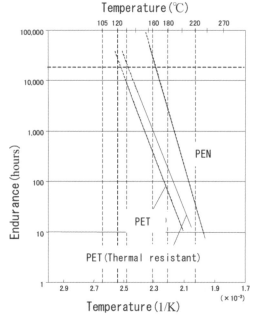

Fig. 10.25 Thermal endurance of insulating film

Table 10.5 Typical properties of main insulating film

		Unit	Standard	PET 50 μm Biaxially stretched	Heat resistant PET 50 μm Biaxially stretched	PEN 50 μm Biaxially stretched	PPS 25 μm Biaxially stretched	Aramid Paper 50 μm
Density		g/cm³	JIS K6760	1.40	1.40	1.36	1.35	0.72
Melting point		°C	DSC	260	260	269	285	Not melt
Second order phase transition temperature		°C	DSC	69	69	113	92	-
Continuous heat-resistance temperature	Mechanical	°C	UL746B	105	130	160	160	220
	Electrical	°C	UL746B	105	130	180	200	220
Tensile strength	MD	MPa	JIS C2318	260	292	274	294	75
	TD	MPa		268	311	265	245	42
Elongation at break	MD	%	JIS C2318	164	167	90	60	8
	TD	%		161	146	90	80	7
Dielectric strength		kV/mm		224	226	250	250	18.1
Volume resistivity		$10^{17}\Omega$cm		2	10	10	5	0.1
Permittivity		(@1kHz)		3.3	3.3	2.9	3	2.6
Dissipation factor		% (@1kHz)		0.44	0.46	0.50	0.06	0.014
Amount of oligomer		wt%		1.04	0.37	0.10	0.24	0.07

ing used in Japan are now permanent magnet synchronous motors. Their stators are dominantly concentrated-winding type, which helps reduce the coil end volume.

Most of the concentrated-winding stators are fitted with an insulator ring like the one shown in Fig. 10.26, which serves both as a winding frame and as electrical insulation.

Modern insulators are produced by plastic molding, which, compared to traditional film-type insulation, involves the use of a significantly larger volume of organic material. Therefore, more careful evaluation and control than before should be provided about possible material elution from these insulators. Commonly used insulator materials are now advanced engineering plastics such as PPS (polyphenylene sulfide), LCP (liquid crystal polymer), and PBT (polybutylene terephthalate).

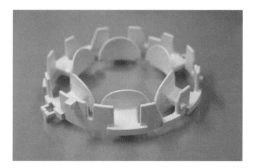

Fig. 10.26 Insulator

10.3.4 Electrical steel sheets

Motor cores are produced by punching out and laminating a number of electrical steel sheets. The laminated sheets are joined together by riveting, crimping, welding or gluing. There are two types of electrical steel sheets, one is grain-oriented (JIS C 2553) type and the other is non-oriented (JIS C 2552[8]) type, the latter of which is shown in Table 10.6 and is the type used for rotating equipment. Besides standard JIS products, many steel manufacturers also offer specially developed electrical steel sheets with original specifications.

The crystal structure of iron has, as shown in Fig. 10.27, specific directionality where magnetic flux passes through the metal more readily in certain directions than in other directions.[9] A grain-oriented steel sheet is a material where the crystal grains are oriented in a specific rolling direction so that magnetic flux will pass more easily in that direction, and a non-oriented sheet is one that is manufactured so that its crystal grains are randomly oriented.

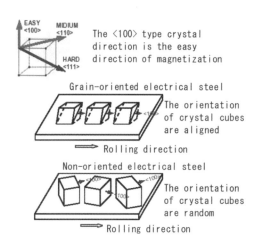

Fig. 10.27 Grain-oriented and non-oriented electrical steel[9]

Table 10.6 Non-oriented electrical steel (based on JIS C2552[8])

Type	Thickness mm	Density kg/dm³	Core loss W/kg $W_{15/50}$	Magnetic Flux Density T B_{50}	Number of bends
35A210	0.35	7.60	≦2.10	≧1.60	2
35A230		7.60	≦2.30	≧1.60	2
35A250		7.60	≦2.50	≧1.60	2
35A270		7.65	≦2.70	≧1.60	2
35A300		7.65	≦3.00	≧1.60	3
35A360		7.65	≦3.60	≧1.61	3
35A440		7.70	≦4.40	≧1.61	3
50A230	0.5	7.60	≦2.30	≧1.60	2
50A250		7.60	≦2.50	≧1.60	2
50A270		7.60	≦2.70	≧1.60	2
50A290		7.60	≦2.90	≧1.60	2
50A310		7.65	≦3.10	≧1.60	3
50A350		7.65	≦3.50	≧1.60	5
50A400		7.65	≦4.00	≧1.61	5
50A470		7.70	≦4.70	≧1.62	10
50A600		7.75	≦6.00	≧1.65	10
50A700		7.80	≦7.00	≧1.68	10
50A800		7.80	≦8.00	≧1.68	10
50A1000		7.85	≦10.00	≧1.69	10
50A1300		7.85	≦13.00	≧1.69	10
65A800	0.65	7.80	≦8.00	≧1.66	10
65A1000		7.80	≦10.00	≧1.68	10
65A1300		7.85	≦13.00	≧1.69	10
65A1600		7.85	≦16.00	≧1.69	10

Notes:
1. Density is used to calculate the sectional area of the test piece.
2. $W_{15/50}$ indicates the core loss at 50 Hz and 1.5 T. B_{50} indicates the magnetic flux density at 5000 A/m.
3. Method of grading steels:
   (Thickness)(Brand name)(Core loss at 1.5 T/50 Hz)
   Example of 35A210: 35 means thickness 0.35 mm, A is material code for non-oriented electrical steel strip and sheet, and 210 means maximum core loss 2.1 W/kg at 1.5 T/50 Hz.

Due to increasingly greater focus on energy conservation to control global warning, various engineering efforts are being made to improve electric motor efficiency, and iron loss reduction is one of the important aspects of such motor efficiency improvement. For maximum comfort and energy saving, most modern air conditioners use inverter driving, which, however, accompanies carrier-induced high frequency iron loss in addition to the fundamental-frequency iron loss.

Figure 10.28 shows the relationship between frequency and iron loss. Iron losses can be divided into eddy current loss and hysteresis loss. The higher the frequency is, the greater the ratio of eddy current loss to the total iron loss will be. To reduce such eddy current loss, modern electrical steel sheets are being made thinner than conventional ones. It is now quite common that electrical steel sheets used in a compressor motor are thinner than 0.35 mm, for example in the 0.2 to 0.3 mm range.

10.3.5 Permanent magnets

Modern permanent magnets that are commonly used for industrial purposes include alnico (Al-Ni-Co), ferrite and samarium-cobalt (Sm-Co) magnets, rare earth based magnets such as neodymium-iron-boron (Nd-Fe-B), and bonded magnets that contains crushed ferrite or rare-earth magnet. Figure 10.29 shows relationships between residual flux density and coercive force of the most commonly used types of magnets.

Table 10.7 shows global permanent magnet production

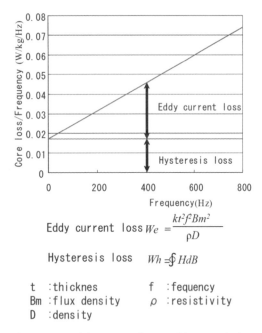

Eddy current loss $W_e = \dfrac{kt^2 f^2 B_m^2}{\rho D}$

Hysteresis loss $W_h = \oint H dB$

t : thicknes
Bm : flux density
D : density
f : fequency
ρ : resistivity

Fig. 10.28 Eddy current loss and hysteresis loss vs. frequency

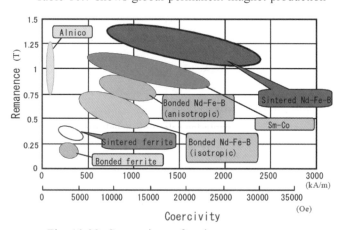

Fig. 10.29 Comparison of various magnet types

Table 10.7 Production volume of permanent magnets in 2007(tons/year)

	Alnico	Sintered Ferrite	Sintered Nd-Fe-B	Bonded Nd-Fe-B	Flex. ferrite	Rigid ferrite	Total	
Japan	300	32,700	10,100	600	3,970	9,800	57,470	8.0%
China	2,000	370,000	16,900	3,200	64,000	4,950	461,050	63.8%
Southeast Asia	800	21,600	-	1,000	10,700	6,800	40,900	5.7%
USA	300	20,000	-	160	26,500	3,800	50,760	7.0%
Europe	600	18,000	800	150	41,000	1,300	61,850	8.6%
Other	-	40,000	-	170	9,000	900	50,070	6.9%
Total	4,000	502,300	27,800	5,280	155,170	27,550	722,100	100%
	0.6%	69.6%	3.8%	0.7%	21.5%	3.8%	100%	

data, listing major magnet producing countries and regions and their production volumes of various types of magnets. Figure 10.30 shows Japan's yearly magnet production records in terms of monetary value[10]. Japan has been one of the world-class producers of permanent magnets and its magnet engineering capability is still among the world's best.

Ferrite (sintered) is the type of magnet of which global production volume, 502 kilotons, is the largest (69.6%) of all the types of magnets. Bonded magnets, 188 kilotons, are the second most produced type of magnets (26.0%), followed by 27 kilotons of sintered Nd-Fe-B (3.8%). The bonded magnet production volume can be further divided into 155 kilotons of flexible ferrite (21.5%), 27 kilotons of rigid ferrite (3.8%) and bonded Nd-Fe-B (0.7%). The sum of sintered and bonded ferrite magnet production is approximately 685 kilotons, making the total production of ferrite-based magnets 95% of the total global magnet production.

Permanent magnets that are used for refrigeration and air conditioning purposes are mostly sintered ferrite and Nd-Fe-B rare earth magnets. Due to greater demand for energy saving, use of rare earth magnets is increasing for their high efficiency capability.

(1) Ferrite magnets

Ferrite magnet provides relatively weak magnetic force, however it can be produced most cheaply of all the types of magnets and is currently used in the greatest quantity in the world. The reason for the low production cost is, in addition to the abundant supply of iron oxides, which are the main production ingredients, ferrite magnets can be produced from a relatively simple manufacturing process of metallic powder sintering. Another advantage of ferrite magnet is its chemical stability, where the metal will not rust or decompose.

(2) Sintered neodymium-iron-boron (Nd-Fe-B) magnets

Nd-Fe-B magnet was invented in 1982 and is currently the strongest known magnet ever produced. Due to the exceptionally high performance of 320 to 440 kJ/m^3 and also assisted by the easy availability of the two primary ingredients, neodymium and iron, the use of Nd-Fe-B magnet is rapidly increasing and now is being spread into diverse applications; not just into air conditioners but into HDDs for computers and DVD units, medical MRIs, mobile phones, handheld video recorders, hybrid vehicles, electric power steering systems (EPS), elevators and power-assisted bicycles.

Figure 10.31 shows a ferrite magnet-based and a rare earth magnet-based rotors. Figures 10.32 and 10.33 show a rare earth magnet tablet and a ferrite magnet tablet, respectively, both used in air conditioners. Due to the difference in their inherent magnetic forces, a ferrite magnet is required approximately ten times as large in terms of volume and four times as large in terms of surface area as a rare-earth magnet, to obtain the same amount of magnetism for air conditioner operation. This yields significant difference in

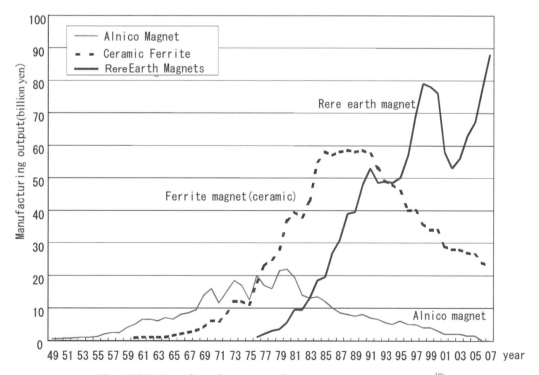

Fig. 10.30 Manufacturing output of permanent magnets in Japan[10]

Compressors for Air Conditioning and Refrigeration

	Rare-earth magnet IPMSM	Ferrite magnet IPMSM
BH_{max}	290 (kJ/m³)	37 (kJ/m³)
Br	1.25 (T)	0.44 (T)
Coercivity	940 (kA/m)	318 (kA/m)
Structure	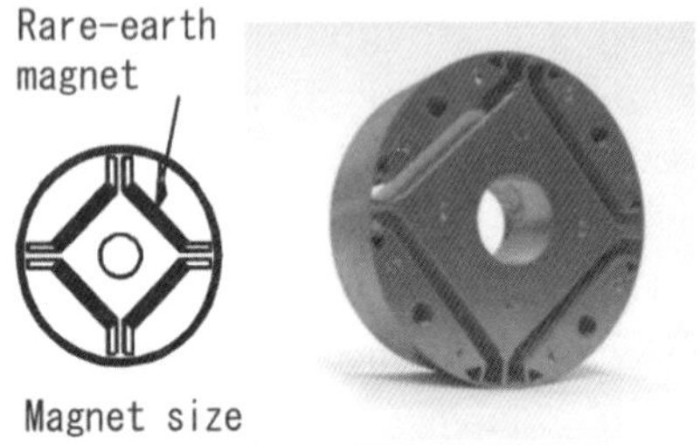 Rare-earth magnet Magnet size Volume (cm³): 10.1 Surface area (cm²): 32.1	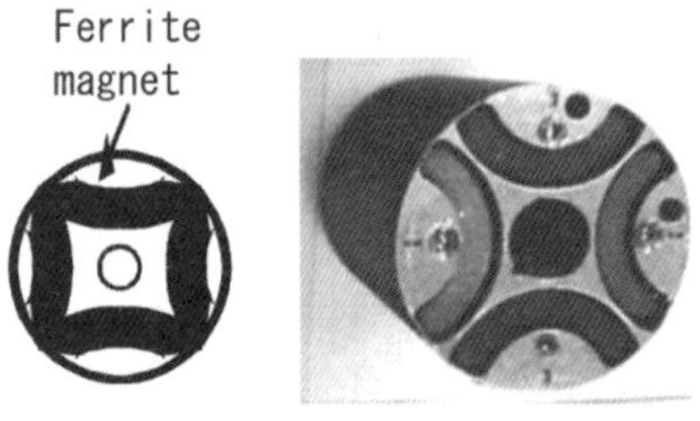 Ferrite magnet Magnet size Volume (cm³): 99.6 Surface area (cm²): 140.7

Fig. 10.31 Rotors with rare-earth(left) and ferrite(right) magnets for air conditioner compressors

Fig. 10.32 Rare-earth magnet used in an air conditioner

Fig. 10.33 Ferrite magnet used in an air conditioner

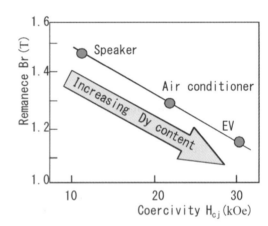

Fig. 10.34 The effect of Dysprosium on remanence Br and coercivity H_{CJ}

the motor size.

As part of the engineering efforts to further improve the coercive force and temperature tolerance of sintered Nd-Fe-B magnet, partial replacement of neodymium with dysprosium or other elements is being researched and implemented. Figure 10.34 shows the performance data of a rotor produced by such alternative technology, where magnetic crystalline anisotropy of the $Nd_2Fe_{14}B$ compound phase is introduced to the rare-earth magnet. This produces a magnet that can function under high temperature conditions.

10.4 Inverter Structure and Control

10.4.1 Inverter classifications and characteristics

Inverters can be largely divided into voltage-source type and current-source type.

Voltage-source inverters supply a variably controlled voltage to the motor as the drive power source. They are commonly used in residential and commercial air-conditioners. On the other hand, current-source inverters supply a variably controlled current to the motor as the drive power source. Current-source inverters are used for larger scale air conditioning systems. Voltage-source inverters are much more commonly used for air conditioning and refrigeration applications. The term "inverter" also usually refers to a voltage-source type inverter. Therefore, the remaining part of this section mainly focuses on the voltage-source inverters.

10.4.2 Inverter structure and components

Figure 10.35 shows the printed circuit board of an inverter for residential air conditioner and its electrical diagram. The alternating current from the commercial power supply first passes through the large-capacity power factor correction. Afterward, the current is fed to the converter's rectifier diode and then smoothed by the electrolytic capacitor, thus converted into a direct current before being fed to the inverter. The direct current is then converted through the inverter module into a three-phase alternating current having a specific frequency and voltage before being supplied to the motor. Conversion into the alternating current is achieved by a switching element, which typically is an insulated gate bipolar transistor (IGBT). The switching element is controlled by a microcontroller to generate desired voltage and frequency so as to obtain a variably controlled three-phase alternating current waveform to control various types of motors. In a system that uses permanent magnets, position sensing elements such as Hall element will not be used and the rotational position of the motor is determined from motor current and voltage ("sensorless position control"). Most inverter circuits being used in a residential air conditioner look like the one shown in Fig. 10.35, where almost all the circuit elements excluding the inductor are directly mounted or printed on the circuit board. Circuit elements that dissipate greater heat, such as the inverter module switching element and the converter diode, may be additionally cooled by an aluminum cooling fin.

(1) Converter (diode-based rectifier)

As commonly used power supply control methods for refrigeration and air conditioning systems, single-phase full-wave or voltage-doubling rectification is done on single-phase power supply and three-phase full-wave rectification is done on three-phase power supply. Figure 10.36 shows how each type of rectification system works.

A single-phase full-wave rectifier circuit is shown in 10.36 (a). A single-phase voltage-doubler rectifier circuit, shown in Fig. 10.36 (b), gives an output voltage with a peak value is approximately twice that of the input voltage. A three-phase full-wave rectifier, shown in Fig. 10.36 (c), provides a relatively smooth output waveform. The output from a diode-rectifier is connected to an electrolytic capacitor to smooth out the pulses that remains after rectification so that an almost perfectly linear direct current will be supplied to the inverter. The input current waveform from the commercial power supply usually contain distortions and harmonics that are detrimental to the power factor. Therefore, an inductor is provided on the input side for phase

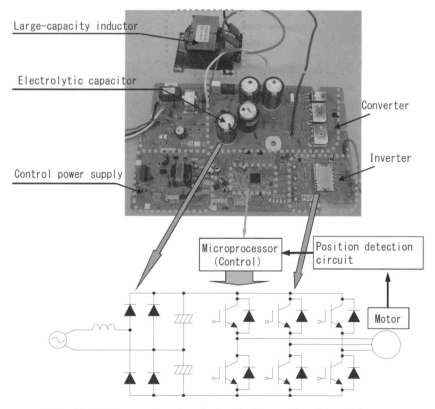

Fig. 10.35 Inverter board and circuit diagram for air conditioner

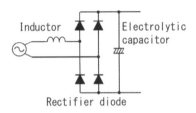

(a) Single-phase full-wave rectifier circuit

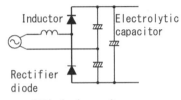

(b) Single-phase voltage doubler rectifier circuit

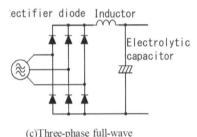

(c) Three-phase full-wave rectifier circuit

Fig. 10.36 Diode rectifier circuits

compensation and harmonics suppression for power factor improvement.

(2) Inverter (switching circuit)

The inverter module provides high speed control of the direct current supplied from the rectifier so as to output power with a specific frequency and voltage. As the inverter load is usually a three-phase motor, the inverter circuit typically has a three-phase full-bridge configuration. Three switching devices, one for each of the three phases (U, V, W), are provided on the positive side, and another three switching devices are similarly provided on the negative side. Thus, one three-phase inverter circuit comprises a total of six switching devices.

To prevent the positive side and the negative side of a given phase from being turned on at the same time, which will cause short-circuiting of the direct current supply, a no-overlapping period, usually referred to as "dead time", is provided for phase compensation.

(3) Control/power supply system

The control/power supply system generates control signals to generate the desired output waveform from the inverter module. If the load is an induction motor, constant V/f (voltage/frequency ratio) control will be provided. If the load is a permanent magnet motor, the control system determines the motor rotation position and provide control accordingly.

The drive frequency for a motor varies depending on its number of poles. In the case of a two-pole motor, which is common in induction motors, the drive frequency is the same as the motor shaft rotation speed. For example, 50 Hz frequency is input to obtain 50 rps rotation. In other types of induction motors and permanent magnet motors where the number of poles is other than two, the drive frequency must be one that specifically corresponds to such number of poles. In the case of a four-pole motor, 120 Hz must be input to obtain a shaft rotation of 60 rps. With a six-pole motor, 180 Hz must be input to obtain a rotation of 60 rps.

(4) Components

(a) Power device

Figure 10.37 shows the history of inverter switching element technology[11]. Traditionally, various devices such as thyristors (SCR), GTO (gate turn off thyristor), power transistors, MOS-FET (metal-oxide-semiconductor field effect transistor) and IGBT (insulated gate bipolar transistor) have been used as the switching element in an inverter. Currently, IGBT is dominantly used in a wide range of inverters from small-scale air conditioning and refrigerators to larger scale systems. IGBT is a bipolar transistor which has an equivalent gate structure as MOS-FET and therefore is capable of providing a voltage control equivalent to that by a MOS-FET. IGBT is advantageous in that its drive circuit can be configured much more simply than other conventional bipolar transistors but the amount of loss incurred is as low as that of other bipolar transistors while providing much higher dielectric strength. Figure 10.38 shows an IGBT module that is used in a small-size air conditioner[12]. As modern inverter circuit integration is being increased, some inverters contain rectification, drive power and protection circuits as well as IGBTs in a single package.

(b) Microcontroller

In a commonly used air conditioner, a single-chip microcontroller centrally controls air conditioning and inverter operations based on user temperature inputs and various sensor information. A single-chip microcontroller is a package that integrates the computation unit, program storage ROM (read only memory), RAM (random access memory) that temporarily stores required data, input/output terminals for external device control, AD converter that receives external analog signal inputs and converts them into digital signal, and various counter/timer units that are used for generating later-described waveforms, into a single chip.

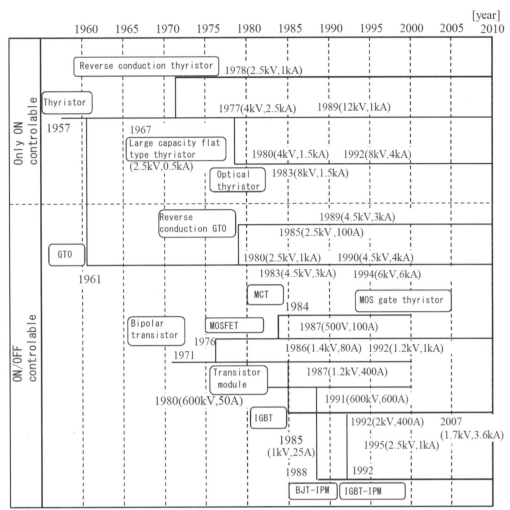

Fig. 10.37 History of switching devices[11]

Fig. 10.38 IGBT module[12]

Figure 10.39 shows the internal structure of a typical microcontroller.[13] The performance of modern microcontrollers is being rapidly improved, achieving greater computation performance (faster computation) and larger ROM capacity and also integrating waveform generation circuits.

(c) Inductor

Figure 10.40 shows an inductor unit. The inductor comprises a copper wire coil would around an iron core which is built by laminating thin electrical steel sheets. Depending on the power supply type and other design considerations, the inductor may be positioned on the alternating current side or the direct current side of the rectifier diode. An inductor has an effect to level the electrical current that passes through it to a constant level and is therefore employed to smooth an electric current before it is fed to the converter. Inductors for small-capacity, single-phase power supply, such as ones for residential air conditioners, are usually positioned on the alternating current side of the diode. On the other hand, inductors for larger capacity, three-phase power supply, such as these for industrial air conditioning systems, are usually positioned on the direct current side. As an inductor is subjected to massive current input, copper loss will be incurred in the windings and iron loss will be incurred in the iron core, both heating the inductor. Design caution is required so that electrical insulation and other surrounding materials will not be subject to excessive heating by the inductor beyond their temperature tolerance.

Compressors for Air Conditioning and Refrigeration

(d) Electrolytic capacitor

An electrolytic capacitor has a cylindrical form as shown in Fig. 10.41.[14] The electrodes are immersed in an electrolytic solution that fills the cylinder. The capacitor, by storing electrical energy, tries to level electrical voltage to a constant level. Capacitors are often employed in an inverter to smooth electrical pulses remaining after the incoming alternating current is converted into a direct current by the rectifier diode, so as to output a perfectly constant-voltage direct current. As an electrolytic capacitor operates based on the chemical reaction in its electrolytic bath, the characteristics of an electrolytic capacitor may change depending on its internal temperature and also with the lapse of time (aging). When an electrolytic capacitor stores or discharges electrical energy in order to smooth out current pulses, a ripple current runs through the capacitor to generate heat. In addition to internal heat and aging, the performance and service life of an electrolytic capacitor are also significantly affected by external heat received from the surrounding environment. Operating temperature and installation environment must be carefully considered in the design of an electrolytic capacitor. It is generally believed that a 10°C temperature increase will reduce the electrolytic capacitor life by half. It is one of the components that require a well-thought cooling feature.

(e) Noise filter

Noise filters are employed in an inverter unit to prevent its internally generated noises (EMI; electromagnetic interference) from leaking outside and also to block externally originated noises, such as those caused by static charge or lightning (EMS; electromagnetic susceptibility), from getting inside the inverter to cause malfunction. Noise filters consisting of a small coil and a capacitor are often provided at the input point of the inverter circuit board. Alternatively, a noise filter may be formed by winding the output wire from the inverter circuit board around a ferrite core. Internally generated noises must be blocked from leaking outside in compliance with the standards and regulations described in Section 10.6. The filter constant must be set

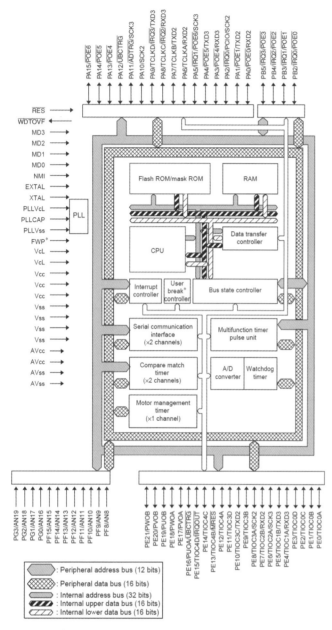

Fig. 10.39 Internal configuration of one-chip microprocessor[13]

Fig. 10.40 Inductor

Fig. 10.41 Electrolytic capacitors[14]

to satisfy these requirements. Externally originated noises must also be blocked in a similar manner, by designing filters that will provide sufficient noise arresting capability to minimize detrimental effects. Control of both outgoing electromagnetic interference (EMI) and incoming electromagnetic susceptibility (EMS) is called "electromagnetic compatibility" (EMC).

10.4.3 Principle of voltage-based inverters
(1) PWM control

PWM, or pulse width modulation, is a control technique to provide virtual control of the output voltage by transforming the output waveform, with the use of a high frequency signal called "carrier signal", into square waves with a specific proportion of ON time, or "pulse width". The output voltage can be virtually controlled by changing this pulse width. Figure 10.42 illustrates how PWM control works. The carrier signal, which is a high frequency triangular waveform signal, is compared to the target waveform to be obtained, so as to send control signals only in the regions where the target waveform is higher than the carrier signal waveform. This results in the generation of a cyclic series of constant-height, variable-width pulses. This signal pattern obtained is referred to as "PWM signal". A PWM inverter unit outputs a direct current by tuning the inverter output on only when the PWM signal is at the high level and turning it off when the signal is at the low level. This would provide, for example, a voltage output to the motor that pulses between 0 V and 280 V. Note that the modulation method may be called differently depending on the type of the target waveform. When the target waveform is sinusoidal, the modulation is called "sinusoidal modulation", When the target waveform is square waves, it is called "square wave modulation". Square wave modulation can be further divided into different types depending on in how much of the half-rotation (180°) cycle the motor will be energized, such as 120° modulation where energization will occur only in a 120° range of the half rotation cycle and will be suspended in the remaining 60° range (non-energization time), or 180° modulation where energization will continue throughout the 180° range. In general, power supply waveforms obtained by sinusoidal modulation are considered to accompany less low-order harmonic components and therefore more efficient and less noisy than those obtained by square wave modulation. Therefore, sinusoidal modulation is widely used for driving air conditioning and other compressors as well as fan motors. Figure 10.43 provides a comparison of motor drive voltage and current waveforms obtained by different modulation methods. As shown in the figure, the waveforms obtained by square wave modulation are distorted with harmonic components while those obtained by sinusoidal modulation are nearly sinusoidal and less distorted.

(2) PAM control

With residential and commercial air conditioners, the input power factor must be improved so as to obtain the maximum air conditioning capacity within the limited current breaker. Although increasing the compressor rotational speed helps to maximize air conditioning capacity, it can

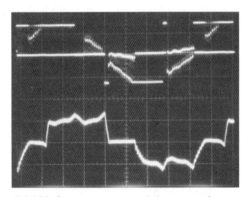

(a) 120° square wave drive waveform

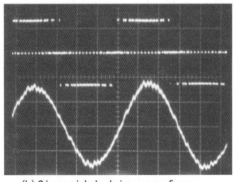

(b) Sinusoidal drive waveform

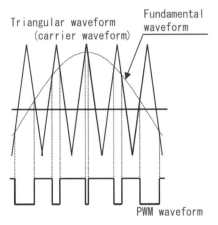

Fig. 10.42 PWM modulation waveform

Fig. 10.43 Inverter output waveforms

only be increased to a certain level due to power voltage limitations. However, use of a voltage step up mechanism will allow further increasing the rotational speed. Furthermore, current harmonics may occur on the commercial power supply side of a system using a diode rectifier circuit, detrimentally affecting other load equipment connected to the same supply. Various industrial and engineering standards have provisions regulating such current harmonics (described in more details in Section 10.6). PAM, or pulse amplitude modulation, has been developed for the purpose of stepping up the direct current voltage and also reducing the effect of current harmonics on the power supply system.

PAM uses a chopper circuit or other features to regulate the output voltage amplitude and also to improve the power factor and reduce current harmonics on the power supply side. PAM may also be referred to as PFC, or power factor control, and can be further divided into different methods. In the air conditioning field, the DC switching method, or the active method, and the AC switching method are the ones most widely used.

(a) DC switching method

Figure 10.44 shows how the DC switching-method PAM works. With a step-up chopper circuit provided in the direct current section of the circuit, control is provided to make the power factor on the power supply side approximately 1 and also to reduce the current harmonics running there. In addition, voltage in the direct current section can be increased when air conditioning load is increasing and the compressor motor needs to run faster, thus providing a greater refrigeration capacity.

The DC switching method is high-performance but may be disadvantageous in that the high frequency switching action needs to be supported by an inductor with superior high frequency characteristics and a fast operating switching element, which increases the production cost. Also, the fast switching operation and the resultant greater loss may lead to efficiency reduction.

(b) AC switching method

Figure 10.34 shows how the AC switching method PAM works. Instead of perform switching in the direct current section like the DC switching method, the AC switching method provides a switching element in the alternating current section upstream of the diode rectifier. As illustrated in Fig. 10.46, one to several control pulses are sent every half rotation cycle to execute switching, so that the current will run on the power supply side for a longer period. The AC switching method does not provide as much improvement in the power factor and current harmonics as the active method does, but is less costly than the active method as the switching speed is slower and does not require expensive fast switching elements. The AC switching method also provides higher efficiency than the active method.

10.4.4 Principle of motor control
(1) Induction motors

The induction motor is driven by supplying the required alternating current which is generated by controlling the switching element based on the target output frequency and voltage determined by the PWM signal, which in turn is controlled by the rotational speed command given from a higher controller on the basis of constant V/f (voltage/frequency) control.

The rotational speed of an induction motor is determined by the combination of the drive frequency output from the inverter, the number of motor poles and the amount of "slip". Therefore, the motor speed can be variably controlled by changing the drive frequency from the inverter. Constant V/f control refers to a control method where the inverter output voltage is controlled together with the drive frequency so as to provide variable motor speed control without reduc-

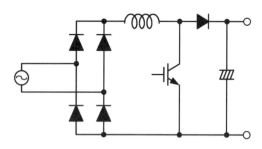

Fig. 10.44 Circuit configuration of active type

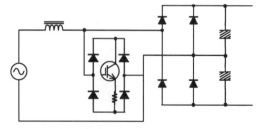

Fig. 10.45 Circuit configuration of partial switching type

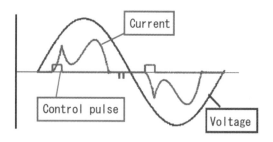

Fig. 10.46 Control waveform of partial switching type

ing motor efficiency and power factor. As indicated by the name, "Constant V/f" control maintains a constant ratio between voltage and frequency. Constant V/f control has been widely implemented for decades as it is a relatively simple yet efficient induction motor control method. It still is commonly employed for various motor control purposes. When designing a motor control system, the start-up voltage (V_0), the V/f line inflection point (V_1, f_1) and the maximum voltage point (V_2, f_2), each shown in Fig. 10.47, are the decisive parameters to be determined.

(2) Permanent magnet synchronous motors

While brushed permanent magnet synchronous motors are still commonly used for toys and other simple applications, the so-called "brushless DC motors", that are the type of motors now dominantly used in air conditioners, are permanent magnet synchronous motors where the conventional brushes and commutator are replaced with the combination of an electronic motor positioning sensor and an inverter[15].

The rotor positioning sensor accurately determines the magnetic pole position (angular position) of the rotor so as to cause the inverter to energize the motor winding in the best timing for the given rotor position. With this, the motor can be driven efficiently to generate greater torque. If the winding energization timing is not optimized for the rotor's magnetic pole position, not only the required torque cannot be efficiently obtained but also, in worst cases, motor operation may be interrupted as the motor steps out of phase. To prevent such situation, high-precision rotor magnetic pole sensing is required. However, in the high-temperature high-pressure environment inside the compressor, conventional position sensing devices like the Hall element cannot be used. To address this difficulty, various rotor position determination techniques without using a position sensing element have been developed and utilized. Blower fan motors, which are not subject to severe environment, still commonly use the Hall element for rotor position sensing. The following paragraphs explain various techniques where the rotor position is determined without using a magnetic pole position sensing device.

(a) Back electromotive voltage detection method

A permanent magnet synchronous motor is driven by an inverter to generate torque, while it also generates electric power by its own rotation. Such electric power generated by a rotating permanent magnet synchronous motor is called "back electromotive force". Figure 10.49 shows how the back electromotive force detection system works. When the inverter drives a motor, the back electromotive voltage will appear on the motor terminal during the preset non-energization time, as shown in Fig. 10.50. The rotor position can be determined by finding the zero voltage point for the back electromotive voltage detected. The back electromotive voltage detection method is disadvantageous in that it needs to have a non-energization time to work, which restricts the drive waveform and also makes it impossible to move the current phase across the non-energization time. However, by taking advantage of simplicity and easiness of the method, it is used as a common solution for less demanding applications such as driving an SPMSM (surface permanent

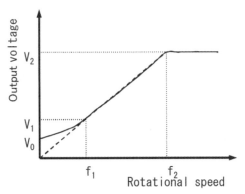

Fig. 10.47 Example of V/f pattern

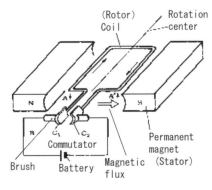

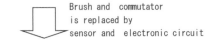

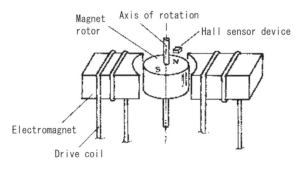

Fig. 10.48 Detection method of rotor position

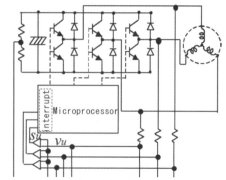

Fig. 10.49 Example of back electromotive voltage detection

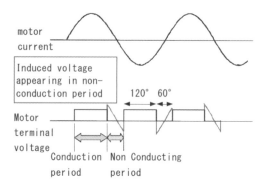

Fig. 10.50 Back electromotive voltage detection waveform

magnet synchronous motor).

(b) Motor model computation method

The motor model computation method employs a mathematical motor model inside the microcontroller program and runs estimating computation based on the motor current and voltage detected, so as to determine the internal motor state from the mathematical model to calculate the motor position. One drawback is that the complex estimating computation needs to be supported by a high performance microcontroller. With that requirement satisfied, however, the motor model computation method will be capable of efficiently driving various types of permanent magnet synchronous motors. Another advantage is that the method does not need any special circuit to be added other than the current and voltage sensing mechanism. This method therefore is becoming increasingly common for air-conditioners where there is stronger demand for high performance.

(c) Other motor drive control techniques
i) Start-up

When a permanent magnet motor is being started up from a stationary state, where no back electromotive force is obtained, either of the above-described techniques will by itself be incapable of determining the rotor position. To overcome this difficulty, it is necessary to either provide a means to determine the rotor position in a stationary state or employ a start-up system that will work regardless of the rotor position. The latter of these two methods is more commonly employed. It starts up the motor by receiving an alternating current input from a 0 Hz state regardless of where the rotor is located. The rotational speed will then be gradually increased to several Hz where synchronization becomes possible. When the rotor has generated a detectable amount of back electromotive voltage by its own rotation, one of the above two methods, the back electromotive voltage detection method or the motor model computation method, will be applied to commence a sensorless position control operation. When designing an inverter control system, the voltage to be given at the start of synchronization, the upper limit frequency and the acceleration rate are the decisive factors to be determined.

ii) Flux-weakening control

As previously explained, motor rotation generates a back electromotive voltage. The faster the motor rotates, the greater the amount of this back electromotive voltage will be. However, an inverter without a PAM circuit cannot output a voltage higher than that of the commercial power supply provided. When the motor is rotating very fast, there comes a point where the inverter output voltage to the motor cannot be increased above a certain level to control the motor. With a motor that has a magnetic field created by coils, field-weakening control, where the current that runs through the coils is reduced to weaken the magnetic field in order to restrict the back electromotive force from increasing too much. With motors that employ permanent magnets, however, the intensity of the magnetic field cannot be adjusted in the same way. To overcome this difficulty, IPMSMs, where the magnets are embedded inside the rotor, can have the apparent magnetic flux reduced by appropriately controlling the energization timing in the rotor magnetic pole position, thus restricting the amount of back electromotive voltage generated. With this, the motor is allowed to rotate faster under inverter control. This technique, where the current phase that runs through the motor is controlled so as to reduce the motor's back electromotive voltage, is called "flux-weakening control". The flux weakening control is widely used for driving IPMSMs.

iii) Torque control

Figure 10.51 shows changes in the compression torque of a single-cylinder rotary compressor[16]. As shown in the figure, the more advanced the rotation angle gets, the greater the compression torque will become, before reaching the discharge pressure and starting to decrease again. Thus, the amount of torque required to run a compressor varies

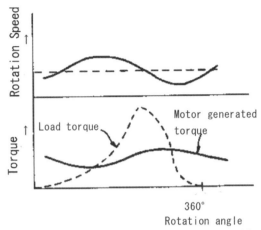

Fig. 10.51 Variation of compression torque[16]

significantly during a single compressor rotation. If the compressor motor is run at a constant voltage and current, it will output a constant torque. In this situation, compressor rotational speed will significantly fluctuate due to the difference between the motor output torque and the compression load torque. This will generate greater amounts of noise and vibration from the compressor. A method to reduce such vibration is to detect changes in the rotational speed and to provide feedback control so that the amount of changes detected will be minimized. Or, anticipated variation patterns can be stored in controller memory to provide feedforward control accordingly. "Torque control" refers to a control technique to reduce compressor vibration by way of motor torque compensation. The former method, feedback-type control, will be capable of providing highly accurate compensation under any operating conditions, but it requires a sensing device to detect vibration changes during each rotation and must be supported by complex data processing capability. Although the latter method, feedforward-type control, will not be able to provide as accurate a control as that by the feedback-type control due to limitation in the number of rotation patterns that can be memorized and how accurate they can be made, it can work without special sensing devices or additional data processing and is capable of providing practically good enough compensation performance. Therefore, the feed-forward type of control is widely used in small-capacity air conditioners.

10.5 Electrical Characteristic Measurements (with Inverter Control)

10.5.1 Types of loss and measurement methods

As shown in Fig. 10.35, the electrical circuitry of an air conditioner inverter unit consists of such components as input filters, converter, inductor, inverter module (such as IPM) and control circuit. These circuit components generate various losses.

Losses incurred in the input filter includes capacitor induction loss and coil resistance loss. Converter related losses may include diode losses that accompany forward-direction voltage drop and the smoothing electrolytic capacitor induction loss. The amount of induction loss is influenced by harmonics contained in the output current from the rectifier circuit. The greater the amount of such harmonics contained, the larger the induction loss will be. When PAM control, which works to bring the power supply side power factor as close to 1 as possible so as to reduce current harmonics, is implemented, switching losses will be incurred.

Losses incurred in the inductor, which is one of the main circuit components, include direct current resistance loss and iron loss. The direct current resistance loss increases as the converter input current increases. Iron loss is related to input voltage and current harmonics and will be larger when greater amounts of harmonic components are contained.

Losses incurred in the inverter potentially include steady-state losses of the parasitic diodes and switching elements, and also switching losses that occur when circuit elements are switched on and off. Both types of losses increase as the inverter output current increases. The latter type of loss, or switching loss, is related to the frequency of switching actions, and may therefore increase to a detrimental level during high speed switching operations. Switching losses are also influenced by the type of PWM modulation method. When three-phase modulation, which has all of the three phases (U, V, W) switched and two-phase modulation, which has only two of the phases (for example, U and V) switched, are compared, the two-phase modulation incurs smaller loss as it accompanies fewer switching actions per unit time.

Losses in main circuit components such as the input filter, converter, inductor and inverter module (such as IPM) may decrease or increase depending on the inverter's input and output situation. It is well known that the load of an inverter-driven air conditioner compressor increases as the operating frequency increases, leading to greater main circuit losses.

The control circuit comprises a microcontroller, an IC and a power module, each of which consumes a largely constant amount of power. The control circuit power consumption is to some degrees related to the inverter operating status through the switching element drive circuit. An ideal PWM inverter will always have one of the pair of top and bottom arms on. That is, ON signals will always be given to three of the six switching elements (this is the case of a 180° sinusoidal modulation system which is most commonly used as the start-up mechanism of modern compressors; in a 120° modulation system, two of the six elements will be on), where power consumption by the drive circuit does not change. Therefore, the amount of loss in the control circuit hardly changes with the inverter operating status. Figure 10.52 shows the composition of inverter power consumption factors, or losses. This data represents a situation where a residential air conditioner is operated under light duty condition with an inverter output of approximately 200 W. In the total inverter loss, switching element losses account for 47%, converter loss 33%, and power consumption by the control circuit and other components 20%.

The above paragraphs explained about losses in and related to inverter circuitry, loss of motor driven in an inverter-driven motor will increase or decrease depending on the amount of harmonic components contained in the inverter output current.

The use of an inverter driven compressor accompanies not only the above-described losses in the inverter circuitry, but also it induces loss in the inverter-driven motor depending on the amount of harmonic components contained in the inverter output current. In general, the higher the switching frequency is set, the less the amount of inverter-harmonics-induced motor losses will be. That is, the amount of motor loss induced by inverter harmonics is inversely proportional to the amount of switching losses, indicating the need for comprehensive design optimization for the entire motor-inverter package. Inverter losses can be investigated by running the motor or compressor with a wattmeter connected between the above-mentioned inverter components.

Figure 10.53 shows a motor characteristics evaluation unit that is used for motor duty adjustment. In addition to running motor characteristics investigation, these evaluation units, which are designed and built for perfectly constant loading, can be used for measuring inverter circuit board losses.

10.5.2 Types of noises and measurement methods

Audible noises generated by an air conditioning compressor includes electromagnetic noises, that are generated by causes related to the motor's electrical characteristics, and mechanical noises, that are caused by the movement of compressor components and also by fluid compression and expansion actions. When an air conditioning compressor is driven by an inverter for variable speed control, harmonics-related noises will be added to the above types of noises.

Audible noises from inverter drive operation can be divided by their frequency range. One type of noise is "carrier noise", which is high frequency noise related to the inverter's PWM control operation. Another type of noise is "sideband noise", of which frequency is m times higher or lower than the fundamental frequency produced by the motor drive operation of which center frequency is n times the carrier frequency. If the carrier frequency is denoted as f_c and the motor drive frequency as f_m, the noises generated will include frequency f_c noise and frequency $(n \cdot f_c \pm m \cdot f_m)$ noise. Figure 10.54 shows an actual noise frequency distribution. As inverter noises are distributed over a very wide frequency range, resonance with the compressor's and the motor's natural frequencies may cause a specific frequency noise to increase significantly. Careful consideration should be given to both the design of mechanical components and the carrier frequency setting.

JIS B 8346 specifies compressor noise measurement methods. In general, noise measurement should be taken in an environment where there is smallest possible amount of

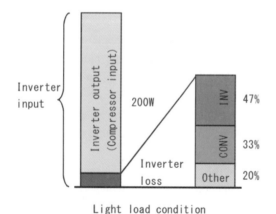

Fig. 10.52 Inverter loss distribution (at light load operation)

Fig. 10.53 Evaluation apparatus of motor characteristics

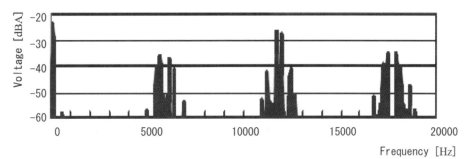

Fig. 10.54 Spectrum of switching frequency component (voltage frequency analysis results)

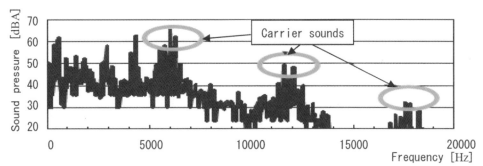

Fig. 10.55 Analysis example of acoustic noise characteristic (noise frequency analysis results)

sound reflection except that from the floor and where the background noise level is more than 10 dB lower than when the target noise is present. In the case of a compressor for residential or commercial air conditioner where the motor is built into the compressor, noise measurements should be taken at a point 1 m away from the compressor surface, in each of the four directions radiating from the compressor. The measurement height should be approximately 1/2 of the compressor unit height.

Noise evaluation may be done based on overall noise analysis, frequency analysis or octave analysis. Figure 10.55 shows the result of a noise characteristics analysis. This analysis was done on a compressor, installed in a commercial air conditioner with a 14 kW heating capacity and operated by a triangular-wave-comparison PWM inverter. The output voltage and noise characteristic analysis results are plotted. The carrier frequency of this inverter is 6 kHz. The diagram shows that noise components caused by PWM inverter switching actions stand out.

The above data shows that an inverter's carrier frequency significantly affect noise generation. In an air conditioning compressor, the inverter carrier frequency is usually set to several kHz depending on the performance characteristics of the sound insulation used.

Since more than a decade ago, PWM inverter control has been introduced to most fan motors that are fitted in the outside unit (condenser unit), where noise reduction is achieved by setting the carrier frequency out of the human audible range.

10.6 Standards and Regulations

10.6.1 Power supply voltages, standards and regulations in major countries

Table 10.8 lists the single-phase power supply voltages and frequencies in major countries of the world. Table 10.9 shows household outlet socket shapes being used around the world. As the power voltage, frequency and outlet socket shape differ from country to country, electrical equipment should adopt an electrical circuit and an outlet socket that are suited to the country or region where it is intended to be used. Even within Japan, multiple types of household outlet sockets are used in 100V/200V range. The most appropriate type of outlet socket must be provided for an air conditioning or refrigeration equipment based on its operating voltage and current.

Regulations concerning electromagnetic compatibility (EMC) are being incorporated into the IEC (International Electrical Commission) standards primarily under the European Union leadership. However, local standards and regulations slightly differ between countries and regions. Major EMC-related standards and regulations implemented around the world are listed in Table 10.11.

The European Union adopts a series of EN (European Norm) standards as the unified regional engineering standards. The EN standards regulating EMC issues are based on the international standards originally published by CIS-

Table 10.8 Worldwide single-phase power supply voltage and frequency

Country	Voltage(V)	Frequency(Hz)	Outlet shape
Japan	100	50/60	A
Korea	110/220	60	A
China	220	50	A
India	220/230	50	B
US	120	60	A
Canada	120	60	A
Australia	240	50	D
New Zealand	230/240	50	D
England	240	50	B1
France	127/220/230	50	C
Germany	220	50	C
Egypt	220	50	C
Russia	127/220	50	C
Italy	220	50	C
Saudi Arabia	127/220	50/60	A
South Africa	220/230	50	B

Table 10.9 Outlet shape of the worldwide

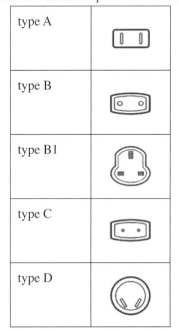

Table 10.10 Japan outlet shape

[image]	[image]
125V 15A	125V 15A
[image]	[image]
125V 20A	250V 20A

PR (International Special Committee on Radio Interference) and TC77 (an IEC committee in charge of electromagnetic compatibility), adopted with some modifications. EU member countries are required to make all their electrical and electronic equipment compliant with the EMC directive that governs electromagnetic interference issues. All such compliant products must carry a CE marking.

Many non-EU countries and regions are starting to adopt a similar regulatory frameworks by issuing their own EMC

Table 10.11 International EMC regulatory status

		Japan	Europe	China	Australia	United States
EMC overall	Decree name	—	EMC directive (CE marking)	—	—	—
	Application range	—	All electrical & electronic equipment	—	—	—
Radio noise	Decree name	Electrical appliances and material safety act	—	①Standardization and product quality laws ②Compulsory product certification manage-ment regulations	Electromagnetic compatibility framework	—
	Standard name	Electrical appliance technical standards	①EN55014 ②EN55022	GB4343-1	①AS/NZS CISPR14 ②AS/NZS CISPR22	FCC 47 CFR Part 15
	Application range	Motor input $\leqq 7$ kW	①Home electric equipment and the like ②Information technology equipment	②Cooling capacity $\leqq 21000$ kcal/h	①Household electrical device and the like equipment ②Information technology equipment.	Radio frequency equipment or technical standards for those components
	Decree name	—	—	①Standardization and product quality laws ②Compulsory product certification management regulations	—	—
Harmonic	Standard name	①JISC61000-3-2 ②Specific customer guidelines	①EN61000-3-2 ②EN61000-3-12	GB17625-1-1998	AS/NZS 61000.3.2 (non-mandatory standard)	IEEE519-1992
	Application range	①Home appliances and general purpose products ($\leqq 20$ A) ②—	①Electrical & electronic equip. ($\leqq 16$ A) ②Electrical & electronic equip. (>16A & $\leqq 75$A)	②Cooling capacity $\leqq 21000$ kcal/h	Electric and electronic equipment ($\leqq 16$ A)	Regulations for specific customers
Voltage variation (Flicker)	Decree name	—	—	Standardization and product quality laws	—	—
	Standard name	—	①EN61000-3-3 ②EN61000-3-11	GB17625.2	—	—
	Application range	—	①Electrical & electronic equip. ($\leqq 16$ A) ②Electrical & electronic equip. (>16A & $\leqq 75$A)	Rated $\leqq 16$ A	—	—

standards, most of which are largely based on CISPR or TC77 standards except for slight modifications. In Japan, electromagnetic interference issues are regulated by the Electrical Appliances and Materials Safety Act, which primarily governs household electrical products.

References

1) Y. Takeda, N. Matsui, S. Morimoto, Y. Honda: "Design and Control of Interior Permanent Magnet Synchronous Motor", Ohmsha, Tokyo (2001). (in Japanese)

2) IEEJ Investigating R&D Committee on Precisely Small Motor: "Small Motors", Corona Publishing Co., Ltd., Tokyo (1991). (in Japanese)

3) Y. Sugii, Y. Kojima, K. Hamamoto: "Technique and Application of High Efficiency Motor", IEEJ, JI-ASC2000, 1, pp. 93-98 (2000). (in Japanese)

4) M. Kawahira: "Hermetic Refrigerating Machinery", Japanese Association of Refrigeration, Tokyo (1981). (in Japanese)

5) IEEJ Investigating R&D Committee on Drive Systems of the Electromagnetic Actuator for Control: "Drive systems of the electromagnetic actuator for control", IEEJ Tech. Report, 614, (1996). (in Japanese)

6) K. Hirose, H. Haitani: "Design of electric machines outline (revision 4)", IEEJ, (2007). (in Japanese)

7) Sumitomo Electric Wintec, Inc.: "MAGNET WIRE TECHNICAL BOOK", p. 11. (in Japanese)

8) Japanese Industrial Standards: "JIS C2552".

9) Y. Arita, C. Kaido, T. Kubota: "Development of Thin Gauge Non-oriented Electrical Steel Sheet", IEEJ, RM-03-46, pp. 73-78 (2003). (in Japanese)

10) N. Ishigaki, H. Yamamoto: "Overview of Permanent Magnets and Their Markets", Magnetics Japan, Journal of the Magnetics Society of Japan, 3 (11), Tokyo (2008). (in Japanese)

11) M. Igarashi, Y. Onozawa, T. Goto, H. Miyashita, M. Watanabe: "Power Device IGBT Basic and Applications", CQ Publishing, Tokyo (2011). (in Japanese)

12) Mitsubishi Electric Corporation: "DIPIPM PS21963-A", August (2007).

13) Renesas Electronics Corporation: "SH7046 Group Hardware Manual", Rev. 4.00, p. 3, December (2005).

14) NICHICON CORPORATION: "ALUMINUM ELEC-TROLYTIC CAPACITORS LQR", p. 341.

15) K. Tanikoshi: "DC Brushless Motor and Control Circuit", Sogodenshi Publishing, Tokyo (1984). (in Japanese)

16) Publication of Patent Applications: "H04-006349".

Chapter 11 Testing

11.1 Performance Tests

Performance of a refrigeration system is determined by the flow rate of refrigerant that is forced to circulate inside the system by the compressor and the amount of heat that is exchanged between the refrigerant and the external heat media by the heat exchanger. Tests to determine performance of the refrigeration system must be based on measuring its refrigeration capacity, which is governed by the refrigerant flow characteristics in the compressor and the heat exchange characteristics in the heat exchanger unit. Also, compressor efficiency can be obtained by measuring the amount of compressor-driving energy (electrical input etc.) to provide the refrigeration capacity.

Table 11.1 lists major JIS standards concerning performance tests to be conducted on a compressor that uses refrigerant as the working medium. The following sections explain about commonly implemented compressor performance tests.

11.1.1 Main measurement items
(1) Refrigeration capacity

Refrigeration capacity can be obtained by multiplying the difference of refrigerant specific enthalpies, which is the difference between specific enthalpy of the saturated liquid refrigerant corresponding to the discharge pressure at the compressor outlet measurement point and that of the suction gas refrigerant at the compressor inlet measurement point, by the measured refrigerant mass flow rate of the compressor. The specific enthalpy values at the above-mentioned measurement points should be obtained based on the refrigerant's thermodynamic property table or the equation of state. Measurement of the refrigerant mass flow rate should be done by one of the methods described in Section 11.1.2.

(2) Volumetric efficiency

Volumetric efficiency is the ratio of the actual volumetric flow rate sucked into the compressor to the compressor's ideal flow rate based on the compressor displacement volume. The actual volumetric flow rate, which changes with various factors such as internal leakage, suction gas pressure loss and suction gas heating, should be measured in the position specified by JIS B 8606-4.3.2.

(3) Input power

In the case of an open type compressor, compressor input power is the amount of driving power received by its shaft. In the case of a hermetic compressor, compressor input power is the amount of electrical input received at its motor terminal. Note that compressor input power should also include the amount of power consumed by auxiliary devices, such as the lubrication pump, that are necessary to support the compressor's operation. In the case of an inverter-driven compressor, compressor input power should also generally include the amount of power consumed by its inverter.

(4) Coefficient of performance (COP)

Coefficient of performance, or COP, is the ratio of the

Table 11.1 JIS standard (performance test of refrigerant compressors)

Number	Standard name
JIS B 8606	Testing of refrigerant compressors
Appendix A	Calibration and uncertainty of measurement instruments
Appendix B	Analysis of uncertainty
Appendix C	Quantity symbol
Appendix D	Measurement method of oil circulation rate
Appendix E	Measurement method of oil circulation rate with attached oil separator
Appendix F	References
JIS B 8600	Standard conditions of rating temperature for refrigerant compressors
JIS B 8603	Performance testing and inspection methods by air for open type reciprocating refrigerant compressors
JIS B 8608	Refrigerating systems−Test methods−General requirements

Compressors for Air Conditioning and Refrigeration

refrigeration capacity to the compressor input power. COP represents the compressor efficiency and its value determined from measurements under actual or close-to-actual operating conditions of the refrigeration system is an important energy saving index of the compressor.

(5) Oil circulation ratio

Oil circulation ratio is the ratio of the lubrication oil mass contained in the refrigerant-lubrication oil mixture circulating through the refrigeration cycle to the total mass of the mixture. An excessively high oil circulation ratio will reduce the system's heat exchange efficiency and will also detrimentally affect the compressor's reliability due to resulting lubrication oil depletion in the compressor.

11.1.2 Test methods

Although the actual performance test procedure should be determined based on the compressor specifications, rated operating conditions for compressor performance tests are provided in JIS B 8600, as shown in Table 11.2.

To determine the refrigeration capacity of a system under specified conditions, at least four consecutive measurements should be taken with a 15 to 20 minute interval between each measurement, and the average of four consecutive measurements obtained should be used as the measurement value, based on which the refrigeration capacity is calculated. This is for the purpose to assure that all the measurements are taken under the same stabilized conditions. Variables that are measured in a compressor performance test and the allowable steady-state variation of each variable

Table 11.2 JIS B 8600 Rated temperature condition of refrigerant compressors

Class	Suction saturated temperature or dew-point temperature [℃]	Discharge saturated temperature or dew-point temperature [℃]	Suction superheated vapor temperature [℃]			Refrigerant liquid temperature or undercooling degree			
						Undercooling refrigerant liquid temperature [℃]	undercooling degree of refrigerant liquid [K]		
			I	II	III	I	II	III	
A1	7.2	37.7	-	18.3	-	-	0.0	8.3	
A2	7.2	54.4	-	18.3	35.0	-	0.0	8.3	
B	7.0	55.0	15.0	18.0	-	-	0.0	5.0	
C	5.0	55.0	13.0	18.0	-	-	0.0	5.0	
D	4.4	40.6	-	18.3	-	-	0.0	5.0	
E1	0.0	40.0	8.0	18.0	-	-	0.0	5.0	
E2	0.0	50.0	8.0	-	-	-	0.0	5.0	
F	-5.0	45.0	3.0	18.0	-	-	0.0	5.0	
G	-6.7	54.4	-	-	35.0	-	0.0	8.3	
H	-10.0	45.0	-2.0	18.0	-	-	0.0	5.0	
I	-12.2	54.4	-	-	32.2	32.2	0.0	-	
J1	-15.0	35.0	5.0	-	-	-	0.0	5.0	
J2	-15.0	40.0	5.0	-	-	-	0.0	5.0	
J3	-15.0	40.6	-	18.3	-	-	0.0	-	
J4	-15.0	54.4	-	18.3	-	-	0.0	-	
J5	-15.0	55.0	-7.0	18.0	-	-	0.0	5.0	
K1	-23.3	40.6	-	-	32.2	32.2	0.0	-	
K2	-23.3	54.4	-	18.3	32.2	32.2	0.0	-	
L	-25.0	55.0	-	-	32.2	32.2	0.0	-	
M	-28.0	50.0	-20.0	18.0	-	-	0.0	5.0	
N1	-30.0	35.0	-10.0	-	-	-	0.0	5.0	
N2	-30.0	40.0	-10.0	-	-	-	0.0	5.0	
O	-35.0	40.0	-27.0	18.0	-	-	0.0	5.0	
P1	-40.0	35.0	-20.0	-	-	-	0.0	5.0	
P2	-40.0	40.0	-20.0	18.0	-	-	0.0	5.0	
P3	-40.0	40.6	-20.0	18.3	-	-	0.0	5.0	
P4	-40.0	45.0	-20.0	18.0	-	-	0.0	5.0	

Note:
1. Select one of class A-P, suction superheated vapor temperature I-III and refrigerant liquid temperature or undercooling degree I-III, depending on the type of refrigerants and type, capacity and use of compressors.
2. In case of single component or azeotropic mixed refrigerant, the temperature of its suction pressure or discharge pressure is indicated by the saturated temperature. In case of zeotropic mixed refrigerant, the temperature is done by the dew point temperature.
3. If the rated temperature condition of the above table cannot be applied by depending the type of refrigerants and the use of compressors, the changed condition shall be specified.

242

Chapter 11 Testing

(operation stability criterion) are specified in Table 11.3. Accuracy requirements for the test instruments are listed in Table 11.4. As specified in JIS B 8606-4.2, performance test should, whenever possible, be concurrently conducted by two different methods. However, one of such test methods may be skipped if the accuracy of the other test method is verified and guaranteed.

(1) Refrigeration capacity measurement

Refrigeration capacity of a refrigerating system should be calculated based on the refrigerant mass flow rate obtained by one of the nine measurement methods listed in JIS B 8606-8 to -15 and the specific enthalpy of the refrigerant obtained from its thermodynamic property table or the equation of state. As an example, the following paragraphs explain about the secondary refrigerant calorimetry method described in JIS B 8606-8.

Figure 11.1 illustrates how secondary refrigerant calorimetry works[1]. Secondary refrigerant calorimetry is a technique to measure the refrigeration capacity of a compressor in a refrigerating system consisted of a compressor,

Table 11.3 JIS B 8606 Maximum allowance tolerance at static state

Item	Variation against base	Tolerance
Power voltage and frequency	Difference between rated value and measured value	Power voltage: ±3% Frequency: ±1%
Suction temperature at compressor	Difference between saturated pressure at test condition and measured pressure	±1% or ±2 kPa*
Suction temperature at compressor	Difference between test condition temperature and measured temperature	±3K
Discharge pressure at compressor	Difference between saturated pressure at test condition and measured pressure	±1% or ±10 kPa*

*Only when suction or discharge pressure is low.

Table 11.4 JIS B 8606 Accuracy of instruments

Segment	Measurement item	Accuracy Standard deviation
Temperature	Heat medium temperature in calorimeter or cooling temperature in calorimeter	0.06 K
	Other temperature	0.3 K
Pressure	Suction pressure (abs)	1.0%
	Other pressure (abs)	2.0%
Flow rate	Refrigerant liquid and colling water	1.0%
	Refrigerant gas	2.0%
Time	Measurement time	0.1%
Speed	Motor rotation speed, Shaft rotation speed	0.75%
Torque	Motor, shaft	2.5%
Measurement instrument	Integrated data	1.0%
	Current data	1.0%

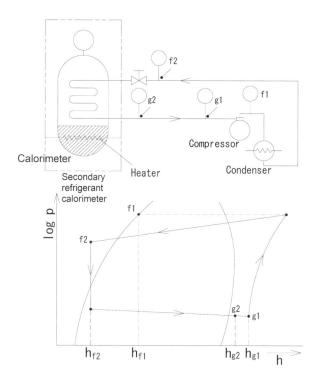

Fig. 11.1 Secondary refrigerant calorimeter method JIS B 8606

a condenser, an expansion valve and a dry-type evaporator. Specifically, it measures the amount of heat received by the evaporator from outside. A calorimeter that contains a sealed volume of secondary refrigerant maintained at saturation pressure serves as the system's dry-type evaporator during the measurement. The heater is immersed in the secondary refrigerant bath at the bottom of the calorimeter. Based on the operating state where thermal equilibrium between the evaporator and the heater is maintained, the refrigerant mass flow rate q_{mf} of the compressor is expressed by Eq. (11.1).

$$q_{mf} = \frac{P + \phi_1}{h_{g2} - h_{f2}} \quad (11.1)$$

P : Electrical power supplied to the calorimeter heater
h_{g2} : Specific enthalpy of refrigerant at the calorimeter outlet
h_{f2} : Specific enthalpy of liquid refrigerant at the expansion valve inlet

Here, thermal leakage ϕ_1 can be expressed, using a predetermined thermal leakage coefficient G, by Eq. (11.2).

$$\phi_1 = G(t_a - t_p) \quad (11.2)$$

t_a : Average ambient temperature
t_p : Refrigerant saturation temperature corresponding to the secondary refrigerant pressure

Based on the above equation, refrigeration capacity of the compressor, ϕ_e, can be calculated from Eq. (11.3).

$$\phi_e = q_{mf}(h_{g1} - h_{f1}) \tag{11.3}$$

h_{g1} : Specific enthalpy of the refrigerant at the compressor inlet

h_{f1} : Specific enthalpy of saturated liquid refrigerant corresponding to the compressor discharge gas pressure

(2) Input power measurement methods

Input power measurement is done by one of the two methods described in JIS B 8606-17.1, the direct method and the indirect method.

The direct method measures the compressor's drive shaft torque directly by using an appropriate measuring equipment, and calculates the amount of input power from the measured torque. The average input power must be obtained within an error tolerance of ±2.5% using the average torque.

The indirect method uses an electric motor with known characteristics to drive the compressor. Electrical input power at the motor terminal is measured to obtain the amount of driving power supplied to the compressor. When a belt transmission system is used, belt transmission loss must be considered in the calculation.

(3) Oil circulation rate measurement

Oil circulation rate should be obtained by the measurement method specified in JIS B 8606 Appendix D or E.

(4) Information to be included in the measurement record

Where applicable, the following information should be included in the measurement record:

- Measurement date, location and weather condition
- Type of refrigerant and the thermodynamic property table used
- Test method, measurement instrument and measurement conditions
- Average readings from the measurement
- Relevant calculation results

JIS B 8606 Appendix B specifies an estimated error analysis method for values calculated from the measurement results.

11.2 Noise Tests

A compressor repeats a cycle of suctioning, compressing and discharging the refrigerant. This accompanies pressure pulsation and load torque variation, which generates discrete noise over a relatively wide frequency range. This may in some cases detrimentally affect human occupancy environment. Therefore, noise test is an important element of compressor performance evaluation. The following sections explain the basic knowledge necessary to conduct a noise test. For more details about noise measurement techniques, measurement equipment and sound power levels, refer to JIS standards listed in Table 11.5..

Table 11.5 JIS standard (Sound level of refrigerant compressors)

Number	Standard name
JIS B 8346	Fans, blowers and compressors − Determination of A-weighted sound pressure level
JIS C 1508	Electroacoustics − Random incidence and diffuse field calibration of sound level meters
JIS C 1509-1	Electroacoustics − Sound level meters − Part 1: Specifications
JIS C 1509-2	Electroacoustics − Sound level meters − Part 2: Pattern evaluation tests
JIS C 1512	Level recorders for recording sound level and/or vibration level
JIS Z 8731	Acoustics − Description and measurement of environmental noise
JIS Z 8732	Acoustics − Determination of sound power levels of noise sources using sound pressure − Precision methods for anechoic and hemi-anechoic rooms
JIS Z 8733	Acoustics − Determination of sound power levels of noise sources using sound pressure − Engineering method in an essentially free field over a reflecting plane
JIS Z 8734	Acoustics − Determination of sound power levels of noise sources using sound pressure − Precision methods for reverberation rooms
JIS Z 8736-1	Acoustics − Determination of sound power levels of noise sources using sound intensity − Part 1 : Measurement at discrete points
JIS Z 8736-2	Acoustics − Determination of sound power levels of noise sources using sound intensity − Part 2 : Measurement by scanning
JIS Z 8736-3	Acoustics − Determination of sound power levels of noise sources using sound intensity − Part 3: Precision method for measurement by scanning

11.2.1 Measurement conditions and equipment

(1) Background noise control

Noise measurement should be conducted in a room (anechoic room, soundproof room etc.) where measures are taken to minimize the effects of other noise sources. Figure 11.2 shows a compressor noise measurement setup in a hemi-anechoic room, where the floor is a reflective surface and the other surfaces are composed of a series of sound-absorbing wedges that do not reflect sound. A number of microphones are positioned in multiple positions around the compressor. In a steady-state noise measurement, noise meter readings with the target noise (A dB) and without the target noise (background noise only; B dB) should indicate at least a 10 dB difference (A - B) for the background noise to be considered ignorable. If the difference is less than 10 dB, the background noise influence cannot be considered ignorable. In that case, a compensation value indicated in Table 11.6[2)] may be added to reading A to estimate the noise level when only the target noise is present.

(2) Measurement equipment

In general, a noise measurement system like the one illustrated in Fig. 11.3 should be used. To accurately determine a noise source and plan countermeasures, it is necessary to fully understand the frequency characteristics of the

Fig. 11.2 Sound power level measurement in hemi-anechoic room

Table 11.6 Difference between measurement noise and background noise [Unit:dB]

Difference between measurement noise and background noise (A-B)	4	5	6	7	8	9
Correction value	−2	−2	−1	−1	−1	−1

Fig. 11.3 Sound level instruments

target noise by using a frequency analyzer. The frequency analysis result is automatically stored and recorded in a data recorder unit or other similar recording device. When using measurement equipment such as microphones and noise meters, their operating temperature and humidity ranges should be checked.

(3) Microphone

Figure 11.4 shows a typical noise measurement microphone. Where windage effect is a concern, use a windscreen like the one shown alongside the microphone in Fig. 11.4. Microphones should be positioned sufficiently away from the wall so as to minimize sound reflection effects. Also, measures should be taken to prevent vibration from being directly transmitted to the microphone. Signal cables should be routed so as to avoid electric and magnetic interferences.

(4) Noise meter

A noise meter is an instrument that quantifies the loudness of audible noise. Noise meters have an audibility correction circuit so as to be able to selectively apply A- or C-weighted frequency responses. A-weighting is generally used for noise measurement regardless of the loudness level, while C-weighting may be used in some cases, for example when measuring an impact noise.

(5) Frequency analyzer

Frequency analyzers are generally divided into two groups depending on the bandwidth to be analyzed. One group of analyzers is the constant ratio bandwidth analyzers including the octave-band analyzer, where the ratio of

Fig. 11.4 Microphone and windscreen

Compressors for Air Conditioning and Refrigeration

the highest and the lowest frequencies of the bandwidth is kept constant regardless of where the bandwidth is located or how wide it is. Therefore, the bandwidth will be wider at higher frequencies. Constant ratio bandwidth analyzers have a high level of correlation with human perception and therefore are suitable to evaluate human occupancy environment. Another group of analyzers is the constant frequency bandwidth analyzers, most typically the FFT analyzer. A constant frequency bandwidth analyzer works with a fixed-width frequency band. Higher frequency resolution can be obtained by setting a narrower band. Constant frequency bandwidth analyzers are suitable for noise source determination and other similar investigation.

(6) Recording devices

Noise meter readings and frequency analysis results should be stored and recorded in a data recorder or other similar recording devices. Check the relationship between the recorder's dynamic range and other range settings and the anticipated target noise level variation. Have the noise meter calibration signal recorded before and/or after the measurement session.

11.2.2 Measurement methods

Noise measurement of the compressor on its own should be conducted by mounting the compressor to a unit noise measurement bench equipped with its own refrigeration cycle. Run the compressor at the predetermined pressure and temperature conditions to measure the noise level and carry out frequency analysis. Noise-generating components of the refrigeration cycle should be located outside the measurement room. If noise due to system piping or support structure vibration is suspected, provide appropriate vibro-isolation (vibro-isolating support or lagging, vibration damper installation) and sound insulation (sound absorbing or insulating material, vibration damper installation) measures. The following sections describe other noise test requirements.

(1) Measurement position

Select a position that best represents the overall compressor noise (for example, 1 m away from the center of the compressor side and at a level 1/2 of the compressor side height). Provide multiple measurement points as appropriate.

(2) Variables to be measured

As a general rule, A-weighted sound pressure level (noise level) should be measured, but non-frequency-weighted

pressure levels may also need to be measured to investigate interaction between the noise source and other physical phenomena.

(3) Reading the instrument indication

If the fluctuation of the indicated value is small, visually take an average reading at the center of the fluctuation range and record such reading. If the indicated value cyclically or intermittently fluctuates, take a sufficient number of fluctuation peak readings and record the average of such readings. If considered necessary, description of the observed fluctuation pattern (period, frequency etc.) should be included in the record report. If the indicated value exhibit erratic and significant fluctuations, investigate the cause of such fluctuation first.

(4) Information to be included in the measurement record

Where applicable, the following information should be included in the measurement record:
- Measurement date and weather condition
- Measurement location and position
- Equipment operating conditions (temperature, humidity, pressure, shaft rotation speed)
- Type and model of the measurement instrument
- Sensory determination etc.

11.3 Vibration Tests

Vibration of a compressor may be caused not only by its pressure pulsation and load torque fluctuations but also by component assembly problems. Vibration measurement therefore can help evaluate compressor assembly accuracy. The following sections explain the basic knowledge necessary to conduct a vibration test. For details about vibration measurement techniques and measurement equipment, refer to JIS standards listed in Table 11.7.

11.3.1 Measurement conditions and equipment

(1) Background vibration control

The ratio of vibration readings with and without the target vibration should ideally be smaller than 1/3 (corresponding to approximately -10 dB of vibration acceleration). If the ratio is larger than 1/3, measures should be taken to reduce the amount of vibration when the target vibration is not present (for example by adding a vibro-isolating support to the equipment base).

Table 11.7 JIS standard (vibration of refrigerant compressors)

Number	Standard name
JIS B 0906	Mechanical vibration − Evaluation of machine vibration by measurements on non-rotating parts − General guidelines
JIS B 0907	Mechanical Vibration of Rotating and Reciprocating Machinery − Requirements for Instruments for Measuring Vibration Severity
JIS B 0908	Methods for the calibration of vibration and shock pick-ups − Basic concepts
JIS B 0909	Shock and vibration measurements − Characteristics to be specified for seismic pick-ups
JIS B 0910	Mechanical vibration of non-reciprocating machines − Measurements on rotating shafts and evaluation criteria − General guidelines
JIS B 0911	Mechanical vibration − Susceptibility and sensitivity of machines to unbalance
JIS C 1510	Vibration level meters
JIS C 1512	Level recorders for recording sound level and/or vibration level
JIS C 1513	Octave-band and third-octave-band analyzers for sounds and vibrations
JIS Z 8735	Methods of Measurement for Vibration Level

(2) Measurement equipment

The measurement equipment to be used for vibration test is the same as the noise measurement equipment described in Section 11.2.1, with the microphone replaced with a vibration pickups and the noise meter replaced with a vibrometer. Similarly to the noise measurement, vibration measurement equipment should be selected by checking their operating temperature and humidity ranges and also the measurable frequency range.

(3) Vibration pickup

Figure 11.5 shows a typical vibration pickup. Normally, a pickup like this one is touched to the compressor body to measure vibration. A commonly used pickup is designed to detect vertical vibration only, but some models are capable of simultaneous three-axis (vertical and horizontal) measurements. When measuring a rotating component or other objects or areas where a contact type pickup cannot be used, a non-contact-type pickup is used. Non-contact type vibration pickups include vortex current type, laser-based type, electrostatic capacitance type and laser-Doppler type. Of these, the vortex current type is the simplest and the most inexpensive.

Vibration pickups can be divided into two groups. One is charge output type where vibration is converted into electric charge, and the other is voltage output type where vibration is converted into and output as voltage. Note that each type requires a different type of vibrometer (amplifier). When using a vibration pickup, be careful that vibration measurement will not be influenced by such factors as the pickup's own weight and installation rigidity, signal cable vibration, wind, and electric and magnetic interferences. Make sure that the pickup's maximum sensitivity line is aligned with the target vibration direction.

(4) Vibrometer (amplifier)

As explained in the vibration pickup section, a charge-output type pickup requires a charge-amplifier type vibrometer and a voltage-output type pickup requires a sensor-amplifier type vibrometer.

11.3.2 Measurement methods

Similarly to noise measurement explained in Section 11.2.2, vibration measurement of the compressor on its own should be conducted by mounting the compressor to a unit vibration measurement bench equipped with its own refrigeration cycle. Run the compressor at the predetermined pressure and temperature conditions to measure the vibration level and carry out frequency analysis. During the measurement, check also the piping sections, the mounting base and other support structures for excessive noise and vibration which could be induced by resonance with the compressor body. The following sections describe other vi-

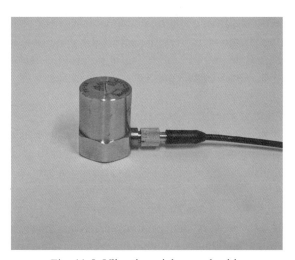

Fig. 11.5 Vibration pickup and cable

bration test requirements.

(1) Measurement position

Select a position that best represents the overall compressor vibration. To fully understand the directional characteristics of vibration, it is desirable to carry out a three-axis measurement.

(2) Variables to be measured

Vibration variables that can be measured in a vibration test include vibration displacement, vibration velocity and vibration acceleration. In the case of most machines, vibration velocity is the variable that best characterizes the machine's vibration in a wide operating speed range. However, in such situations where low frequency vibration is dominant, vibration displacement should be measured. If high frequency vibration is dominant, vibration acceleration should be measured.

(3) Reading the instrument indication

If fluctuation of the indicated value is small, visually take an average reading at the center of the fluctuation range and record such reading. If the indicated value cyclically or intermittently fluctuates, take a sufficient number of fluctuation peak readings and record the average of such readings. If considered necessary, description of the observed fluctuation pattern (period, frequency etc.) should be included in the record report. If the indicated value exhibit erratic and significant fluctuations, investigate the cause of such fluctuation first to avoid potential danger.

(4) Information to be included in the measurement record

Where applicable, the following information should be included in the measurement record:

- Measurement date and weather condition
- Measurement location, measurement position and direction
- Equipment operating conditions (temperature, humidity, pressure, shaft rotational speed)
- Type and model of the measurement instrument
- Sensory determination etc.

11.4 Reliability Tests

Reliability is defined in JIS Z 8115 as "the capability of an item to perform its required functions for a given period of time and under given conditions". Here, the term "item" refers to any or all of components, devices, systems or elements for which reliability is sought. "Reliability test" refers to testing operations to evaluate the reliability characteristics of a subject item in order to assure that they satisfy the required criteria.

11.4.1 Types of reliability tests
(1) Laboratory reliability test

Laboratory reliability tests are conducted under specified and controlled conditions to measure, determine and classify the reliability-related characteristics and properties of a subject item. Laboratory reliability tests must be carried out under appropriately selected and specified operational and environmental conditions with the actual user site stress taken into consideration.

(2) Durability test

A durability test is conducted over a specified period of time to investigate how a specific stress and its continuous or repetitive impression will affect the characteristics of the subject item. When such test is continued until the item fails, the test may also be called "life test".

(3) Accelerated test

To evaluate the item's response to a stress in a shortened time or to see an increased response within a given period of time, an accelerated test applies an amount of stress that is far beyond what is expected during use within that period of time. An important requirement in such accelerated test is that the degree of acceleration must be quantified in advance through physical and chemical evaluation.

(4) Safety margin test

A safety margin test determines how much safety margin can be expected of a subject item in resisting a given stress. It is important that a safety margin test is implemented by correctly identifying and reflecting anticipated variations. Testing should start from the actual use condition, after which the degree of stress should be progressively increased until a specified lifetime point is reached.

(5) Field test

Field tests are implemented to determine the reliability characteristics of a subject item under actual use condition which may not be adequately evaluated in a laboratory environment. Field tests are typically run at the user site. It is critical to fully record the circumstantial information related to the test, including the operating conditions, the environmental conditions, usage conditions, operating hours and the exact situation when a failure is observed during the test.

11.4.2 Reliability test implementation

Reliability testing of an item should be started by running general tests such as safety margin test and accelerated test, mainly focusing on newly introduced structural or material features, so as to identify any potential reliability issues the item might have. Then, based on the findings from such initial testing, plan and implement more customized tests specifically adopted to the item's reliability determination needs, using carefully designed testing conditions including the sample quantity.

Tested samples that failed must be thoroughly investigated to determine the reliability failure factors and develop countermeasures, before repeating the test again to evaluate the effectiveness of the countermeasures. Samples that did not fail must also be checked by disassembly investigation afterward, so as to precisely measure and evaluate any changes observed and to see if any sign of potential failure is present.

To assure the design lifetime of an item, reliability test should ideally be continued through to the end of its design life. If that is not possible due to development timeline and other limitations, such verification that the item does not exhibit any progressively worsening deterioration of reliability-related parameters should be made in place of end-of-life testing.

Failure rate assurance should be implemented by determining the number of accumulated component operating hours that corresponds to the target failure rate and testing the appropriate number of samples accordingly.

11.4.3 Reliability data analysis

"Reliability data analysis" refers to a statistical technique to obtain quantified reliability estimation of a subject item from its reliability test data.

Statistical analysis must be based on a statistical theory and requires complex and time-consuming computations. To facilitate the analysis, graphical analysis method using a specially ruled "probability paper" is often used. One of the most convenient and widely used probability papers is the Weibull paper.

Effective use of a Weibull paper will allow not only estimating the average life and reliability of the item but also predicting its anticipated failure modes (initial failure, accidental failure, wear-induced failure) to help determine where the future reliability assurance activities should be focused.

11.4.4 Reliability evaluation

Reliability evaluation should be implemented by analyzing the data obtained from the above-described reliability tests, including a detailed investigation of disassembly inspection results and looking into any progressively worsening tendency.

Samples that failed must be scientifically analyzed to determine the failure cause. Findings from such analysis must be fed back to the development project.

11.5 Pressure Tests

Compressor operation entails high pressure inside the casing which potentially leads to casing rupture or other dangers. In addition to complying with all laws and standards listed in Table 11.8, manufacturers must carry out exhaustive safety verification tests to assure the safety of a compressor product. Working fluid leakage must be checked by practical testing.

Pressure test procedures are specified in details by JIS B 8620 and B 8240. The following sections provide an overview of the pressure test procedures. For more details, refer to the JIS standards.

11.5.1 Pressure resistance test

Pressure resistance test must be conducted as a 100% test by the following method:

1) Test pressure must not be less than 1.5 times the design pressure.

2) The test must be a liquid pressure test with air inside completely removed. Pressure must be gradually increased to the test pressure and held there for least one minute. Afterward, check the compressor for damage, leakage and any unacceptable deformation.

3) If the design pressure is less than 0.2 MPa, gas pressure test may be conducted as an alternative to a liquid pressure test. Air or other inert gas must be used for such pressure test. Provide sufficient safety measures to assure

Table 11.8 Laws and standards concerning refrigerant compressors (in Japan)

High pressure gas safety acts	Act No. 204
[Safety ordinances]	Refrigeration safety ordinance and standards concerned with refrigeration safety ordinance
[Testing standards]	Testing standards for refrigeration equipment
[Facilities standards]	Facilities standards for refrigeration equipment (KHKS-0302 series)
JIS B 8620	Safety code for small refrigerating equipment
JIS B 8240	Construction of pressure vessels for refrigeration

the safety of people and equipment during the test.

If manufacturing quality control is good enough so that a fully consistent product quality can be relied upon, the following strength test on one sample may be used to omit the 100% pressure resistance test for all the remaining production pieces belonging to the same lot.

11.5.2 Strength test

On the condition that a fully consistent product quality can be relied upon, the following strength test may be done on a randomly taken sample from the production run.

1) One test sample should be taken to represent a single lot, where all the production pieces belonging to that lot are manufactured in the same production plant, by the same production method using the same material, and have exactly the same shape and structure.
2) The test pressure must not be less than three times the design pressure.
3) The test must be a liquid pressure test with air inside completely removed. Pressure must be gradually increased to the test pressure and held there for least one minute. Afterward, check the compressor for damage, leakage and any unacceptable deformation.
4) The tested sample must not be used for production.

11.5.3 Airtightness test

Airtightness test must be conducted as a 100% test by the following method:

1) Test pressure must not be less than the design pressure.
2) Charge air or inert gas into the compressor to gradually increase the pressure to the test pressure. With the test pressure achieved and held, place the compressor sample in water or apply a bubbly solution over the surface. No bubbles should be observed. Alternatively, use a gas leak detector to verify that the sample is not leaking.

11.6 Other Tests

To assure that a compressor product performs fully satisfactorily, many verification tests other than the above-described ones must be conducted.

Table 11.9 lists specific examples of such tests. As some of those tests do not have a unified industrial standard, an optimized test procedure must be developed according to the customer's actual use conditions, which needs to be done based on careful discussion with the customer prior to testing.

Table 11.9 Other confirmation test of refrigerant compressors

Test name	Contents
Drop test	Test for confirming the effects on the function of the compressor by dropping impact
Transportation test	Test for confirming the effects on the function of the compressor by vibrations during such as shipment transportation
Weather resistance test	Test for confirming the influence by sunlight, weather, temperature and humidity, salt
Electric test	Test of starting current, winding resist, insulation resistance, dielectric strength Electrical Appliances and Material Safety Act (Japan)

References

1) JAR : "JAR Handbook 5 Edition", Vol.II, p.545, (1993).
2) M.Kabira : "Enclosed Refrigerating Machine", p.84, (1981).

Chapter 12 Measurement Technologies

12.1 Measurements of Basic Physical Quantities

There are many physical quantities to be measured, such as length, angle, shape, electric current, voltage, power, light, sound, chemical properties, human five senses and so on. However, it is practically impossible to cover all measurement technologies about these quantities in this chapter. Considering the rapid technological progress, there is a risk that any information contained herein soon becomes obsolete. Therefore, this chapter will focus only on most essential measurement technologies that are employed in the research and development of refrigeration and air conditioning compressors. The basic measurement activities performed in the field of compressor engineering are divided into three groups: 1. Measurements of basic physical quantities such as temperature, pressure, vibration, noise, flow rate, electric current, voltage and power; 2. Tribological measurements (friction force, oil film thickness, presence of oil etc.) that involves more advanced, indirect type of measurement; and 3. The determination of the amount of lubrication oil that discharged from the compressor with refrigerant.

Most of the basic measurement technologies and techniques about the above mentioned quantities are incorporated in national standards, such as JIS of Japan and international standards, such as ISO, IEC and others. About these standardized techniques, this textbook offers only a minimum amount of information, citing the standard numbers to be referred to. JIS Z 8103-2000 "Glossary of terms used in measurement" provides terms and their definitions related to basic measurement technologies and techniques.

On the other hand, efforts are made to offer as much information as possible about more advanced and compressor-specific measurement techniques, citing actual examples where they have been employed for compressor measurements.

For more details about compressor performance tests and noise and vibration tests, refer to the respective chapters.

12.1.1 Temperature measurement

Temperature measurement techniques are roughly classified into contact-based and non-contact-based measurements. Contact-based measurement refers to a technique, such as thermocouple measurement, that determines the temperature of an object by touching a measuring device to it. On the other hand, non-contact-based measurement is a technique like radiation thermometer that determines the temperature of an object based on its thermal radiation.

Standards for general (non-compressor-specific) temperature measurement techniques are provided in JIS Z 8710-1993 "Temperature measurement - General requirements", which describes the application ranges and characteristics of various types of thermometers. Figure 12.1 shows the regular measurement range for a contact-based thermom-

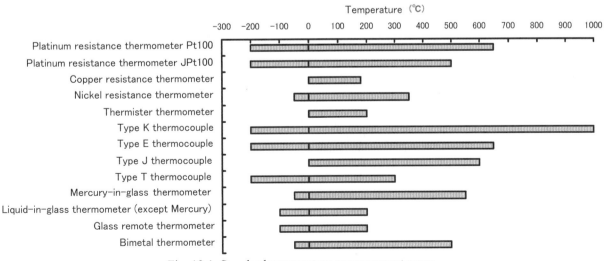

Fig. 12.1 Standard temperature measurement range

eter commonly used for compressor measurement.

JIS Z 8703-1983 "Standard atmospheric conditions for testing" provides a set of criteria for the standard environmental condition for a temperature measurement location. For each of the standards mentioned herein, the latest version of the relevant standard should be referred to whenever a temperature measurement is required, to assure that measurement will be done using an appropriate method and in an appropriate environment.

Although radiation thermometers are not commonly used for compressor-related temperature measurements, JIS C 1612-2000 "Test methods for radiation temperatures" provides criteria for temperature measurements using a radiation thermometer, which should be referred to as required.

12.1.2 Thermocouples and resistance thermometer sensors

Thermocouples and resistance thermometer sensors are often used for compressor temperature measurements. The reasons for this may be that many measuring equipment to which thermocouples and resistance thermometer sensors can be connected are commercially sold and easily available, and that these devices allow the measurement results to be electronically recorded and stored easily.

JIS Z 8704-1993 "Temperature measurement - Electrical methods" provides criteria concerning the use of thermocouples and resistance thermometer sensors.

(1) Thermocouples

The operating principle of a thermocouple is based on the Seebeck effect, where a temperature difference between two contact points of dissimilar metals (hot and cold contact points) produces an electromotive force. Thermocouples are classified in various types listed in Table 12.1 according to the metallic materials that are used to make the contact elements.

When the surface temperature of a compressor housing needs to be measured, thermocouple elements are usually joined by electric spot welding or being twisted and soldered together. Alternatively, commercially available thin-film thermocouples can be used.

Attachment of a thermocouple to the measurement surface is often done by attaching the thermocouple to the surface either by spot welding or gluing with a thermal adhesive (aluminum nitride or other thermally conductive metal mixed into the base glue) and covering the joint with an insulation material. If the hot contact point needs to be electrically insulated, an adhesive that contains ceramic slurry is used.

Table 12.1 Type of thermocouple

Type code	Anode	Cathode
B	Pt, Rh:30%	Pt, Rh:6%
R	Pt, Rh:13%	Pt
S	Pt, Rh:10%	Pt
N	Ni, Cr, Si	Ni, Si
K	Ni, Cr (Chromel)	Ni (Alumel)
E	Ni, Cr (Chromel)	Cu, Ni (Constantan)
J	Fe	Cu, Ni (Constantan)
T	Cu	Cu, Ni (Constantan)

Measurement inside a piping section or inside the compressor housing can be done by a sheath type thermocouple. The sheath type thermocouple comprises a pair of thermocouple elements housed in a rigid sheath. Figures 12.2 and 12.3 show a grounded-sheath type and an electrically insulated-sheath type thermocouples, respectively. The thermocouple sheath is typically austenitic stainless or heat-resistant alloy. In the common temperature measurement range that involves refrigeration compressors (up to 250°C), no significant measurement error will result from direct insertion into the measurement atmosphere due to relatively small thermal radiation effect. However, use in the proximity of a surface that is heated to 500°C or above (for example, areas near a heater unit) requires the sheath type thermocouple to be protected by a shielding feature that blocks heat radiation.

To minimize measurement errors, compensating lead wires are used for connection between the thermocouple unit and the measuring equipment. Note that the compensating lead wire is compatible with the type of thermocou-

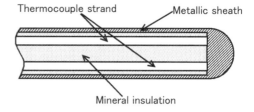

Fig. 12.2 Grounded sheath type thermocouple

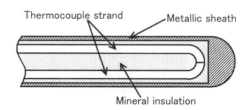

Fig. 12.3 Insulated sheath type thermocouple

Chapter 12 Measurement Technologies

ple used. Compensating lead wires have a specific polarity, which must also be checked for correct connection at the time of installation.

(2) Resistance thermometer sensors

Resistance thermometer sensors are a device that detects changes in electrical resistance caused by temperature change. A platinum element is used as the resistance thermometer sensor. There are three types of resistance thermometer sensor wiring configurations:

- Two-wire configuration

In a two-wire configuration, the resistance thermometer sensor unit and the measuring equipment are connected with two wires. An advantage of the two-wire configuration is that it works with the simplest wiring, but it requires compensation for lead wire resistance which is added to the sensing element resistance. In many cases, resistance variations that occur due to temperature changes in the lead wires cannot be adequately compensated, resulting in inferior measurement accuracy compared to other wiring configurations. Due to these disadvantages, two-wire configuration is generally not used.

- Three-wire configuration

This is the most commonly used wiring configuration for industrial resistance thermometer sensors. If used with a measuring equipment that is designed for full compatibility with three-wire configuration, it is possible to reduce lead wire resistance effects to a practically ignorable level. In order to actually reduce wiring resistance to such ignorable level, installation must be done in such a way to assure that all the three wires will be at the same temperature.

- Four-wire configuration

A system designed for four-wire configuration has two voltage terminals and two current terminals, enabling measurement without being influenced by lead wire resistance. Disadvantages of four-wire configuration are complex wiring and higher sensor cost. Use of this type of wiring configuration is limited to special applications where precision temperature measurement is required.

The following engineering standards apply to thermocouples and resistance thermometer sensors. These standards provide calibration procedures and other detailed information that are not covered in this textbook. The latest version of the relevant standard should be referred to when actually using thermocouples or resistance thermometer sensors for temperature measurement.

- JIS C 1602-1995: Thermocouples
- JIS C 1604-1997: Resistance thermometer sensors
- JIS C 1605-1995: Mineral insulated thermocouples
- JIS C 1610-1995: Extension and compensating cables for thermocouples
- JIS C 1611-1995: Thermistor for temperature measurement

(3) Other devices

Thermo label: A single use indicator that can be used to measure the surface temperature of a compressor. Thermo labels may be selected when only the maximum temperature within a given period of time needs to be known at the lowest cost. Thermo labels are divided into melting-point-based irreversible type and liquid-crystal-based reversible type. Each type should be selectively used depending on the purpose of use.

Thermo viewer: A radiation thermometry system that captures infrared rays radiating from the surface of an object and converts them into graphic temperature information using imaging elements. Some thermo viewers are capable of combining the temperature information with visible light camera images. The most important characteristic of a thermo viewer is the capability to record temperature distribution as a surface information. Additional image processing will allow displaying highest and lowest temperature regions and other special display functions. When a glass pane is installed between the measurement object and the thermo viewer camera, germanium glass is commonly used for its infrared absorption characteristics, or a thick quartz glass pane may be used if it also needs to serve as a pressure tight partition. In the latter case, tests must be run with and without the glass pane to establish a calibration curve before running the actual measurement.

12.1.3 Pressure measurement
(1) Aneroid pressure gauges

Like temperature, pressure is a property that needs to be measured more often. The most commonly used pressure measuring device is the aneroid pressure gauge (Bourdon tube gauge) shown in Fig. 12.4.

"Aneroid pressure gauge" is a collective term that refers to all the types of pressure measuring devices that use an elastic pressure-sensing element to convert pressure into displacement and mechanically enlarge that displacement for display. "Bourdon tube gauge" refers to a type of aneroid pressure gauge where a Bourdon tube is used as the sensing element. The aneroid pressure gauge is advantageous in that it can convert pressure into mechanical displacement in a simple manner without requiring external power supply, but the measurement data is not very compatible with electronic measurement data management. Some types of Bourdon tube gauge are combined with a switch mechanism so as to trigger the switch to activate when an

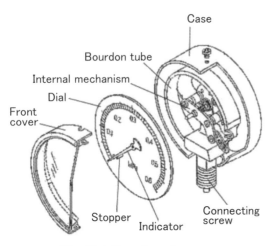

Fig. 12.4 Bourdon tube gauge

upper or lower pressure limit is reached.

JIS B 7505-1-2007 "Aneroid pressure gauge, Part 1: Bourdon tube pressure gauge (explanation included)" provides criteria about aneroid pressure gauges.

(2) Pressure transducers

A pressure transducer is a pressure-sensing device that converts the displacement of its elastic sensing element into electrical signal. The term "transducer" usually refers to a module in which the pressure sensing and conversion mechanism is combined with a signal amplifier circuit. Conversion of displacement into electrical signal is generally achieved using a strain gauge.

When using an aneroid pressure gauge or a pressure transducer for static pressure measurement in a space where fluid flow is present, it is important to position the device with its opening at the right angle to the direction of the flow, which helps minimize measurement errors. When the device setup is slanted as shown in Fig. 12.5, the reading will be influenced by dynamic pressure and will therefore be different from true pressure value P_0. To accurately compensate for this error, a calibration curve proportional to the flow rate, like the ones shown in Fig. 12.6, must be established. If the T-shaped pressure tap in the flow path is tilted toward the downstream side, calibration curve "a" will be obtained. If the tap is mounted the other way around so that it is tilted toward the upstream side, curve "b" will

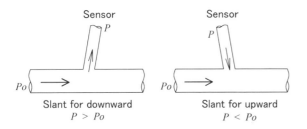

Fig. 12.5 Effect of connecting tube slant on pressure

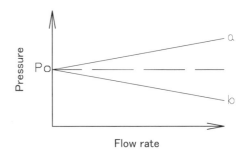

Fig. 12.6 Calibration curve method

be obtained. Compensation of measurement values will be implemented on the assumption that true pressure P_0 will be approximately at the center of these two calibration curves. (In reality, there is no guarantee that curves "a" and "b" will be symmetrical about true pressure P_0. Therefore, this technique should be used as a tool to roughly determine the degree of dynamic pressure influence).

Turbulent flow in a pipe contributes to measurement errors. To minimize such in-pipe turbulence, there should desirably be straight pipe sections both immediately upstream and downstream of the measuring point, whose length is five to ten times larger than the diameter of the pipe.

(3) Pressure sensors

There are various types of pressure sensors, including ones that use a quartz or potassium sodium tartrate element with large piezoelectric effect, or other semiconductor-based types where the p-n junction band width indicates changes in electrical resistance when pressure is applied.

Figure 12.7 shows a piezoelectric quartz pressure sensor. These quartz sensing elements are used in many high precision pressure sensors as quartz exhibits a good linear proportionality to pressure input and the effect of temperature could be made extremely low given a favorable crystal grain orientation. As characteristics of the sensor, all the raw pressure readings from a piezoelectric quartz sensor are AC-based and therefore DC compensation is required for conversion into absolute values. As the piezoelectric action produces only an extremely small amount of electrical charge, the measurement system impedance must be kept high. Therefore, it is necessary to carefully control the

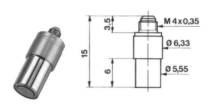

Fig. 12.7 Quartz pressure transducer

cleanliness of electrical contacts.

On the other hand, semiconductor-based pressure sensors are advantageous in that they produce greater changes in electrical resistance in response to pressure changes and therefore does not require an expensive processing system. However, semiconductor-based sensors exhibit significant output shift with temperature changes and also their output response to pressure changes includes nonlinearity.

When using a piezoelectric quartz pressure sensor for compressor pressure measurement, make sure that the sensor's acoustic resonance frequency, that is governed by the volume and length of the pressure lead tube and sensor adapter and the type of fluid used, will not fall in the frequency range that needs to be measured. Figure 12.8 shows a compression chamber measurement setup on a scroll compressor where multiple sensors are positioned on the fixed scroll end plate[1].

As each compression chamber in a scroll compressor moves from the outer edge of the scroll toward the center, a series of sensors must be provided so as to be able to track pressure changes during the compression cycle from suction to discharge by combining output signals from these sensors. Screw compressors, which entail compression chamber movement like scroll compressors, also require multiple sensors to track pressure changes in the compression chamber. On the other hand, a reciprocating compressor only needs a single sensor per cylinder to track pressure changes over the entire compression cycle from suction to discharge. In the case of rotary compressors, two sensors per cylinder are used, one for suction and the other for compression to discharge, to track pressure changes.

Figure 12.9 shows data obtained from the scroll compressor compression chamber measurement shown in Fig. 12.8, where outputs from seven pressure sensors are combined and plotted on a single diagram. The measurement result can be expressed as a P-V diagram, shown in Fig. 12.10, which describes the relationship between volume V

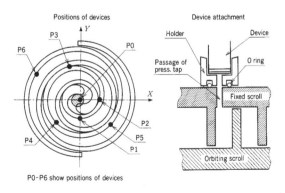

Fig. 12.8 Installation of pressure sensor[1]

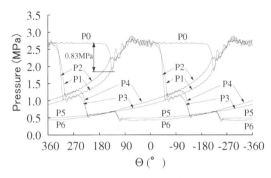

Fig. 12.9 Measurement result by using pressure sensor[1]

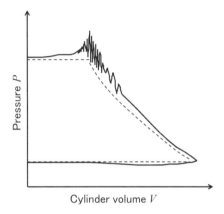

Fig. 12.10 PV diagram[1]

and pressure P. The area surrounded by the P-V curve corresponds to the amount of compression work done, W.

The ratio of the power obtained by multiplying the amount of work represented on the P-V diagram with the number of compressor rotations to the amount of power that would be consumed to compress the refrigerant through a fully isentropic process from the compressor inlet temperature and pressure state, or "theoretical power", is called "indicated efficiency". The indicated efficiency value is commonly used for compressor loss analysis.

12.1.4 Flow rate measurement

To obtain the volumetric efficiency of a positive displacement compressor used for refrigeration or air conditioning purpose, refrigerant flow measurement is required. Flow rate measurement of supplied lubrication oil is also required to investigate the lubrication condition inside the compressor.

Figure 12.11 provides the classification of various types of flowmeters. Flowmeters can be largely divided into two groups; volumetric and mass flowmeters. Volumetric flowmeter is further divided into direct volume measurement, where a measuring container with a known calibrated volume is used for measurement, and transduction-based measurement where a physical quantity (frequency, differential

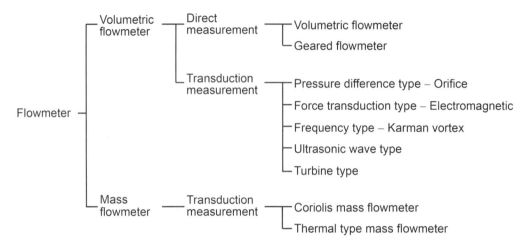

Fig. 12.11 Classification of flowmeter

pressure, force, etc.) that is proportional to volumetric flow rate is measured and converted into a flow rate.

The most commonly used type of flowmeters are those that works on differential pressure (orifice plate, nozzle, etc.). Differential pressure flowmeter is commonly used to measure the amount of water or air flow. However, use of a differential pressure flowmeter to measure refrigerant flow would require checking if the Reynolds number range where the device is calibrated is appropriate for the proposed use condition. Also, such use must be accompanied by accurate pressure and temperature measurements both upstream and downstream of the flowmeter. In addition, care must be taken to thermally insulate the measurement area from the outside and also to evaluate and control the degree of thermal conduction across the flowmeter. Due to these difficulties, use of orifice plate-type flowmetry, which is a very common measurement technique for air and water flow measurement, is usually avoided in refrigerant flow measurement. Almost no instance of orifice plate flowmetry for refrigerant measurement is cited in academic papers.

As a simpler technique to measure refrigerant flow, volumetric turbine flowmeters, which count the number of turbine blade rotations proportional to the flow speed, are commonly used. One particular area, where the volumetric turbine flowmeters are often used, is the flow measurement in a refrigerant line that injects liquid refrigerant into the cylinder for the purpose of lowering the discharge gas temperature of the compressor ("liquid injection line"). These volumetric turbine flowmeters have much less pressure loss than the later described Coriolis flowmeters, and therefore their use will not affect the refrigerant circuit characteristics as much as Coriolis flowmeters do.

Other types of flowmetry that can be used for refrigerant flow measurement include the detection of the Karman vortex frequency that is proportional to the flow speed. This technique measures volumetric flow rate based on the principle that the Karman vortex frequency is directly proportional to both the Strouhal number and the flow speed and is also inversely proportional to the width of the vortex-producing object. The vortex frequency will be detected by piezoelectric or ultrasonic sensing.

On the other hand, mass flowmeters are dominantly Coriolis force-based, which measures the amount of Coriolis force and converts it into a flow rate. Coriolis flowmeters will be described later in details.

Other types of mass flowmeters include thermal type meters, which however are inherently not suited for measurements in a refrigeration system where localized temperature distribution often induces pressure changes. There is almost no published example where a thermal type mass flowmeter is used for refrigeration compressor measurement.

Flow rate readings obtained from a Coriolis or turbine flowmeter would include lubrication oil dissolved in the refrigerant. To obtain the refrigerant-only flow rate, flowmeter readings must be corrected by deducting the amount of lubrication oil contained in the refrigerant. To determine the refrigerant-and-lubrication oil solubility that is required for such correction, a sampling cylinder with shutoff valve is commonly used.

A procedure of sampling cylinder method to measure the refrigerant solubility in the lubrication oil is basically the same as that to measure the lubrication oil solubility in the refrigerant. Oil-in-refrigerant measurement requires more accurate measurement as the amount of oil that remains in the cylinder is only several percent of the total amount. The sampling cylinder-based solubility measurement procedure is given below:

(a) Evacuate the sampling cylinder thoroughly before use, and then weigh the cylinder.
(b) Connect the cylinder to the refrigeration circuit, and open the shutoff valve to suck liquid refrigerant containing lubrication oil into the cylinder, together.
(c) Close the valve, and weigh the cylinder.
(d) Slightly open the valve so that refrigerant will slowly vaporize and escape to the outside, which will remain lubrication oil only in the cylinder.
(e) Evacuate the cylinder again to remove any residual refrigerant dissolved in the oil, and then weigh the cylinder again.
(f) The ratio of the last remaining lubrication oil weight to the initial sampling weight is the solubility.

(1) Coriolis mass flowmeters

Coriolis mass flowmeters are commonly used for refrigerant mass flow rate measurements. Figure 12.12 shows the structure and the operating principle of a Coriolis mass flowmeter.

Coriolis force arises when fluid passes through a vibrating U-shaped piping section. As the direction of Coriolis force is dependent on the direction of the flow, a U-shaped flow generates a twisted force. The degree of this twist angle is proportional to the mass flow rate. Therefore, measuring the twist angle enables calculating the mass flow rate. The vibration frequency should be set so that it coincides with the natural frequency determined by the mass of the U-shaped pipe and the mass of the fluid inside. As the mass and the internal volume of the U-shaped pipe is known, fluid density can be calculated from the resonance frequency, which enables to output the volumetric flow value. This measurement technique can be applied to a gas-only or liquid-only phase flows, while it is difficult to apply this technique to a gas-liquid two-phase flow.

The Coriolis flowmeters are divided into single pipe type and double pipe type. The double pipe type is advantageous in that pipe inertia can be canceled out by vibrating the two pipes in mutually opposed phases and that any external vibration can also be canceled. However, condition in the two pipes must be in an exactly identical state. Also, greater error will occur when measuring a gas-and-liquid mixed phase flow. The double pipe type is generally more expensive. On the other hand, the single pipe type is inexpensive but is more easily influenced by external vibration. Therefore, care must be taken to assure good installation. To obtain larger Coriolis force, flow speed through the flowmeter must be increased. This is often done by designing the vibrating pipe with a relatively small cross section. This, however, may increase the amount of pressure loss through the flowmeter.

JIS B 7555-2003 "Method of flow measurement by Coriolis meters (mass flow, density and volume flow measurement)" provides criteria concerning the use of a Coriolis flowmeter.

(2) Relevant standards

The following industrial standards apply to flow measurements:
- JIS B 7554-1997: Electromagnetic flowmeters
- JIS B 7555-2003: Method of flow measurement by Coriolis meters (mass flow, density and volume flow measurement)
- New JIS Z 8762-1-2007: Measurement of fluid flow by means of pressure differential devices inserted in circular cross-section conduits running full − Part 1: General principles and requirements
- New JIS Z 8762-2-2007: Measurement of fluid flow by means of pressure differential devices inserted in circular cross-section conduits running full − Part 2: Orifice plates
- New JIS Z 8762-3-2007: Measurement of fluid flow by means of pressure differential devices inserted in circular cross-section conduits running full − Part 3: Nozzles and Venturi nozzles
- New JIS Z 8762-4-2007: Measurement of fluid flow by means of pressure differential devices inserted in circular cross-section conduits running full − Part 4: Venturi tubes
- JIS Z 8765-1980: Method of flow measurement by turbine meters
- JIS Z 8766-2002: Vortex flowmeters − Methods of flow measurement
- JIS Z 8767-2006: Measurement of gas flow by means of critical flow Venturi nozzles

12.2 Other Measurement Technologies and Applications

When a measurement inside the compressor is required, it is often difficult to directly measure the target physical

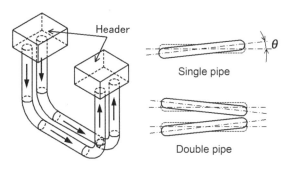

Fig. 12.12 Coriolis mass flowmeter

quantity due to restrictions regarding the sensor size or the presence of refrigerant atmosphere. In such case, the target physical quantity must be converted into another measurable physical quantity, while a relationship between the two quantities must be correlated before actual measurement. Load and torque in particular are often obtained from strain measurements, examples of which are cited in Reference 6). In these types of measurements, additional post-measurement analyses such as separating the effects of temperature and pressure, evaluating crosstalk components with misprediction of loading direction and so on greatly influence the measurement reliability.

Most of applied measurements are related to tribology (oil film pressure and thickness at journal bearing, refrigerant solubility, etc.). Measurements inside the compressor are often required for the purpose of assurance of compressor reliability despite the difficulty to use commercially available sensors. Therefore it is quite common that an original sensor or a new measurement method needs to be developed to carry out a specific measurement inside the compressor. Examples of such development are cited in References 3) to 5) and 7) to 13).

An effective internal measurement technique is the visualization of the inside of the compressor. It used to be that the inside of the compressor can only be viewed from a sight glass. Now, recording with a high speed camera and analyzing the images obtained make it possible to quantitatively evaluate the flow conditions inside the compression chamber.

Reference 2) cites a measurement setup, shown in Fig. 12.13, where 24 sight glasses are provided on the compressor housing to record the movement of refrigerant gas and lubrication oil droplets inside the housing with high speed cameras. Processing of the images obtained enabled the calculation of velocity vectors as shown in Fig. 12.14 for quantitative evaluation.

Concerning measurement of refrigerant solubility in the lubrication oil, Figs. 12.15 and 12.16 show sensors cited in Reference 5) that are employed for the capacitance-based measurement of refrigerant concentration in refrigeration oil. These sensors, developed by the authors of Reference 5) themselves, are designed to work based on the difference in electric permittivity between refrigerant and refrigeration oil.

Figure 12.17 shows an experimental measurement setup, cited in Reference 6), where the drive torque of a rotary compressor is directly measured. A strain gauge is installed to the shaft section between the motor rotor and the compression mechanism, and the gauge's output signals are

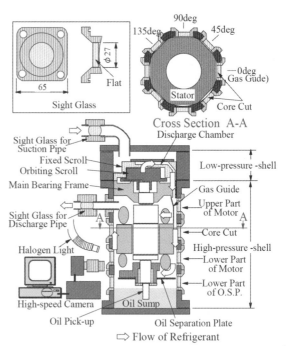

Fig. 12.13 Visualization of the inside of Compressor[2]

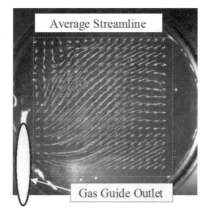

Fig. 12.14 Velocity vector of compressor internal flow[2]

extracted to the outside through a slip ring at the top end of the shaft. For this measurement to work, the relationship between torque and the amount of strain must be established in advance. As long as that requirement is satisfied, this technique is quite advantageous in that the sensor can be installed in a narrow space just where it is required and that torque can be directly measured inside an actually operating compressor.

Other load measurement examples include the ones cited in References 8) and 9) and are shown in Figs. 12.18 and 12.19, both are related to thrust bearing friction load measurement inside a scroll compressor. Both setups are part of scientific research programs where vertical and horizontal loadings are measured at the same time to determine the

Chapter 12 Measurement Technologies

Fig. 12.15 Cylindrical sensor[5]

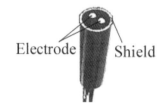

Fig. 12.16 Shielded needle sensor[5]

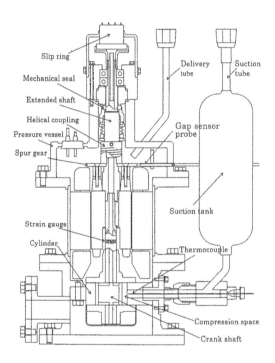

Fig. 12.17 Measuring equipment of driving torque for rotary compressor[6]

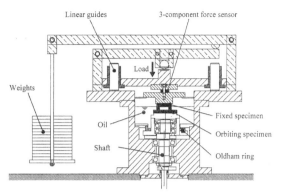

Fig. 12.18 Measuring equipment of thrust load for scroll compressor (1)[8]

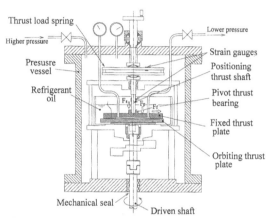

Fig. 12.19 Measuring equipment of thrust load for scroll compressor (2)[9]

thrust load will induce a significant displacement that influences the motion behavior of the orbiting scroll, possibly resulting in the detection of an incorrect frictional force that differs from what would be measured from a normally operating compressor.

Component separation error (crosstalk error) refers to a situation where a horizontal load component is detected even when only vertical load is given. In addition to possible sensor errors, such component force detection may also be caused by inaccurate equipment assembly or heat distortion. To cope with these types of errors to assure measurement reliability, a calibration curve must be established in advance so as to be able to compensate the readings afterward.

As an example of oil film thickness and oil film condition investigation in sliding parts, Reference 11) cites an oil film measurement conducted on the bushes of a swing rotary compressor. Figure 12.20 shows a displacement sensor embedded inside the bush. Figure 12.21 shows the position relationship between the bush and the blade. The bush-blade distance is measured by small displacement sensors embedded inside the bush. Since the distance to be measured is

friction coefficient. The equipment is designed to simulate the orbiting movement of the scrolls. The eccentric pin at the end of the shaft drives a member corresponding to the orbiting scroll into an orbiting motion.

In Fig. 12.18 setup, a balance weight load and two load components orthogonal to that make up a total of three load components, which are measured by a single piezoelectric quartz sensor. In Fig. 12.19 setup, frictional force is measured by a strain gauge fitted to the thrust load shaft.

Critical factors to be controlled in these load measurements include sensor rigidity and component separation errors (crosstalk errors). If the sensor rigidity is insufficient,

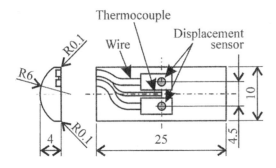

Fig. 12.20 Displacement sensor of swing-bush [11]

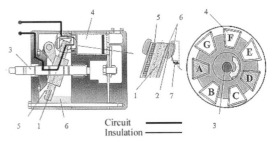

1: Swashplate   2: Shoe   3: Shaft
4: Piston   5: Insulating coat   6: oil
7: Spring

Fig. 12.22 Circuit of inside compressor [12]

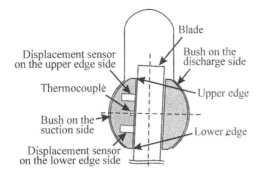

Fig. 12.21 Sensor arrangement for oil film thickness measurement [11]

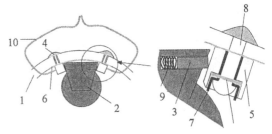

1. Compressor shell   2. Piston   3. Brush
4. Bond   5. Electrode   6. PET film
7. Heat contraction tube   8. Bolt   9. Spring
10. Lead wire

Fig. 12.23 Bush conduction [12]

extremely small around 30 μm, the resulting output voltage changes will be in the range of 0.2 - 0.3 V. This range of the output changes is approximately on the same order as that resulting from temperature changes. To address this issue, a temperature-sensing thermocouple is embedded near the sensor to provide temperature compensation.

Similarly, Fig. 12.22 shows a conduction-based measurement of oil film formation condition between the shoe and the swash plate in a swash plate compressor, cited in Reference 12). Conduction-based measurement is a measurement technique that works based on the relationship between oil film formation and changes in electrical resistance, where electrical resistance will be greater when oil film is present and will be smaller when metal-to-metal contact occurs.

Figure 12.23 shows how electricity is supplied to the shoe. This conduction-based measurement requires complete electrical insulation so that no electrical current will run through components and areas other than the ones to be measured. In the case of a compressor composed of metallic components, an insulation layer needs to be formed with surface coating or bonding. When electricity is supplied to moving components, the electrical terminal must be made fully compliant with the component movement to eliminate contact resistance in order to assure accurate measurements. This would require the electricity supply section to be held in position with a spring.

Figure 12.24 shows the basic electrical circuit required for a conduction-based measurement. In actual measurement, a resistor-formed bridge is often used to measure electrical potential difference.

Another example of refrigerant-lubrication oil mixture measurement other than the capacitance-based method from Reference 5) is cited in Reference 13), where a sonic speed-based measurement is implemented. Figure 12.25 shows the shape of the ultrasonic element, and Fig. 12.26 shows the measurement setup. This technique works based on the principle that the sonic speed will vary depending on the refrigerant-oil ratio in the mixture.

What is described in the above paragraphs are currently published measurement methods that have been actually implemented on compressors and refrigeration systems.

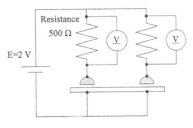

Fig. 12.24 Electric circuit for conductive measurement [12]

Chapter 12 Measurement Technologies

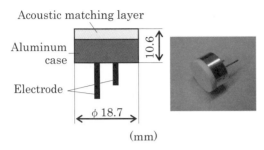

Fig. 12.25 Piezoelectric ceramic ultrasonic element[13]

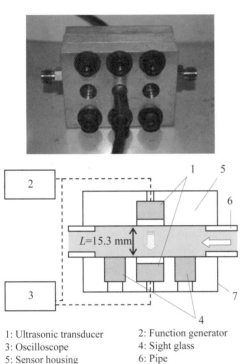

1: Ultrasonic transducer 2: Function generator
3: Oscilloscope   4: Sight glass
5: Sensor housing   6: Pipe
7: O-ring

Fig. 12.26 Measuring equipment of oil-refrigerant mixing ratio[13]

Considering the rapid ongoing advances in sensor technology, it is important to continuously obtain the latest knowledge from academic publications and other information sources.

References

1) T. Itoh, M. Fujitani, Y. Sakai: "Analysis of Discharge Pressure Pulsation of Scroll Compressor", Trans. of the JSME (Series B), 67 (661), pp. 2166-2173 (2001). (in Japanese)
2) T. Toyama, H. Matsuura, Y. Yoshida: "Visual Techniques to Quantify Behavior of Oil Droplets in a Scroll Compressor", Proc. of 17th Int. Comp. Eng. Conf. at Purdue, West Lafayette, C026 (2006).
3) N. Matsumura, H. Kobayashi, H. Machida, T. Hirano, T. Nakahara, Y. Hori: "Measurement of Micron-Size Gap in Rotary Compressor for Air Conditioners", Trans. of the JSRAE, 24 (2), pp. 127-138 (2007). (in Japanese)
4) S. Suzuki, Y. Fujisawa, S. Nakazawa, M. Matsuoka: "Measuring Method of Oil Circulation Ratio Using Light Absorption", Trans. of the JAR, 8 (1), pp. 25-34 (1991). (in Japanese)
5) M. Fukuta, T. Yanagisawa, Y. Ogi, J. Tanaka: "Measurement of Concentration of Refrigerant in Refrigeration Oil by Capacitance Sensor", Trans. of the JSRAE, 16 (3), pp. 239-248 (1999). (in Japanese)
6) M. Matsushima, T. Nomura, N. Nishimura, H. Iyota, K. Inaba: "Study on Efficiency Improvement of Hermetic Rotary Compressors (Direct Torque Measurement of Compressors Using the Strain Gauge)", Trans. of the JSRAE, 18 (1), pp. 39-50 (2001). (in Japanese)
7) H. Hattori, Y. Ito, T. Hirayama, K. Miura: "Mixed Lubrication Analysis for Journal Bearings in Rotary Compressors —Solid Contact Characteristics in Oil Film at Low Speed Operation—", Trans. of the JSRAE, 25 (4), pp. 337-346 (2008). (in Japanese)
8) H. Sato, T. Itoh, H. Kobayashi: "Frictional Characteristics of Thrust Bearing in Scroll Compressor", Trans. of the JSRAE, 25 (4), pp. 347-354 (2008). (in Japanese)
9) N. Ishii, T. Oku, K. Anami, T. Tsuji, T. Ozasa, K. Sawai, T. Morimoto, N. Iida: "An Experimental Study of Lubrication at Thrust Slide-Bearing of Scroll Compressors —Effect of Thickness and Inside Form of Thrust Plate—", Trans. of the JSRAE, 25 (4), pp. 355-364 (2008). (in Japanese)
10) H. Nakao, H. Maeyama, N. Hattori, T. Takayama: "Wear-Reducing Technologies for Rotary Compressors Using CO_2 Refrigerant", Trans. of the JSRAE, 25 (4), pp. 365-374 (2008). (in Japanese)
11) S. Tanaka, S. Zuo, A. Hikam, T. Toyama: "Measurement of Oil Film Thickness between Bush and Blade in Swing Type Refrigerant Compressor", Trans. of the JSRAE, 25 (4), pp. 375-382 (2008). (in Japanese)
12) H. Suzuki, M. Fukuta, T. Yanagisawa: "Measurement of Lubricating Condition between Swashplate and Shoe in Swashplate Compressors under Practical Operating Conditions", Trans. of the JSRAE, 25 (4), pp. 383-390 (2008). (in Japanese)
13) M. Fukuta, T. Suzuki, T. Yanagisawa: "Concentration Measurement of Refrigerant/Refrigeration Oil Mixture by Sound Speed", Trans. of the JSRAE, 25 (4), pp. 391-400 (2008). (in Japanese)

Chapter 13 About This Book

Refrigeration and air conditioning systems are indispensable to people's lives today, to offer a wide variety of food products, to provide personal comfort, to maintain steady and efficient production operations, and to support commercial activities. Various machines and systems, including refrigerators, residential and automotive air conditioners, refrigerating showcases, frozen and cold food storages, freezer trucks and ships, constant temperature rooms and hot water systems, need a refrigeration system to work. The refrigerant-based refrigeration system is the core of all such refrigeration and air conditioning functions. The most commonly used type of refrigeration system is the vapor compression system, the heart of which is the refrigerant compressor.

Refrigerant compressors influence the refrigeration system performance and reliability significantly. Thanks to decades of research and development efforts in academic and industrial sectors, refrigerant compressors have made progressive improvements to higher performance and better reliability. However, to improve performance and reliability furthermore, we need to educate and train engineers and researchers for a deep understanding of scientific and technical knowledge concerning refrigerant compressors.

Businesses and organizations that are involved in the research, development or manufacture of refrigerant compressors are training their engineers in refrigeration and air conditioning fields using commercially available literature or preparing compressor textbooks by themselves. However, there are few commercially available publications that offer systematic knowledge about compressor technology. Most compressor knowledge available from commercial literature is fragmented descriptions contained as part of refrigeration engineering or automotive air conditioning learning. This is far from sufficient as an educational material for future compressor researchers and engineers. The only one existing Japanese publication that offers comprehensive knowledge about refrigerant compressor technology was "Hermetic Refrigerating Machinery", authored by Mutsuyoshi Kawahira and published from the Japanese Association of Refrigeration (JAR), which was the predecessor to the publisher of this textbook, the Japan Society of Refrigerating and Air Conditioning Engineers (JSRAE).

Having been published in 1981, however, part of the information contained in the Kawahira's textbook is now obsolete as the book does not cover modern refrigerant compressor technology. In addition, the book is currently out of print and is not available. While private businesses and organizations individually strive to prepare textbooks that are good enough for in-house training, it is getting difficult, considering the accelerating technological advancement and intensifying competition for better performance and higher technology in the industry, for individual companies to provide a textbook that comprehensively covers the highly advanced and diversified knowledge related to modern refrigerant compressor technology.

To address this situation, the Compressor Technology Subcommittee (now Compressor Technology Committee) of JSRAE launched a project to prepare an original compressor textbook by collecting academic and engineering knowledge about refrigerant compressor technology from various sources to broadly share the knowledge and thereby help aspiring refrigeration and air conditioning engineers and researchers. One of the greatest difficulties anticipated toward bringing the project to success was to have researchers and engineers belonging to competing manufacturers and suppliers to join forces to prepare the compressor textbook. Luckily enough, a team of university researchers provided a core leadership under which engineers belonging to the member companies of JSRAE became actively committed to the project.

Since the Compressor Technology Subcommittee meeting in April 2006 when the textbook preparation project was first proposed, discussions took place in subsequent meetings about what information should be contained in the textbook, how it should be structured and whom to ask to contribute the chapters. An important decision about textbook structure was whether the chapters should be organized on the basis of scientific and technical categories or by the types of compressors. For the best convenience of the readers, a medium between these two approaches was adopted. Considering the situation where businesses and organizations individually work on their largely proprietary compressor technologies to provide respectively unique compressors and related products, it has been agreed that all such existing technologies cannot be completely covered in this single textbook. Therefore, this book primarily focuses on common and universal technologies that need to be understood while covering any specially interesting or important trends and developments.

Based on the above project policy, this textbook offers a wide variety of academic and engineering knowledge about refrigerant compressors in the following chapter structure:
- Introduction (purpose of the compressor and its history)
- Basic theory (compression, mechanism, lubrication, efficiency and refrigerant)
- Variety of compressors (reciprocating, rotary, scroll, twin screw, single screw, and automotive types)
- Refrigeration oil
- Motors and inverters
- Tests
- Measurement technologies

In a basic way, this textbook has been prepared by having members of the Compressor Technology Committee and the employees of its member companies contribute individual chapters, but external contribution was obtained in some parts. For example, Chapter 9 "Refrigeration Oil" was prepared through the cooperation of a lubrication oil manufacturer.

Actual writing by individual authors started from the middle of 2008. As all the authors were busy engineers and researchers who had to spare time from their main corporate or organizational duties for this textbook, and also as a significant amount of coordination was necessary to assure consistency between chapters, this textbook was finally completed after much delay. However, we are confident that the many years we spent to complete this book also allowed to us to carefully review and improve its content before publication, in such a way that would be impossible if the book was completed in a short period of time. Finally, the Japanese version textbook was completed and issued in 2013. Based on the Japanese version, the English version textbook was published in 2017.

This textbook is primarily intended to help younger engineers and researchers who are involved in or are aspiring to be involved in refrigeration and air conditioning fields to acquire academic and technical knowledge about refrigerant compressors in a systematic manner. However, experienced engineers can also benefit from this textbook by reviewing or reconfirming their knowledge or by finding new inspirations for future engineering breakthroughs. To pursue the goals of global warming reduction and building of a low-carbon society, greater social focus is being placed on energy saving in refrigeration and air conditioning operations not only based on user needs but also as part of national governmental strategy. We hope that this textbook will also be useful in that respect by accelerating development of energy efficient refrigerant compressors.

The refrigerant compressor technology described in this textbook reflects its current state today, which is expected to be further improved and developed to attain even higher performance and better efficiency, to offer more effective control of noise and vibration and further reduction of size and weight, and to realize wider operating ranges, oil-free operation and novel mechanisms. Needless to say, refrigerant compressor development efforts will help not only to improve the performance of the compressor on its own but also to make the refrigeration system more efficient, more reliable and better functioning. For example, reduction of oil discharge from compressor to refrigeration circuit, application of liquid refrigerant injection, multiple-stage compression and utilization of waste energy at the expansion process will lead to improvement of comprehensive performance of the refrigeration system, reduction of power requirement during peak load operation and better characteristics during partial load operation.

The publisher of this textbook, the Compressor Technology Subcommittee (renamed to Compressor Technology Committee in 2011) of the Japan Society of Refrigerating and Air Conditioning Engineers (JSRAE) is a body that acts to promote human networking and information sharing among compressor engineers and researchers belonging to various businesses and organizations involved in the manufacture of refrigerant compressors. In addition to having member meetings several times a year, the committee is responsible for organizing compressor-related sessions and seminars as part of the JSRAE annual conference. The committee members have contributed to the publication of this textbook, for sparing their valuable time from everyday duties at their respective employment, with the hope that the completed textbook will be a great help to those who need it. They are happy if their hard work is finally being rewarded.

Lastly, we offer our sincere appreciation to JSRAE and its secretariat and publication management committee, for their consistent support and cooperation throughout this textbook project including the permission to use drawings and other quotes in this textbook.

Appendix

The first edition of this textbook in Japanese was published in 2013. The textbook has been welcomed by Japanese researchers and engineers, and the second print of the first edition was issued in 2016.

As many foreigners were interested in this textbook and requested to publish its English version, the Compressor Technology Committee started to translate the Japanese

Compressors for Air Conditioning and Refrigeration

version into English one. The final English textbook was published in January 2018.

Index

A

AB (Alkyl benzene) 190, 191
Absolute pressure 7
Absolute work 9
Accelerated test 248
Addendum 106
Additive 197, 200
Adiabatic compression 6, 9, 10
Adiabatic efficiency 29
Annealing 212
Anti-rotation mechanism 90, 166
Aramid paper 221
Arc-spot welding 82
ASHRAE 31
Asymmetric profile 101, 107
Asymmetric wrap 88
Auxiliary winding 214
Average flow model 19
Azeotropic 30, 31

B

Back electromotive voltage detection 233
Back pressure chamber 91
Background noise 237, 245
Background vibration 246
Backlash 107, 140
Balance piston 103, 117, 122
Balance type 43, 116
Balance weight 55, 209
Balance ratio 116
Ball coupling 90, 178
Ball joint 50, 159
Barrel polishing 41
Big end 38
Blowhole 106, 110, 148
Bosch type 180
Built-in compression ratio 151
Built-in volume ratio 88, 103, 121, 143
Bypass piston 179

C

Capacitance-based measurement 258
Capacity control 70, 96, 156
Capacity control mechanism 44, 119, 146, 169, 179
Capacity control valve 170, 179
Carnot cycle 6
Carrier noise 236
Caulking 82, 211
Center distance 109
Centrifugal pump 39, 76
Centrifugal separation 76, 148, 183
Centrifugal type 5
CFC (Chloro fluoro carbon) 29, 30
Check valve 172
Closed gap 41
Closed system 9
Closure delay 58, 108
Compressor efficiency 27, 63
Concentrated-winding stator 209, 212
Condenser 1, 7, 155
Conduction-based measurement 260
Constraint force 16, 59, 122
Contact angle 148
Contact theory 19, 20
Continuous capacity control 121, 146
Converter 227, 235
COP (Coefficient of performance) 6, 8, 241
Copper loss 63, 207
Critical speed 125
Crosstalk error 259
Current phase 215
Cusp 110, 112
Cycle 1, 6, 152

D

D'Alembert's principle 17
DC motor 53, 205, 206, 233
Dedendum 106
Die casting 212
Differential pressure lubrication 67, 76
Differential pressure oil supply 118
Direct current drive 206
Displacement volume 54, 87, 111, 149, 181
Distributed-winding stator 209, 210, 212
Drop test 250

Compressors for Air Conditioning and Refrigeration

Dysprosium 226

E

Economizer 128, 152
EHL (Elasto-hydrodynamic lubrication) 27
Ejector 76
Electric test 250
Electrical insulation 195, 221
Electrical steel sheet 208, 222
Electrolytic capacitor 227, 229
Electromagnetic clutch 157
Electromagnetic noise 80, 236
Electrostatic capacitance 247
EMC (Electromagnetic compatibility) 231, 237
Envelope curve 86
Equation of energy 12, 18, 47, 61, 95, 124
Evaporator 1, 8, 155
Excitation force 18, 47, 62, 77, 124
Expansion valve 1, 7, 156
Externally controlled valve 170, 172

F

Fan angle 150
Fanno flow 14
Fatigue life 198
Female rotor 102
Ferrite magnet 209, 225
FFT analyzer 246
F-gas regulation 32
Field test 248
Fixed scroll 85
Fleming's rule 213, 214
Flowmeter 255
Fluid diode 76
Fluid noise 80, 136
Flux-weakening control 234
Force of inertia 17, 45
Four-wire configuration 253
Friction tester 198

G

Gas cooler 66
Gas injection 71
Gate rotor 106, 145, 149

Gate rotor support 145
Germanium glass 253
Grain-oriented electrical steel 223
GWP (Global warming potential) 29

H

HCFC (Hydro chloro fluoro carbon) 30
Heat pump 2
Helix centrifugal pump 76
Hermetic compressor 4, 38, 52, 125
Hermetic motor 217
Hermetic type 4, 199
HFC (Hydro fluoro carbon) 30
High pressure casing 73, 92, 183
Hydraulic efficiency 28

I

IGBT (Insulated gate bipolar transistor) 227, 228
Immiscible oil 190
Indicated efficiency 255
Indicated loss 63
Induced electromotive force 213
Induction motor 205, 213
Inductor 229
Industrial work 9
Inertia force 17, 47, 59, 97, 168
Inertia moment 46, 78, 95, 164
Inertial supercharging 42
Infinitely long bearing 22
Infinitely short bearing 24
Injection port 70, 96, 113, 152
Input power 63, 241, 244
Inserter 212
Insoluble oil 119, 130
Insulation coating 210
Insulation material 218, 221
Insulator 210, 221
Intermediate pressure 50, 70, 74, 152, 179
Internal leakage 12, 151
Internal leakage passage (or path) 110, 121
Internally controlled valve 171
Inverter 205, 226, 227, 228, 231, 235
Inverter control 235
Involute angle 87

266

Involute curve ·········· 86, 98, 177

IPMSM (Interior permanent magnet synchronous motor)
·········· 206, 214

Iron loss ·········· 63, 208

Irreversible cycle·········· 7

J

Journal bearing ·········· 21

K

Kinetic type·········· 3

L

Latent heat ·········· 7

LCP (Liquid crystal polymer)·········· 210, 222

Lead error ·········· 139

Lead wire ·········· 252, 253

Lead wiring·········· 210

Lip seal ·········· 43, 159

Liquid (refrigerant) injection ·········· 70, 96, 128, 152

Low pressure casing ·········· 93

Lubrication system ·········· 39, 67, 74, 76, 177

Lysholm profile ·········· 107

M

Magnet torque ·········· 215

Magnet wire ·········· 210, 218

Male rotor ·········· 102

Material compatibility·········· 201

Mechanical efficiency ·········· 18, 47, 61, 95, 124

Mechanical loss ·········· 16, 27, 63

Mechanical seal ·········· 42, 103, 115, 146

Miscibility ·········· 193

Mixture viscosity ·········· 194

MO (Mineral oil)·········· 190

Modified Reynolds equation ·········· 19

Moment of inertia ·········· 17, 45, 61

Montreal protocol ·········· 30

Motor core ·········· 222

Motor efficiency ·········· 27

Motor loss ·········· 63, 208

Motor model computation method ·········· 234

Motor rotor ·········· 208, 209

Motor stator·········· 213

Multiple-stage compression ·········· 131

N

Natural frequency ·········· 19, 126, 174

Nd-Fe-B magnet ·········· 225

Neodymium·········· 209, 224

Noise filter ·········· 230

Noise meter ·········· 246

Non-oriented electrical steel ·········· 208, 223

Non-sinusoidal (pulsed) power source ·········· 206

Nozzle·········· 12, 14, 256

O

ODP (Ozone depletion potential) ·········· 29

Oil circulation rate ·········· 241, 244

Oil circulation ratio·········· 75, 242

Oil cooler ·········· 118, 152

Oil film pressure ·········· 22, 27

Oil film thickness ·········· 19, 27, 251

Oil injection ·········· 105, 113, 152

Oil mixture ·········· 242, 260

Oil ring ·········· 41

Oil separator ·········· 77, 105, 148, 183, 209

Oil supply system ·········· 117

Oil-free ·········· 101, 104, 118, 131

Oldham ring ·········· 89, 90, 94, 198

Open system ·········· 9

Open type ·········· 4, 105, 145

Open type compressor·········· 4

Orbiting radius ·········· 86, 178

Orbiting scroll·········· 85

Order of rotation ·········· 174

Output characteristics ·········· 215

Output power ·········· 63, 217

Over-compression ·········· 28, 42, 58, 127

P

PAG (Polyalkylene glycol) ·········· 190, 191

PAM (Pulse amplitude modulation) ·········· 231

PBT (Polybutylene terephthalate) ·········· 210, 222

PEN (Polyethylene naphthalate)·········· 221

Permanent magnet synchronous motor ·········· 206, 214, 233

Permanent magnet ·········· 206, 209, 224

PET (Polyethylene terephthalate) ·········· 201, 221

Petroff's equation ·······21
PFC (Power factor control)······· 232
P-h diagram·······6, 66, 70, 74, 129
Pilot valve ······· 146
Pipe friction······· 13, 15
Pitch circle ······· 106, 123, 160
Pitch error ······· 137
POE (Polyol ester) ······· 190, 192
Polyester film ······· 210, 221
Polyvinyl formal resin······· 218
Positive displacement compressor ·······3, 28, 255
Positive displacement type ······· 3
PPS (Polyphenylenesulfide) ······· 210, 222
Pressurized oil supply system······· 117
Profile coefficient ······· 111
Profile modification······· 107
Punching ······· 210
P-V diagram ·······9, 64, 127, 179, 255
PVE (Polyvinyl ether) ······· 190, 191
PWM (Pulse width modulation) ······· 231

Q

Quality·······7, 15
Quantity of state ······· 1

R

Radial bearing······· 103, 114, 159
Radial load ······· 113, 122, 146
Radiation thermometer ······· 251
Rare earth magnet ······· 209, 224
Ratio of specific heats ······· 11
Real contact area······· 20
Reed valve ·······40, 160
Refrigeration capacity······· 6, 241, 243
Refrigeration cycle ······· 1, 6
Refrigeration effect ······· 6
Refrigeration oil ······· 190
Reluctance motor ······· 206
Reluctance torque ······· 207, 215
Resonance ·······80, 125, 164
Reversed Carnot cycle ······· 6
Reversible process ······· 6
Reynolds equation ·······19
Rolling bearing ·······26, 113, 146

Rolling piston ·······4, 52, 56, 60
Rotary type ······· 4, 116
Rotary vane ·······4, 52
Rotary vane compressor ······· 3, 154, 180
Rotating magnetic field ······· 213
Rotational balance ······· 52, 68, 164
Rotor ······· 180
Rotor bouncing ······· 135
Rotor core ······· 208
Rotor profile ······· 106
Rotor tooth ······· 148

S

Safety margin test ······· 248
Sampling cylinder ······· 256
Scooping type ······· 180
Scratchproof wire ······· 219
Screw rotor ······· 101, 111, 136, 142, 145
Seal ring ·······41, 182
Sealing line ······· 110
Sealing mechanism ·······91, 159
Secondary conductor ······· 209
Secondary refrigerant calorimetry ······· 243
Seizuer ······· 199
Semi-hermetic (type) ······· 4, 105, 145
Sensible heat ······· 7
Shaft power·······28, 120, 126
Shaft seal·······42, 115, 146
Shaft seal(ing) mechanism ·······42, 159
Shock loop ·······39
Shoe·······38, 162
Simultaneous-drive capacity control ······· 147
Single-phase ······· 214, 227, 237
Single-stage compression ······· 131
Sinusoidal modulation ······· 231
Sinusoidal power source ······· 206
Slide valve ······· 103, 119, 143, 146
Sliding bearing ······· 21, 60, 114
Sliding vane (type) ·······4, 52
Slip ······· 214, 232
Slip ratio ·······15
Slip ring ······· 258
Small end ·······38
Solid-borne noise ·······80

Solubility ·················· 191, 194, 256
Sommerfeld number ················ 21, 25
Sound power level ··················· 244
Specific enthalpy ······················· 8
Specific entropy ························ 7
SPMSM (Surface permanent magnet synchronous motor)
······························ 206, 214
Squirrel-cage ························ 213
Starting current ····················· 215
Starting torque······················ 215
Start-up (winding) ··············· 214, 234
Stator ····················· 157, 209, 213
Stepped capacity control·················· 146
Stepped-height wrap ·····················89
Suciton gas bypass (release) ···············71
Swedish steel ······················ 41, 58
Swing compressor ·····················68
Swinging angle ·······················38
Symmetrical circular arc··················· 107
Synchronous motor··············· 206, 214, 233

T

Temperature tolerance class ············· 217
Temperature tolerance life ············· 221
Thermo label ······················· 253
Thermo viewer ····················· 253
Three-phase···················· 214, 227
Three-wire configuration ··············· 253
Through-vane type ··················· 180
Thrust bearing·············· 24, 90, 114, 159, 178
Tilting pad bearing ················ 26, 114
Timing gear···················· 101, 104
Tip seal ······················91, 175
Top clearance ···············36, 44, 64, 161
Torque characteristics ················ 217
Torque control ·····················78, 234
Torque limiter ···················· 167, 173
Torsional resonance ··············· 126, 164
Total efficiency ·······················29
Trailing type ······················· 180
Trans-critical cycle ·····················32
Transportation test ··················· 250
Trochoid pump ·······················40
Turbo compressor ······················ 3

Two-cylinder type ··················· 52, 67
Two-phase separation ················· 193
Two-stage compression (or compressor) ·········· 50, 73
Two-wire configuration ················· 253

U

Ultrasonic element ··················· 260
Unbalance type ···················43, 116
Unbalanced inertia force····················18

V

V/f control ························· 232
Valve lift ··························57
Vane pump ··························40
Vane slot ······················40, 65, 81
Vapor compression refrigeration cycle ··········· 1, 6
Variable cylinder management ···············72
Variable-pitch (variable thickness) wrap ···········88
Varnish ························ 212, 219
Vent hole ······················· 39, 75
Vi variation mechanism ················· 104
Vibration pickup ···················· 247
Vibrometer ························ 247
Viscosity ····················· 13, 16, 195
Viscosity-pressure coefficient ············· 199
Volume resistivity ················· 191, 195
Volumetric efficiency ·················28, 128

W

Water-based varnish ·················· 221
Wear ·························43, 198
Weather resistance test ················ 250
Weibull paper ····················· 249
Windscreen ························ 245
Wrap angle ························ 109

X

Y

York type···························· 180

Z

Zeotropic ························ 30, 31

\<Authors\>

Chapter 1 Introduction

Mitsuhiro Fukuta	Shizuoka University

Chapter 2 Basic Theory

Mitsuhiro Fukuta	Shizuoka University
Masaaki Motozawa	Shizuoka University
Noriaki Ishii	Osaka Electro-Communication University
Kiyoshi Sawai	Hiroshima Institute of Technology
Chun-cheng Piao	DAIKIN INDUSTRIES, LTD

Chapter 3 Reciprocating Compressors

Takahide Ito	Mitsubishi Heavy Industries, Ltd.
Shigeki Miura	Mitsubishi Heavy Industries Thermal Systems, Ltd.
Noriaki Ishii	Osaka Electro-Communication University
Takuya Hirayama	Toshiba Carrier Corporation

Chapter 4 Rotary Compressors

Yoshinori Shirafuji	Mitsubishi Electric Corporation
Masao Tani	Mitsubishi Electric Corporation
Kazuaki Fujiwara	Panasonic Corporation
Noriaki Ishii	Osaka Electro-Communication University
Takuya Hirayama	Toshiba Carrier Corporation
Atsushi Kubota	Hitachi, Ltd.
Kazuhiro Furusho	DAIKIN INDUSTRIES, LTD.

Chapter 5 Scroll Compressors

Kenji Tojo	Hitachi Appliances, Inc. (retired)
Tsutomu Nozaki	Hitachi, Ltd.
Masatsugu Chikano	Hitachi, Ltd.
Noriaki Ishii	Osaka Electro-Communication University

Chapter 6 Twin Screw Compressors

Kouichi Matsuo	MAYEKAWA MFG. CO., LTD.
Harumi Sato	MAYEKAWA MFG. CO., LTD.
Noriaki Ishii	Osaka Electro-Communication University

Chapter 7 Single Screw Compressors

Kaname Ohtsuka	DAIKIN INDUSTRIES, LTD. (retired)
Kazuhiro Furusho	DAIKIN INDUSTRIES, LTD.

Chapter 8 Automotive Air Conditioning Compressors

Hitoshi Azami	Sanden Automotive Components Corporation
Kazuhiko Takai	Sanden Advanced Technology Corporation (retired)
Takahide Ito	Mitsubishi Heavy Industries, Ltd.
Shigeki Miura	Mitsubishi Heavy Industries Thermal Systems, Ltd.
Takashi Morimoto	Panasonic Corporation
Kiyoshi Sawai	Hiroshima Institute of Technology
Noriaki Ishii	Osaka Electro-Communication University

Chapter 9 Refrigeration Oil

Masato Kaneko	Idemitsu Kosan Co.,Ltd.

Chapter 10 Motors and Inverters

Kazuo Ida	DAIKIN INDUSTRIES, LTD.
Akio Yamagiwa	DAIKIN INDUSTRIES, LTD.
Sumikazu Matsuno	DAIKIN INDUSTRIES, LTD.
Yoshinori Shirafuji	Mitsubishi Electric Corporation

Chapter 11 Testing

Takuya Hirayama	Toshiba Carrier Corporation
Yukio Yokomizo	Toshiba Carrier Corporation (retired)

Chapter 12 Measurement Technologies

Takahide Ito	Mitsubishi Heavy Industries, Ltd.
Shigeki Miura	Mitsubishi Heavy Industries Thermal Systems, Ltd.

Chapter 13 About This Book

Tadashi Yanagisawa	Shizuoka University

JSRAE Technical Book Series
Compressors for Air conditioning and Refrigeration
Printed in January, 2018

Copyright © 2018 by JSRAE, All rights reserved.
No part of this publication may be reproduced, stored in retrieval system, or transmitted, in any form or by any means, electric, mechanical, photocopying, recording, or otherwise, without the prior written permission of the publisher.

ISBN 978-4-88967-136-0

Edited by JSRAE Compressor Technology Committee
Published by Japan Society of Refrigerating and Air conditioning Engineers
     Nihonbashi Otomi BLDG., 13-7 Nihonbashiodenmacho, Chuo-ku, Tokyo 103-0011, Japan
Printed in Japan by Adot Works, Inc.
     2-1 Yamashitacho, Naka-ku, Hamamatsu, Shizuoka